AMERICAN MATHEMATICAL SOCIETY TRANSLATIONS

Series 2

Volume 96

Ten Papers on Algebra and Functional Analysis

by

N. D. Filippov   Ju. P. Ginzburg   L. N. Ševrin
A. L. Garkavi   V. V. Rozen   A. A. Vinogradov
            B. M. Šaĭn

Published by the
AMERICAN MATHEMATICAL SOCIETY
Providence, Rhode Island
1970

International Standard Book Number 0-8218-1796-5
Library of Congress Catalog Number A51-5559

TABLE OF CONTENTS

Page

Amer. Math. Soc. Transl.
(2) Vol. 96, 1970

BASIC PROBLEMS IN THE THEORY OF PROJECTIONS OF SEMILATTICES *

UDC 519.47

L. N. ŠEVRIN

Introduction

A *projection* of a semilattice Γ onto a semilattice Γ' is an isomorphism of the lattice of all of the subsemilattices of Γ (the empty set is also considered to be subsemilattice) onto the lattice of all of the subsemilattices of Γ'. The notion of a projection applies to an arbitrary class of algebraic systems. Just as extensively used in this connection is the term "lattice isomorphism." Clearly, the term "projection" is preferable in the case of semilattices and especially of lattices. A projection is also called a projectivity (see, for example, [1]).

Studies have been made of the projections of groups [2,3], semigroups [4], linear manifolds over a field or, more generally, modules over a regular ring [1,5] (in the papers cited, see the corresponding extensive bibliography), as well as certain other systems, for example, Boolean algebras [6]. A study of the projections of semilattices, besides being of independent interest, is needed for the theory of projections of semigroups, in particular, semigroups of idempotents, in view of the very appreciable role which semilattices play in the theory of semigroups; on the one hand, it can be roughly stated that they constitute one of the most important classes of semigroups of idempotents, viz. the class of commutative semigroups of idempotents (for the exact nature of the relationship between these classes see the book [7], Chapter II, or the book [8], §1.8), and, on the other hand, they are closely connected with certain important decompositions of semigroups into bands (concerning this, see [7] and [8], where a relevant bibliography is given). The present paper actually arose in connection with an attempt by the author to systematically study the projections of semigroups of idempotents. It has turned out that in the commutative case the language of semilattices is not simply one of various possible languages, but is essentially required, since for almost all of the notions and proofs arising in the solution of the corresponding problems

---

* Translation of Mat. Sb. 66 (108) (1965), 568–597.

the most natural form is precisely that one in which this language is employed (see the notions of homogeneous subsemilattice, weak isomorphism, etc.
below).

The following (which, like the subsequent problems, is formulated directly for semilatices for the sake of simplicity) should be regarded as a basic
problem in the theory of projections:

A) *The determination of necessary and sufficient conditions under which
two arbitrary semilattices project onto each other; in other words, the determination of all of the projections of an arbitrary semilattice.*

The solution of this problem in a certain sense gives a classification of
semilattices according to their lattices of subsemilattces, which partitions
semilattices into classes of projectively equivalent semilattices. Here problem A) plays the same role as analogous problems in other theories, for example, the problem concerning elementary equivalence in the theory of
modules. Of course, as was correctly noted in the article [9], any problem
concerning the determination of necessary and sufficient conditions for some
property bears a somewhat indefinite character, but, the indefiniteness not-
withstanding, it is customary to consider such a problem as solved if conditions are found which are suitable for the applications. In this sense the do-
minant role of problem A) consists in the fact that its solution facilitates or
makes trivial the solution of other problems connected with projections.

Isomorphic semilattices obviously project onto each other. The converse statement (as in the case of other algebraic systems) is generally not
true. There thus arises the following problem concerning the semilattices for
which the indicated converse statement holds:

B) *A description of all of the semilattices which project only onto isomorphic semilattices.*

Each isomorphism of a semilattice induces a projection of it. The converse statement is generally not true, even if the semilattice satisfies the
condition of problem B). There thus arises the following problem:

C) *A description of all of the semilattices each projection of which is
induced by some isomorphism.*

Complete and partial solutions of these problems are known for various
classes of algebraic systems. For example, problem A) has been completely
solved for finite groups [2]. Many sufficient conditions have been established for systems satisfying the conditions of problems B) and C). We mention, in particular, the first fundamental theorem of projective geometry (see
the book [1]) and its generalization in the paper [10]. Various classes have

been found of groups (semigroups) which are determined by the lattices of
their subgroups (subsemigroups) and so on (for the details concerning the
corresponding results see the papers cited above). In this connection semi-
linear transformations play the role of isomorphisms in the case of modules
over a ring.

The present article is almost entirely devoted to the solution of problem
A) for semilattices. We obtain a complete solution of this problem, which, in
particular, easily permits one to also completely solve problem C). As for
problem B), its solution has required the introduction of certain additional
notions and constructions connected with them. It will be dealt with in another paper.

The article consists of eight sections. In § 1 we present the definitions
of the basic notions used in the paper and some of their simplest properties.
Here, too, we prove a proposition on the connection between projections and
weak isomorphisms, which serves as a starting point for many subsequent constructions. A formulation of the main result, viz. Theorem A, which solves
problem A), is contained in § 8; there we also present a proof of the necessity of the assertion of this theorem and prove Theorem C, which solves problem C). The sufficiency of the assertion of Theorem A is essentially proved
in § 4, where we also give a definition of the notion (involved in the formulation of the basic result) of a semilattice obtained from a given semilattice by
permuting the components of the irreducible decompositions of some of its
saturated subsemilattices into ordinal sums of semilattices. §§ 2 and 3 serve
as preparatory sections for § 4. §§ 5–7 together with §§ 3 and 4 pave the way
for a proof of the necessity of the assertion of Theorem A. The material of
§§ 5 and 6 is of independent interest.

The interdependence of the sections is indicated by the following dia-
gram:

$$1\begin{smallmatrix} \nearrow 2\to3\to4 \searrow \\ \\ \searrow 5\to6\to7 \nearrow \end{smallmatrix}8.$$

A brief announcement of the results of this paper was contained in the
note [11].

§ 1. Basic definitions and notation

By a *semilattice* (lower semilattice [8] or minorant semilattice [12]) is
meant a partially ordered set in which any pair of elements has a greatest
lower bound. The greatest lower bound of two elements a and b is called
their *intersection* and denoted by $a \wedge b$. A *subsemilattice* of a semilattice Γ

is a subset of Γ which is a subalgebra of Γ considered as a universal algebra with respect to the operation of intersection (for other possible meanings
of this term see [13], Chapter IV). The empty set is also considered to be a
subsemilattice. As usual, a *proper* subsemilattice of a semilattice Γ is a subsemilattice that is different from Γ. We will let $\{M\}$ denote the subsemilattice generated by a set M of elements of a given semilattice. The set of
all of the subsemilattice of a semilattice Γ is a complete lattice with
respect to set-theoretic inclusion. In accordance with our semilattice-
theoretic notation, this lattice will be denoted by $\Sigma'(\Gamma)$.

The partial ordering relation in a given partially ordered set and, in particular, in a semilattice will be denoted by $\leq$. As usual, we write $a < b$ if
$a \leq b$ and $a \neq b$. The symbols $\geq$ and $>$ are used in the usual way. The letter σ will denote the binary relation of comparability, i.e. the union of the
relations $\leq$ and $\geq$, while $\bar{\sigma}$ will denote the complementary relation of incom-
parability. If $a \, \sigma \, b \, (a \, \bar{\sigma} \, b)$ then, as usual, we say that a and b are *comparable (incomparable)*. We will say that an element x of a semilattice Γ is
a *proper divisor*[*] of an element $a \in \Gamma$ if $x \neq a$ and there exists a $y \in \Gamma$
such that $y \neq a$ and $x \wedge y = a$, in which case y will also be a proper divisor
of a. An element of a semilattice which has proper divisors is said to be
decomposable; otherwise it is *indecomposable*.

The symbols $\cup$, $\cap$ and $\setminus$ denote the corresponding set-theoretic operations, and $\emptyset$ denotes the empty set.

Let Γ be a semilattice. A subsemilattice $H \subseteq \Gamma$ is said to be *completely isolated* (as in [7] for semigroups) if $\Gamma \setminus H$ is a subsemilattice. From
this definition there immediately follows

Lemma 1.1. *If the set-theoretic sum of a set of completely isolated subsemilattices is a subsemilattice, then it is completely isolated.*

A subsemilattice $H \subseteq \Gamma$ is said to be *homogeneous* if either $H = \Gamma$ or
the following alternative holds for an arbitrary $x \in \Gamma \setminus H$: either

$$x \sigma h \qquad \text{for any} \qquad h \in H, \tag{1}$$

or

$$x \, \bar{\sigma} \, h \qquad \text{for any} \qquad h \in H. \tag{2}$$

From this definition there immediately follows

Lemma 1.2. *Suppose H is a homogeneous subsemilattice of a semilat-
tice Γ, $x \in H$, $y \in \Gamma \setminus H$ and $x \, \bar{\sigma} \, y$. Then $x \wedge y \in \Gamma \setminus H$.*

---

[*] This term is due to the connection with semigroups mentioned in the introduction.

For a homogeneous subsemilattice $H \subseteq \Gamma$ we will let $\Gamma_\sigma(H)$ denote the set of all $x \in \Gamma \backslash H$ satisfying condition (1) and let $\Gamma_{\underline{\sigma}}(H)$ denote the set of all $x \in \Gamma \backslash H$ satisfying condition (2). We have $\Gamma_\sigma(H) \bigcup \Gamma_{\underline{\sigma}}(H) = \Gamma \backslash H$ and $\Gamma_\sigma(H) \cap \Gamma_{\underline{\sigma}}(H) = \emptyset$. A subsemilattice $H \subseteq \Gamma$ is said to be *saturated* if (i) it is a completely isolated homogeneous subsemilattice and (ii) any completely isolated homogeneous subsemilattice $F \subseteq \Gamma$ which strictly contains H satisfies the strict inclusion $\Gamma_{\underline{\sigma}}(F) \subset \Gamma_{\underline{\sigma}}(H)$. Γ itself is a saturated subsemilattice.

The notion of an *ordinal sum* of partially ordered sets is assumed to be known (see [14]). In the present paper, when speaking of the decomposability of a semilattice into an ordinal sum, we will always have in mind an ordinal sum of subsemilattices, often without stating this explicitly; the only exception will occur in part of § 5, where we consider a decomposition into an ordinal sum of subsets, which is precisely specified in the appropriate places. A semilattice having a decomposition into an ordinal sum of proper subsemilattices will be said to be *decomposable*; otherwise it will be said to be *indecomposable*. The terms of a decomposition of a semilattice into an ordinal sum will be called the *components* of this decomposition. From the corresponding definitions there immediately follows

Lemma 1. 3. *Any component of a decomposition of a semilattice Γ into an ordinal sum is a completely isolated homogeneous subsemilattice of Γ.*

If, for two components A and B of a decomposition of a given semilattice into an ordinal sum, $a < b$ for any $a \in A$ and $b \in B$, we will say that A *precedes* B. A natural linear order is established in this way on the set of components of a given decomposition. In the case when this set is finite, we will frequently number the components in accordance with the indicated order (the latter remark is important in regard to our work on the solution of problem B)).

Suppose a semilattice Γ is decomposable into an ordinal sum of some set of semilattices. It is easily seen that the set of components of this decomposition and the rule for ordering them in the sense indicated above uniquely determine Γ to within an isomorphism. If the set of components of some decomposition of semilattice Γ' into an ordinal sum coincides to within an isomorphism with the set of components of an analogous decomposition of Γ (so that to within an isomorphism Γ' differs from Γ by at most only the arrangement of these components), we will say that Γ' *is obtained from* Γ *by permuting the components* of the given decomposition.

It is not difficult to show that an arbitrary semilattice has a unique *irreducible decomposition* into an ordinal sum, i.e. a decomposition whose components are indecomposable. This follows, for example, from the corresponding property of successively annihilating bands in the theory of semigroups (see [7], Chapter VIII), since, in passing to the language of the theory of semigroups, the notion of an ordinal sum for semilattices goes over into the notion of a successively annihilating band. Since a successively annihilating band is a special case of a strong band [15,4], by using the definition of the preceding paragraph and applying Lemma 5 of [16], which asserts that the lattice of subsemigroups of a strong band of semigroups is isomorphic to the direct product of the lattices of subsemigroups of the components of this band, we obtain the following assertion.

Lemma 1.4. *If a semilattice Γ' is obtained from a semilattice Γ by permuting the components of some decomposition of Γ into an ordinal sum, then Γ projects onto Γ'.*

Since no other decompositions of semilattices are considered in the present paper, in place of the words "components of the irreducible decomposition into an ordinal sum of semilattices" we will frequently say simply "components of the irreducible decomposition."

A one-to-one mapping ϕ of a semilattice Γ onto a semilattice Γ' is called a *weak isomorphism* if the following conditions are satisfied for arbitrary $x, y \in \Gamma$:

1) $x \, \sigma \, y \leftrightarrow \phi(x) \, \sigma \, \phi(y)$,

2) $x \, \bar{\sigma} \, y \rightarrow \phi(x \wedge y) = \phi(x) \wedge \phi(y)$.

Condition 1) means that ϕ is an isomorphism of Γ as a σ-structure in the sense of Bourbaki [17]. This condition can obviously be replaced by the following equivalent condition:

1') $x \, \bar{\sigma} \, y \leftrightarrow \phi(x) \, \bar{\sigma} \, \phi(y)$.

Clearly the inverse mapping ϕ^{-1} is a weak isomorphism of Γ' onto Γ.

Consequently we have the condition

2') $\phi(x) \, \bar{\sigma} \, \phi(y) \rightarrow \phi^{-1}(\phi(x) \wedge \phi(y)) = x \wedge y$.

It is easily verified that the set of conditions 1) and 2) is equivalent to the set of conditions 2) and 2'). This permits one to note that the notion of weak isomorphism is similar to the notion of quasi-isomorphism, which is considered for lattices in papers [18] and [19].[*]

---

[*] Specifically: a quasi-isomorphism ϕ of a finite lattice is a one-to-one mapping which satisfies conditions 2) and 2') and takes the unit of a lattice into the unit.

We agree to multiply mappings below by writing down the factors from right to left. We conclude this section by proving a proposition which gives a preliminary solution to problem A) and clarifies the role played in our work by the notion of weak isomorphism.

Proposition 1. 1. *Two semilattices* Γ *and* Γ_1 *project onto each other if and only if they are weakly isomorphic, each projection of* Γ *onto* Γ_1 *being induced by a weak isomorphism of* Γ *onto* Γ_1.

Proof. Suppose ψ is an isomorphism of $\Sigma'(\Gamma_1)$. The image of a single element subsemilattice of Γ relative to ψ is a single element subsemilattice of Γ_1 (the atoms in $\Sigma'(\Gamma)$ and $\Sigma'(\Gamma_1)$ respectively), and therefore ψ induces a one-to-one mapping of Γ onto Γ_1: if $\psi(\{x\}) = \{x_1\}$, we put $\phi(x) = x_1$. Let us show that ϕ is a weak isomorphism. Suppose $x \,\bar{\sigma}\, y$ $(x,\ y \in \Gamma)$. Then the subsemilattice $\{x, y\}$ consists of three elements: x, y and $x \wedge y$, whence the image of it in Γ_1, viz. the subsemilattice $\{\phi(x),\ \phi(y)\}$, also consists of three elements: $\phi(x)$, $\phi(y)$ and $\phi(x) \wedge \phi(y)$, and consequently $\phi(x \wedge y) = \phi(x) \wedge \phi(y)$. On the other hand, if $x \,\sigma\, y$, the condition $\phi(x) \,\sigma\, \phi(y)$ is trivial in the case $x = y$, while in the case $x \neq y$ the subsemilattice $\{x, y\}$ consists of the two elements x and y, whence $\{\phi(x),\ \phi(y)\}$ also consists of two elements, i.e. $\phi(x) \,\sigma\, \phi(y)$. It follows from what has been proved that $\phi(x) \,\sigma\, \phi(y)$ implies $x \,\sigma\, y$. Thus ϕ is a weak isomorphism.

Conversely, suppose ϕ is a weak isomorphism of Γ onto Γ_1. We extend the mapping ϕ (denoting it by the same letter) into a subsemilattice H of Γ by putting, as usual, $\phi(H)$ equal to the set of all elements of the form $\phi(h)$, where $h \in H$. $\phi(H)$ is a subsemilattice of Γ_1. In fact, the condition $\phi(h) \wedge \phi(g) \in \phi(H)$ for arbitrary elements $\phi(h)$ and $\phi(g)$ of $\phi(H)$ is obvious if we have $\phi(h) \,\sigma\, \phi(g)$, while if $\phi(h) \,\bar{\sigma}\, \phi(g)$ then according to condition 2') we have $\phi(h) \wedge \phi(g) = \phi(h \wedge g) \in \phi(H)$, since $h \wedge g \in H$. In the same way we get that $\phi^{-1}(H_1)$ is a subsemilattice of Γ when H_1 is a subsemilattice of Γ_1. Thus ϕ is a one-to-one mapping of $\Sigma'(\Gamma)$ onto $\Sigma'(\Gamma_1)$. Moreover, ϕ is an isomorphism of $\Sigma'(\Gamma)$ onto $\Sigma'(\Gamma_1)$, since, clearly $H \subseteq F$ in Γ if and only if $\phi(H) \subseteq \phi(F)$ in Γ_1.

The proposition is proved.

On account of Proposition 1.1 it is necessary to clarify the "mechanism of action" of a weak isomorphism. This mechanism, as will be shown, is closely connected with decompositions of a given semilattice into an ordinal sum of (saturated) completely isolated homogeneous subsemilattices. The following two sections are devoted to a consideration of those properties of such subsemilattices that will be needed in the sequel.

§ 2. Completely isolated homogeneous subsemilattices

We begin this section by proving three lemmas on homogeneous subsemilattices.

Lemma 2.1. *If H is a homogeneous subsemilattice of a semilattice Γ and $x \in \Gamma_{\bar{\sigma}}(H)$, then $x \wedge y = x \wedge z$ for any $y, z \in H$.*

Proof. By Lemma 1.2, $x \wedge y \in \Gamma \backslash H$. In addition, $x \wedge y < y$. It follows by virtue of the homogeneity of H that $x \wedge y \, \sigma \, z$. Clearly $x \wedge y < z$, since otherwise we would have $z < x \wedge y < x$, i.e. z would be comparable with x, which contradicts the condition $x \in \Gamma_{\bar{\sigma}}(H)$. In exactly the same way we show that $x \wedge z < y$. Hence $x \wedge y = (x \wedge y) \wedge z = (x \wedge z) \wedge y = x \wedge z$. Q.E.D.

Lemma 2.2. *If in a system of homogeneous subsemilattices of a semilattice Γ the intersection of each pair of subsemilattices is not empty, then the set-theoretic sum of all of the subsemilattices of this system is a homogeneous subsemilattice of Γ.*

Proof. Let $[H_\alpha]$ (α ranges over a set of indices A) be a system of subsemilattices satisfying the condition of the lemma. We put $\mathbf{U}_{\alpha \in A} H_\alpha = H$ and take arbitrary elements $x, y \in H$. We have $x \in H_\alpha$ and $y \in H_\beta$ for certain $\alpha, \beta \in A$. If, in addition, one of the cases $x \in H_\beta$, $y \in H_\alpha$ or $x \, \sigma \, y$ holds, then the inclusion $x \wedge y \in H_\alpha \cup H_\beta \subseteq H$ is obvious. Suppose, therefore, that $x \in H_\alpha \backslash H_\beta$, $y \in H_\beta \backslash H_\alpha$ and $x \, \bar{\sigma} \, y$. By assumption, $H_\alpha \cap H_\beta \neq \emptyset$. We take an element $z \in H_\alpha \cap H_\beta$. By virtue of the homogeneity of H_β we have $x \, \bar{\sigma} \, z$, and hence, according to Lemma 1.2, $x \wedge z \notin H_\beta$. Since $x \wedge x < z$, we get, again by virtue of the homogeneity of H_β, that $x \wedge z \, \sigma \, y$ (more precisely, $x \wedge z < y$ on the basis of Lemma 2.1). But $x \wedge z \in H$. Thus $y \, \sigma \, x \wedge z$ and $y \, \bar{\sigma} \, x$, and this contradicts the homogeneity of H_α, which shows that the last case considered is impossible.

Thus $x \wedge y \in H$ in all possible cases, i.e. H is subsemilattice.

Let us prove the homogeneity of H. We assume the contrary: there exist elements $u, v \in H$ and $w \in \Gamma \backslash H$ such that $w \, \sigma \, u$ and $w \, \bar{\sigma} \, v$. Suppose $u \in H_\gamma$, $v \in H_\delta$ ($\gamma, \delta \in A$) and $t \in H_\gamma \cap H_\delta$. Then, by successively making use of the homogeneity of H_γ and H_δ, we get $w \, \sigma \, t$ and $w \, \sigma \, v$. But the latter contradicts our assumption. This completes the proof of the lemma

Lemma 2.3. *The intersection of an arbitrary set of homogeneous subsemilattices of a semilattice Γ, if it is not empty, will again be a homogeneous subsemilattice of Γ.*

The proof easily follows from the definition of a homogeneous subsemilattice.

Lemma 2.4. *The intersection of an arbitrary set of completely isolated homogeneous subsemilattices of a semilattice Γ, if it is not empty, will again be a completely isolated homogeneous subsemilattice of Γ.*

Proof. Let $H = \bigcap_{\alpha \in A} H_\alpha$, where H_α is a completely isolated homogeneous subsemilattice for each $\alpha \in A$. The homogeneity of H is ensured by the preceding lemma. Let us prove that H is completely isolated. Suppose $H \neq \Gamma$ (in the case $H = \Gamma$ there is nothing to prove) and x and y are arbitrary elements of $\Gamma \backslash H$. If $x \, \sigma \, y$ then $x \wedge y$ is equal to one of the elements x or y, and therefore $x \wedge y \in \Gamma \backslash H$. Suppose $x \, \bar{\sigma} \, y$. There exists an $\alpha_1 \in A$ such that $y \notin H_{\alpha_1}$. If $x \notin H_{\alpha_1}$ then by virtue of the fact that H_{α_1} is completely isolated we have $x \wedge y \notin H_{\alpha_1}$, and hence $x \wedge y \notin H$. But if $x \in H_{\alpha_1}$ then Lemma 1.2 implies that $x \wedge y \notin H_{\alpha_1}$ by virtue of the homogeneity of H_{α_1}. Thus $x \wedge y \notin H$ in all cases, so that $\Gamma \backslash H$ is a subsemilattice.

Corollary. *There exists a unique minimal completely isolated homogeneous subsemilattice containing a given set M of elements of a semilattice.*

The subsemilattice indicated in the corollary will be called *the completely isolated homogeneous subsemilattice generated by the set M.*

Lemma 2.5. *The union of an increasing sequence of completely isolated homogeneous subsemilattices of a semilattice Γ is a completely isolated homogeneous subsemilattice of Γ.*

Proof. Let H be the union indicated in the formulation of the lemma. First of all, H is a subsemilattice. Therefore, by virtue of Lemma 1.1, H is completely isolated. In addition, since the conditions of Lemma 2.2 are obviously satisfied for an increasing sequence, H is homogeneous by virtue of this lemma.

Lemma 2.6. *If H is a completely isolated homogeneous subsemilattice of a semilattice Γ while F is a completely isolated homogeneous subsemilattice of H, then F will be a completely isolated homogeneous subsemilattice of Γ.*

Proof. Let x be an arbitrary element of $\Gamma \backslash F$. We have either $x \in \Gamma \backslash H$ or $x \in H \backslash F$. By making use of the homogeneity of H in Γ in the first case, and of the homogeneity of F in H in the second, we conclude that x is either comparable with any element of F or else not comparable with even one of them. Consequently F is homogeneous in Γ. Let x and y be arbitrary elements of $\Gamma \backslash F$. In the case when they belong to one of the sets $\Gamma \backslash H$ or

$H \setminus F$, the element $x \wedge y$ belongs to the same set by virtue of the condition of the lemma, and thus $x \wedge y \in \Gamma \setminus F$. The same conclusion can obviously be drawn when $x \, \sigma \, y$. Suppose $x \in \Gamma \setminus H$, $y \in H \setminus F$ and $x \, \bar{\sigma} \, y$. Then, by Lemma 1.2, $x \wedge \in \Gamma \setminus H$. Consequently $\Gamma \setminus F$ is a subsemilattice, i.e. F is completely isolated in Γ.

Lemma 2.7. *Let H be a completely isolated homogeneous subsemilattice of a semilattice Γ. For any semilattice T there exists a semilattice Γ' such that it contains a completely isolated homogeneous subsemilattice H' that is isomorphic to T, $\Gamma \setminus H = \Gamma' \setminus H'$,* $\Gamma_\sigma(H) = \Gamma'_\sigma(H')$ and any element $x \in \Gamma \setminus H$ which is less (greater) in Γ than any element of H is less (greater) in Γ' than any element of H'.*

If in addition T is weakly isomorphic to H, then every semilattice Γ' with the above properties is weakly isomorphic to Γ.

Proof. 1. We put $\Gamma \setminus H = F$, denote a semilattice that is isomorphic to T and does not have elements in common with F by H', and consider the set $\Gamma' = H' \cup F$. We define a binary relation μ on Γ' in the following way. If a pair of elements $x, y \in \Gamma'$ belongs to one of the semilattices H' or F, we put $x \, \mu \, y$ in Γ' if and only if $x \leq y$ in the corresponding semilattice. But if $x \in H'$ while $y \in F$, we put $x \, \mu \, y$ if and only if y is greater in Γ than each element of H; $y \, \mu \, x$ if and only if y is less in Γ than some element of H.

We prove that μ is a partial ordering relation. The reflexivity and anti-symmetry of μ are obvious. Let us verify its transitivity. Suppose $x \, \mu \, y$ and $y \, \mu \, z$. We must show that $x \, \mu \, z$. If x, y and z belong to one of the semilattices H' or F, the required result follows from the fact that on each of these semilattices μ coincides with the corresponding partial ordering relation. Suppose that among the elements x, y and z there are elements from both H' and F. We consider the possible cases here.

a) One of the elements x, y or z belongs to H' while the other two belong to F.

a1) $x \in H'$ and $y, z \in F$. Since $x \, \mu \, y$, the element y is greater in Γ than every element $u \in H$. But $y \, \mu \, z$ means that $y \leq z$ in F and hence in Γ, which implies that z is greater than every $u \in H$. Therefore $x \, \mu \, z$.

a2) $y \in H'$ and $x, z \in F$. Since $y \, \mu \, z$, the element z is greater in Γ than every element $u \in H$. From $x \, \mu \, y$ it follows that there exists an element $v \in H$ such that $x < v$. From $v < z$ we then get $x < z$, i.e. $x \, \mu \, z$ in Γ'.

---

* An equality of semilattices and not just an equality of sets.

a3) $z \in H'$, and $x, y \in F$. Since $y \mu z$, we have $y < v$ for some $v \in H$. From $x \leq y$ it then follows that $x < v$, which implies that $x \mu z$ in Γ'.

b) One of the elements x, y or z belongs to F while the other two belong to H'.

b1) $x \in F$ and $y, z \in H'$. Since $x \mu y$, we have $x < v$ in Γ for some $v \in H$. But then $x \mu z$ (the condition $y \mu z$ is not used here).

b2) $y \in F$ and $x, z \in H'$. This case is impossible since $x \mu y$ implies that $y > u$ in Γ for all $u \in H$ while $y \mu z$ implies that $y < u$ for some $v \in H$.

b3) $z \in F$ and $x, y \in H'$. Since $y \mu z$, we have $z > u$ in Γ for all $u \in H$. But then $x \mu z$ (the condition $x \mu y$ is not used here).

Thus μ is a partial ordering relation in Γ'. We now show that, in addition, Γ' is a semilattice relative to μ, i.e. for each pair of elements $x, y \in \Gamma'$ there exists a greatest lower bound $\inf(x, y)$. If x and y belong to one of the semilattices H' or F then it follows directly from the definition of μ that $\inf(x, y)$ exists and coincides with the intersection $x \wedge y$ in the corresponding semilattice. Suppose $x \in H'$ while $y \in F$. If $x \mu y$ then $\inf(x, y) = x$; if $y \mu x$ then $\inf(x, y) = y$. It therefore remains to consider the case when we have neither $x \mu y$ nor $y \mu x$. By virtue of the definition of μ this means that $y \in \Gamma_{\bar{\sigma}}(H)$. According to Lemma 2.1 all of the intersections $y \wedge h \ (h \in H)$ are equal to one and the same element, which belongs to F by virtue of Lemma 1.2. We denote this element here by $p(y)$. We have $p(y) < u$ for any $u \in H$ and $p(y) < y$. Therefore $p(y) \mu x$ and $p(y) \mu y$ in Γ'. Suppose $z \in \Gamma'$ and $z \mu x$, $z \mu y$. Since $y \in \Gamma_{\bar{\sigma}}(H)$ it follows from the definition of μ and the condition $z \mu y$ that $z \in F$. In this case $z < y$ in Γ, so that z cannot be greater than any element of H. From this result and the condition $z \mu x$ (implying that $z < v$ for some $v \in H$) it follows that z is less than all of the elements of H. But then $z \leq p(y)$ and $z \mu p(y)$ in Γ'. It follows from what has been proved that $p(y) = \inf(x, y)$ in Γ'.

Thus Γ' is a semilattice. The definition of μ directly implies that H' is a completely isolated homogeneous subsemilattice of Γ' as well as all of the other requirements contained in the first assertion of the lemma.

2. Suppose Γ' is a semilattice with the properties mentioned in the formulation of the lemma and H' is weakly isomorphic to H. Let F denote the semilattice $\Gamma \backslash H = \Gamma' \backslash H'$. Let ϕ denote a one-to-one mapping of Γ onto Γ' which coincides on H with a weak isomorphism of H onto H', and on F with the identity automorphism. We show that ϕ is a weak isomorphism of Γ

onto Γ'. It is clearly only necessary to consider a pair x, y, where $x \in H$ and $y \in F$. If $x \sigma y$ then $y \in \Gamma_\sigma(H)$. But inasmuch as $\Gamma_\sigma(H) = \Gamma'_\sigma(H')$, the elements $\phi(x) \in H'$ and $\phi(y) = y \in \Gamma'_\sigma(H')$ are comparable. If $x \bar{\sigma} y$ then $y \in \Gamma_{\bar{\sigma}}(H)$ and $x \wedge y = p(y)$ (see the notation in the first part of the proof). By definition $p(y)$ is the greatest of those elements of $\Gamma \backslash H$ which are less than y and less than all of the elements of H. It follows by virtue of our assumption that $p(y)$ has the same properties in Γ' (with H replaced by H'), i.e. $p(y) = h \wedge y$ for any $h \in H'$, and in particular $\phi(x) \wedge y = p(y)$. Hence $\phi(x \wedge y) = \phi(p(y)) = p(y) = \phi(x) \wedge y = \phi(x) \wedge \phi(y)$. This proves the second assertion of the lemma.

Remark. The semilattice Γ' in the above lemma is generally not uniquely determined to within an isomorphism. In fact, we could arrive at the same results by defining the relation μ for elements $x \in H'$ and $y \in F$ in part 1 of the proof of the lemma as follows: $x \mu y$ if and only if the element y is greater in Γ than some element of H; $y \mu x$ if and only if the element y is less in Γ than any element of H. We would then arrive at a semilattice which is generally not isomorphic to the semilattice obtained by the first method. But the uniqueness of Γ' to within an isomorphism does occur when H is a saturated subsemilattice. This will be proved in the next section (Lemma 3.3).

Suppose the semilattice T in Lemma 2.7 is obtained from H by permuting the components of some decomposition of H into an ordinal sum. We then say that Γ' *is obtained from Γ by permuting the components of the given decomposition of H into an ordinal sum.* In general, such a semilattice Γ' is not uniquely determined by the corresponding permutation, since, as was just indicated in the above remark, the semilattice Γ' of Lemma 2.7 is not uniquely determined. But for any such semilattices, by successively applying Lemma 1.4, Proposition 1.1 and Lemma 2.7, we obtain

Proposition 2.1. *A semilattice obtained from a semilattice Γ by any permutation of the components of some decomposition into an ordinal sum of an arbitrary completely isolated homogeneous subsemilattice of it is weakly isomorphic to Γ.*

§ 3. Saturated subsemilattices

Lemma 3.1. *If H is a proper saturated subsemilattice of a semilattice Γ, then the following alternative holds for an arbitrary element $x \in \Gamma_\sigma(H)$: either x is greater than all of the elements of H or x is less than all of the elements of H.*

Proof. We assume the contrary: there exists an element $x \in \Gamma_\sigma(H)$ such that $h_1 < x < h_2$ for certain $h_1, h_2 \in H$. Let F denote the set of all elements of Γ contained between h_1 and h_2. Clearly, $H \cup F$ is a subsemilattice. The homogeneity of H implies that $x \, \bar\sigma \, y$ for any $x \in F$ and $y \in \Gamma_{\underset{\sigma}{}}(H)$ (we note that $\Gamma_{\underset{\sigma}{}}(H) \neq \emptyset$, which follows directly from the fact that H is a proper saturated subsemilattice). It is not difficult to conclude from this result that $H \cup F$ is a completely isolated homogeneous subsemilattice with $\Gamma_{\underset{\sigma}{}}(H \cup F) = \Gamma_{\underset{\sigma}{}}(H)$. But since $H \cup F$ strictly contains H the latter equality contradicts the fact that H is saturated. The lemma is proved.

If, as usual, a subset of a partially ordered set which contains, together with any elements a and b, where $a < b$, every element x such that $a \leq x \leq b$, is said to be *convex*, then from Lemma 3.1 we obtain the following

Corollary. *Every saturated subsemilattice is convex.*

Since, by definition, a weak isomorphism of a semilattice takes comparable elements into comparable elements, incomparable elements into incomparable elements, and a subsemilattice into a subsemilattice (see the proof of Proposition 1.1), the corresponding definitions imply

Lemma 3.2. *Under a weak isomorphism of a semilattice the image of a completely isolated homogeneous subsemilattice is a completely isolated homogeneous subsemilattice, and the image of a saturated subsemilattice is a saturated subsemilattice.*

Lemma 3.3. *If the subsemilattice H in Lemma 2.7 is saturated, then, to within an isomorphism, there exists only one semilattice Γ' with the properties indicated in that lemma.*

Proof. Suppose Γ' and Γ'' are two such semilattices, i. e. H' and H'' are isomorphic and saturated (according to the preceding lemma) subsemilattices in Γ' and Γ'' respectively, with $\Gamma' \backslash H' = \Gamma'' \backslash H''$ (we denote this semilattice by F) and $\Gamma'_\sigma(H') = \Gamma''_\sigma(H'')$. Let ϕ denote a one-to-one mapping of Γ' onto Γ'' which coincides on H' with an isomorphsim of H' onto H'', and on F with the identity automorphism. We show that ϕ is an isomorphism. Let x and y be arbitrary elements of Γ'. In the case when x and y belong to one of the subsemilattices H' or F, the equality

$$\varphi(x \wedge y) = \varphi(x) \wedge \varphi(y) \tag{3}$$

follows directly form the definition of ϕ. Suppose $x \in H'$ while $y \in F$. If $y \in \Gamma'_{\underset{\sigma}{}}(H')$, equality (3) is established in the same way as at the end of the proof of Lemma 2.7. If, on the other hand, $y \in \Gamma'_\sigma(H')$, by virtue of Lemma 3.1 the alternative given in the formulation of this lemma holds for y. But then, by the construction of Γ' and Γ'', the corresponding case is satisfied for y as an element of Γ'', i.e., for example, if y is less in Γ' than any element of H' then y is less in Γ'' than any element of H'', and, in particular, $y < x$ and $y = \phi(y) < \phi(x)$, which implies equality (3). In the other case the conclusion is the same. The lemma is proved.

Corollary. *If the subsemilattice H in Lemma 2.7 is saturated, x is an arbitrary element of $\Gamma \backslash H = \Gamma' \backslash H'$, and h and h' are arbitrary elements of H and H' respectively, then x is found to be in the same relation $(<, >, \bar{\sigma})$ with h in Γ as it is with h' in Γ'.*

Lemma 3.4. *Let H be a completely isolated homogeneous subsemilattice of a semilattice Γ. There exists a greatest completely isolated homogeneous subsemilattice $F \subseteq \Gamma$ for which $\Gamma_{\bar{\sigma}}(F) = \Gamma_{\bar{\sigma}}(H)$. It will be saturated.*

Proof. In the case $H = \Gamma$ there is nothing to prove; we therefore assume that $H \neq \Gamma$. In the set $H \cup \Gamma_\sigma(H)$, which is easily seen to be a subsemilattice, we take a maximal completely isolated homogeneous subsemilattice containing H. Such a subsemilattice exists by virtue of Lemma 2.5. We denote it by $Q(H)$. We will show that $Q(H)$ satisfies the requirements of the lemma. Since $Q(H) \cap \Gamma_{\bar{\sigma}}(H) = \emptyset$ and $Q(H)$ is a homogeneous subsemilattice, every element of $\Gamma_{\bar{\sigma}}(H)$ is incomparable with an element of $Q(H)$. But if $x \notin Q(H) \cup \Gamma_{\bar{\sigma}}(H)$ then it is comparable with any element of H and hence with any element of $Q(H)$. This means that $\Gamma_{\bar{\sigma}}(Q(H)) = \Gamma_{\bar{\sigma}}(H)$.

Let F be an arbitrary completely isolated homogeneous subsemilattice for which $\Gamma_{\bar{\sigma}}(F) = \Gamma_{\bar{\sigma}}(H)$. We consider a subsemilattice $Q(F)$ analogous to $Q(H)$. In the same way as above, we get $\Gamma_{\bar{\sigma}}(Q(F)) = \Gamma_{\bar{\sigma}}(F)$, which, from the choice of F, implies

$$\Gamma_{\bar{\sigma}}(Q(F)) = \Gamma_{\bar{\sigma}}(Q(H)), \tag{4}$$

whence, in particular, $Q(F) \subseteq H \cup \Gamma_\sigma(H)$. Let us show that $Q(H) = Q(F)$. To this end, by virtue of the inclusion just proved it suffices to show that $Q(H) \cup Q(F)$ is a completely isolated homogeneous subsemilattice. We take arbitrary elements $x, y \in Q(H) \cup Q(F)$. If x and y belong to one of the subsemilattices $Q(H)$ or $Q(F)$ then, clearly, $x \wedge y \in Q(H) \cup Q(F)$. Suppose $x \in Q(H) \backslash Q(F)$ while $y \in Q(F) \backslash Q(H)$. Since $x \in Q(H)$, we have $x \notin \Gamma_{\bar{\sigma}}(Q(H))$, and hence, on the basis of equality (4), $x \in \Gamma_{\bar{\sigma}}(Q(F))$. From this result and the fact that $x \notin Q(F)$ it follows that x is comparable with any element of $Q(F)$. Consequently the elements x and y are comparable, and therefore their intersection, which is equal to one of them, belongs to

$Q(H) \cup Q(F)$. Thus $Q(H) \cup Q(F)$ is a subsemilattice. By virtue of Lemma 1.1 it is completely isolated. The homogeneity of $Q(H) \cup Q(F)$ follows directly from equality (4). Thus $Q(H) = Q(F)$, and $Q(H) \supseteq F$. By the same token it is shown that $Q(H)$ contains any completely isolated homogeneous subsemilattice F for which $\Gamma_{\bar{\sigma}}(F) = \Gamma_{\bar{\sigma}}(H)$. Let us show, finally, that $Q(H)$ is a saturated subsemilattice. Suppose P is a completely isolated homogeneous subsemilattice such that $Q(H) \subset P$. By virtue of the definition of $Q(H)$ it follows that $P \cap \Gamma_{\bar{\sigma}}(H) \neq \emptyset$, which in turn implies that $\Gamma_{\bar{\sigma}}(P) \subset \Gamma_{\bar{\sigma}}(H) = \Gamma_{\bar{\sigma}}(Q(H))$. This completes the proof of the lemma.

The notation $Q(H)$ introduced in the proof of Lemma 3.4 will be used in the sequel. By viture of the equality $\Gamma_{\bar{\sigma}}(Q(H)) = \Gamma_{\bar{\sigma}}(H)$ we have

Corollary 1. *If H is a saturated subsemilattice, then $Q(H) = H$.*

If $H \neq Q(H)$ then each element of $Q(H) \backslash H$ is comparable with any element of H. We therefore have

Corollary 2. *An arbitrary component of the irreducible decomposition of H into an ordinal sum will be a component of the irreducible decomposition of $Q(H)$ into an ordinal sum.*

This assertion in turn implies

Corollary 3. *If F is a saturated subsemilattice whose irreducible decomposition into an ordinal sum consists of a finite number of components, then there exist only a finite number of completely isolated homogeneous subsemilattices H for which $Q(H) = F$.*

Proposition 3.1. *If H and F are arbitrary distinct saturated subsemilattices of a given semilattice, then either H and F do not intersect or else one of them is strictly contained in a component of the irreducible decomposition of the other into an ordinal sum.*

Proof. Suppose $H \cap F \neq \emptyset$. Then Lemmas 2.2 and 1.1 imply that $H \cup F$ is a completely isolated homogeneous subsemilattice. If neither H nor F is contained in the other, then from the proof of Lemma 2.2 it follows that $\Gamma_{\bar{\sigma}}(H) = \Gamma_{\bar{\sigma}}(F) = \Gamma_{\bar{\sigma}}(H \cup F)$. This contradicts the fact that H and F are saturated. Thus one of them is contained in the other. Suppose, for definiteness, $H \subset F$. If F is indecomposable then there is nothing to prove. Suppose F is decomposable and A is a component of the irreducible decomposition of F which has a nonempty intersection with H. Let us show that $H \subseteq A$. We assume the contrary: $H \cap B \neq \emptyset$ for some component B other than A of the irreducible decomposition of F. Since an arbitrary element $x \in F \backslash H$ is comparable with each element of at least one of the components A or B, it is comparable with an element of H by virtue of the homogeneity of H. Hence $\Gamma_{\underset{\sigma}{}}(H) = \Gamma_{\underset{\sigma}{}}(F)$, and this contradicts the fact that H is saturated. Thus $H \subseteq A$. But it is easily seen that we have the strict inclusion $H \subset A$. For otherwise an arbitrary element of $F \backslash H$ would be comparable with any element of H, which, as was just noted, leads to a contradiction.

The proposition is proved.

Lemma 3.5. *Suppose H and F are completely isolated homogeneous subsemilattices of a given semilattice. If $H \subseteq F$ then $Q(H) \subseteq Q(F)$.*

Proof. Since $H \subseteq Q(H)$ and $F \subseteq Q(F)$, the condition $H \subseteq F$ implies that $Q(H) \cap Q(F) \neq \emptyset$. It follows on the basis of Proposition 3.1 that one of the semilattices $Q(H)$ or $Q(F)$ is contained in the other. Let us assume that $Q(F) \subset Q(H)$. Then $H \neq Q(H)$, which implies according to Corollary 2 of Lemma 3.4 that $Q(H)$ is decomposable and H is the union of a set of components of the irreducible decomposition of $Q(H)$. On the other hand, by virtue of Proposition 3.1, $Q(F)$ is strictly contained in one of the components of the indicated decomposition, and since $H \subset Q(F)$ it follows that H is strictly contained in the same component. The resultant contradiction shows that $Q(H) \subseteq Q(F)$.

Lemmas 3.4 and 3.5 and Corollary 1 of Lemma 3.4 imply

Lemma 3.6. *The operation on the set of completely isolated homogeneous subsemilattices of a given semilattice that takes H into $Q(H)$ is a closure operation.*

§ 4. The canonical weak isomorphism

The aim of the present section is to prove a proposition which generalizes Proposition 2.1 to the case of any set of completely isolated homogeneous subsemilattices of a given semilattice.

It is necessary for us to begin with some preliminary remarks.

First, it is clear that we will suffer no loss of generality by considering only permutations of components of the *irreducible* decomposition of completely isolated homogeneous subsemilattices into an ordinal sum. This will always be borne in mind in the sequel.

Second, by virtue of Corollary 2 of Lemma 3.4 we need only consider permutations of the components of the irreducible decomposition of *saturated* subsemilattices. This requirement is in general even necessary, and here is the reason. A component of the irreducible decomposition into an ordinal sum of a completely isolated homogeneous subsemilattice H which is not saturated can turn out to be a component of the analogous decomposition of a different completely isolated homogeneous subsemilattice F; therefore not all of the permutations of the components of the irreducible decompositions of H and F will be simultaneously possible. But the situation is different for saturated subsemilattices by virtue of Proposition 3.1.

Thus we will show that, for any system $[H_\alpha]$ (α ranges over a set of indices A) of saturated subsemilattices of a semilattice Γ, a semilattice obtained from Γ by arbitrarily permuting the components of the irreducible decomposition of the subsemilattices of $[H_\alpha]$ will be weakly isomorphic to Γ. It is necessary, of course, to substantiate the correctness (i.e., essentially, to give a precise definition) of the notion of such a semilattice. Clearly, in the substantiation we need only consider the case of an infinite system of subsemilattices $[H_\alpha]$, since in the case of a finite system the indicated generalization is obtained simply by successively applying Proposition 2.1. In the general case this substantiation can be carried out in the following manner. For a given system $[H_\alpha]$ we let $[\psi_\alpha]$ denote a corresponding system of permutations: for any $\alpha \in A$, ψ_α is a one-to-one mapping of the ordered set of components of the irreducible decomposition of H_α onto a set of the same components arranged in a new order (among the ψ_α there can be identity maps). We will say that elements $x, y \in \Gamma$ *participate in the permutation* ψ_α if they belong to different components of the irreducible decomposition of H_α. Proposition 3.1 implies the following assertion.

1. *Arbitrary elements $x, y \in \Gamma$ can participate in not more than one permutation.*

We also note some other assertions.

2. *If $x \,\bar\sigma\, y$ then x and y do not participate in any permutation while $x \wedge y$ does not participate in any permutation with either x or y.*

In this assertion the first conclusion is obvious, while the second follows from the fact that a component of a decomposition into an ordinal sum of a completely isolated homogeneous subsemilattice of a semilattice Γ will be a completely isolated subsemilattice of Γ by virtue of Lemmas 1.3 and 2.6.

3. *If $x \,\bar\sigma\, y$ while z is an arbitrary element that is comparable with both x and y, then x and z participate in a permutation if and only if y and z participate in the same permutation.*

For suppose x and z participate in a permutation ψ_α of the components of H_α. Since $x \,\bar\sigma\, y$, either x and y belong to one and the same component or $y \in \Gamma_{\bar\sigma}(H_\alpha)$. But the latter is impossible inasmuch as $z \in H_\alpha$ while by assumption $y \, \sigma \, z$. Consequently the first part of the alternative holds, which means that y and z participate in the permutation ψ_α. The converse follows from the fact that x and y play the same role.

4. *If x and y do not participate in a given permutation and x' and y' are the images of x and y in the semilattice Γ' obtained from Γ by this permutation (see the definition before Proposition 2.1), then x' is found to be in the same relation $(\leq, \geq, \bar{\sigma})$ with y' in Γ' as x is with y in Γ.*

For suppose x and y do not participate in a permutation ψ_α. This means that either x and y belong to a component of the irreducible decomposition of H_α or x and y do not belong to H_α (and in these two cases the assertion is obvious), or one of the elements x or y belongs to H_α while the other does not belong to it. In the last case the assertion follows from the corollary of Lemma 3.3.

We introduce the following binary relation μ on the set Γ.

If x and y do not participate in any of the permutations of $[\psi_\alpha]$ we put $x \, \mu \, y$ if and only if $x \leq y$.

If x and y participate in a permutation ψ_α we put $x \, \mu \, y$ if and only if the image of x in the semilattice obtained from Γ by this permutation ψ_α is less than the image of y.

The uniqueness of μ in the second case follows from assertion 1 cited above.

The reflexivity and antisymmetry of the relation μ are obvious. Let us prove that it is transitive. Suppose $x \, \mu \, y$ and $y \, \mu \, z$. We make all of the permutations of $[\psi_\alpha]$ in which x and y, x and z, and y and z participate. By virtue of assertion 1 there are no more than three such permutations (it can actually be shown that there are no more than two). Applying assertions 1 and 4, we get that the same relations relative to μ hold among x, y and z as among their images in the semilattice obtained from Γ by the indicated permutations. It therefore follows that $x \, \mu \, z$. Thus μ is a partial ordering relation. We denote the corresponding partially ordered set (consisting of the same elements as Γ) by Γ'. Let us show that Γ' is a semilattice. If $x \, \mu \, y$ then $\inf_\mu (x, y) = x$. Suppose x and y are not comparable relative to μ. Then $x \, \bar{\sigma} \, y$ by the definition of μ, and therefore (according to assertion 2) $x \wedge y$ does not participate in any permutation with either x or y. It follows that $x \wedge y \, \mu \, x$ and $x \wedge y \, \mu \, y$. Suppose $z \, \mu \, x$ and $z \, \mu \, y$. We consider the possible permutations in which x and y, x and z, and y and z participate. By virtue of assertions 1–3 there is at most one such permutation. If there is no such permutation, obviously $z \, \mu \, x \wedge y$. If, on the other hand, there is one such permutation, according to assertions 2 and 3 the elements x, y and $x \wedge y$ belong to one of the components of the irreducible decomposition of some subsemilattice H_α while z belongs to another of these components; hence $z \, \mu \, x \wedge y$. This means that $\inf_\mu (x, y) = x \wedge y$.

Thus Γ' is a semilattice. We call it *the semilattice obtained from* Γ *by the permutations* $[\psi_\alpha]$. We note that Γ' is uniquely defined for a given system of permutations $[\psi_\alpha]$. In the case when $[\psi_\alpha]$ consists of only one permutation it is easily seen that Γ' is isomorphic to the corresponding semilattice of Proposition 2.1 (this semilattice is by virtue of Lemma 3.3 unique to within an isomorphism). Since, in addition, it is true that (i) x and y are comparable in Γ if and only if they are comparable in Γ', and (ii) $x \wedge y = \inf_\mu (x, y)$ for incomparable x and y, we have

Proposition 4.1. *If a semilattice* Γ' *is obtained from a semilattice* Γ *by permuting the components of the irreducible decomposition of some of its saturated subsemilattices into ordinal sums then the mapping which takes any* $x \in \Gamma$ *into the same element considered as an element of* Γ' *is a weak isomorphism of* Γ *onto* Γ'.

The weak isomorphism of Proposition 4.1 is called the *canonical* weak isomorphsim of Γ onto Γ'.

Remark. Suppose a semilattice Γ' is obtained from a semilattice Γ by permutations in saturated subsemilattices H_α $(\alpha \in A)$. Since the relation μ coincides with the relation $\leq$ on the set of all elements of Γ not belonging to one of the subsemilattices H_α (this set is obviously a subsemilattice), we essentially get that *the canonical weak isomorphsim of* Γ *onto* Γ' *induces the identity automorphsim on the indicated subsemilattice, which is contained in both* Γ *and* Γ'.

§ 5. N-closure

Let X be a nonempty subset of a partially ordered set T. Put $N_0(X, T) = X$. Proceeding inductively, let $N_{i+1}(X, T)$ denote the set of all elements of T which are incomparable with at least one element of $N_i(X, T)$ (if $N_{i+1}(X, T)$ is empty, the process terminates at this point). The union of all of the $N_i(X, T)$ for $i = 0, 1, \cdots$ is denoted by $N(X, T)$. We call $N(X, T)$ the N-*closure* of X in T. When T remains the same, we will simply say "N-closure" and denote it by $N(X)$. Clearly this operation satisfies the usual axioms for a closure operation; in particular, $N(N(X)) = N(X)$. When X consists of one element x, we will denote the N-closure of this set by $N(x)$ (respectively, $N(x, T)$) and call it the N-closure of the element x.[*] If $N(X) = X$ then the set X is said to be *N-closed* (more precisely, *N*-closed in T). In particular, we obtain the notion of an *N*-closed subsemilattice in a given semilattice. Clearly, X will be *N*-closed in T if and only if every element of $T \backslash X$ is comparable with every element of X. In particular, the components of a decomposition of T into an ordinal sum of partially ordered sets are *N*-closed in T.

Lemma 5.1. *The classes $N(x)$ are the components of the irreducible decomposition of a corresponding partially ordered set into an ordinal sum of partially ordered sets.*

Proof. Each of these classes, as is easily seen, is indecomposable into an ordinal sum. Let $N(x)$ and $N(y)$ be two distinct arbitrary classes. We will show that either each element of $N(x)$ is less than any element of $N(y)$, or each element of $N(y)$ is less than any element of $N(x)$. We assume the contrary: there exist elements $u_1, u_2 \in N(x)$ and elements $v_1, v_2 \in N(y)$ such that $u_1 < v_1$ and $u_2 > v_2$. Since $u_2 \, \sigma \, v_1$, two cases are possible: 1) $u_2 > v_1$ or 2) $u_2 < v_1$. We consider the first. Here $u_1 < v_1 < u_2$, and since v_1 is comparable with any element of the class $N(x)$, this class divides into two nonempty subsets, viz. the set A of all elements of $N(x)$ that are less than v_1, and the set B of elements that are greater than v_1. $N(x)$ is the ordinal sum of A and B, which leads to a contradiction. In the second case we obtain an analogous contradiction from $v_2 < u_2 < v_1$. This proves the lemma.

Lemma 5.2. *In order for a subsemilattice H of a semilattice Γ to be a component of the irreducible decomposition of Γ into an ordinal sum of semilattices, it is necessary and sufficient that H be an N-closed completely isolated semilattice that is indecomposable into an ordinal sum.*

Proof. The necessity follows directly from the corresponding definitions. Let us prove the sufficiency. In the case $H = \Gamma$ there is nothing to prove; we therefore assume that $H \neq \Gamma$. Suppose H satisfies the three conditions indicated in the formulation of the lemma. The first condition means that any element of $\Gamma \backslash H$ is comparable with each element of H. By virtue of the third condition an arbitrary element of $\Gamma \backslash H$ is then either less than each element of H or greater than each element of H (this is shown in the same way as in the proof of Lemma 5.1). Hence, by virtue of the second condition,

---

[*] In terms of the theory of binary relations, we can give the following definition of $N(X)$. Let ν denote the equivalence generated by the relation $\bar{\sigma}$. Then $N(X)$ is none other than the saturation of X with respect to ν (see [17]). In particular, $N(x)$ is the equivalence class with respect to ν which contains x. Thus $N(X) = \mathbf{U}_{x \in X} N(x)$.

which means that $\Gamma \backslash H$ is a subsemilattice, we see that H is a component of a decomposition of Γ into an ordinal sum of semilattices: if A is the subsemilattice of all elements that are less than each element of H while B is the subsemilattice of all elements that are greater than each element of H (at least one of these subsemilattices is not empty), then Γ is decomposable into the ordinal sum of A, H and B. But then by the third condition H will be a component of the irreducible decomposition of Γ into an ordinal sum of semilattices.

Lemma 5.3. *Let X be a generating set of a subsemilattice H of a semilattice Γ. If X is N-closed then so is H.*

Proof. We must show that any element of $\Gamma \backslash H$ is comparable with each element of H. We represent an arbitrary element $h \in H$ in the form of an intersection $h = x_1 \wedge \cdots \wedge x_m$, where $x_i \in X (i = 1, \cdots, m)$. If $y \in \Gamma \backslash H$ then $y \in \Gamma \backslash X$, and, by assumption, $y \, \sigma \, x_i$ for all $i = 1, \cdots, m$. If $y \leq x_i$ for all i then $y \leq h$, while if $y > x_{i_0}$ for some i_0 then a fortiori $y > h$. Thus $y \, \sigma \, h$. Q.E.D.

Let $\overline{N}(X, \Gamma)$ denote the least N-closed subsemilattice of a semilattice Γ which contains a set X. Such a subsemilattice obviously exists. In accordance with the remark made above we will employ below the more concise notation $\overline{N}(X)$. The notation $\overline{N}(x)$ or the more explicit notation $\overline{N}(x, \Gamma)$ is employed analogously. Clearly, $N(X) \subseteq \overline{N}(X)$. Therefore Lemma 5.3 implies

Lemma 5.4. *$\overline{N}(X)$ coincides with the subsemilattice generated by the set $N(X)$.*

Lemma 5.5. *A subsemilattice $\overline{N}(x)$ either coincides with the set $N(x)$ or contains only one element not belonging to $N(x)$, which will be the least element in $\overline{N}(x)$.*

Proof. Suppose $\overline{N}(x) \neq N(x)$ and $y, z \in \overline{N}(x) \backslash N(x)$. By virtue of Lemma 5.4 each of the elements y or z is decomposable and less than a corresponding element of $N(x)$, and hence, by Lemma 5.1, less than any element of $N(x)$. Since $y = x_1 \wedge \cdots \wedge x_m$, where $x_i \in N(x)$ $(i = 1, \cdots, m)$, and $z < x_i$ for all i, we have $z \leq y$. But it can be shown in a completely analogous manner that $y \leq z$. Therefore $y = z$, which proves the lemma.

Corollary. *The element $y \in \overline{N}(x) \backslash N(x)$ generates a single element N-closed subsemilattice.*

The set of all such elements of a semilattice Γ is denoted by $C(\Gamma)$. In other words, $C(\Gamma)$ is the set of all of the decomposable elements of Γ which are comparable with any element of Γ.

The role of the subsemilattices $\bar{N}(x)$ is explained by the following fact.

Lemma 5.6. *In a semilattice* Γ *the subsemilattices* $\bar{N}(x)$ *for* $x \in \Gamma \setminus C(\Gamma)$ *are the components of the irreducible decomposition of* Γ *into an ordinal sum of semilattices.*

Proof. Inasmuch as $\bar{N}(x)$ is an N-closed subsemilattice, we must verify that the two other conditions indicated in Lemma 5.2 are satisfied. We show that $\bar{N}(x)$ is indecomposable into an ordinal sum of semilattices. If $\bar{N}(x) = N(x)$, this is immediately implied by Lemma 5.1. On the other hand, if $\bar{N}(x) \neq N(x)$, it is implied by the following: $N(x)$ must be contained in a component of an arbitrary decomposition of $\bar{N}(x)$ into an ordinal sum of semilattices by virtue of Lemma 5.1, and hence $\bar{N}(x)$ must be contained in the same component by Lemma 5.4.

It remains to show that $\bar{N}(x)$ is completely isolated. We assume the contrary: there exist elements y, $z \in \Gamma \setminus \bar{N}(x)$ such that $y \, \bar{\sigma} \, z$ and $y \wedge z \in \bar{N}(x)$. From the preceding it then follows that y and z are greater than any element of $\bar{N}(x)$, which in turn implies that $y \wedge z \geq u$ for any $u \in \bar{N}(x)$. Thus $\bar{N}(x)$ has a greatest element $y \wedge z$. Hence, if $N(x)$ consists of more than one element, $\bar{N}(x)$ decomposes into an ordinal sum of semilattices: the single element subsemilattice consisting of $y \wedge z$ and the subsemilattice of all of the other elements of $\bar{N}(x)$. But this is impossible. Consequently $\bar{N}(x)$ consists of one element x, and hence $y \wedge z = x$, i.e. the element x is decomposable. But then $x \in C(\Gamma)$, which contradicts our assumption. It follows that $\bar{N}(x)$ is completely isolated.

Thus all of the conditions of Lemma 5.2 are satisfied for $\bar{N}(x)$, which completes the proof.

Since any weak isomorphism of a semilattice takes comparable elements into comparable elements, it also takes every N-closed subsemilattice into an N-closed subsemilattice as well as a minimal N-closed subsemilattice into a minimal N-closed subsemilattice. But the minimal N-closed subsemilattices are the subsemilattices $\bar{N}(x)$ for all possible x. In addition, it is easily seen that a weak isomorphism of a semilattice Γ onto a semilattice Γ' takes the elements of $C(\Gamma)$ into the elements of $C(\Gamma')$. Comparing all of what has been said and taking Lemma 5.6 into account, we obtain the following proposition, which is the basic result of the present section.

Proposition 5.1. *If two semilattices* Γ *and* Γ' *are weakly isomorphic, then under any weak isomorphism of* Γ *onto* Γ' *the images of the components of the irreducible decomposition of* Γ *into an ordinal sum of semilattices are the components of the irreducible decomposition of* Γ' *into an ordinal sum* *of semilattices.*

§ 6. D-closure

Let X be a nonempty subset of a semilattice Γ. Put $D_0(X) = X$. Proceeding inductively, let $D_{i+1}(X)$ denote the set of all proper divisors of elements of $D_i(X)$ (if $D_{i+1}(X)$ is empty then the process terminates at this point). The union of all of the $D_i(X)$ is denoted by $D(X)$ and called the *D-closure* of X. We could employ the notation $D(X, \Gamma)$ to emphasize that the D-closure is taken in the semilattice Γ, but there is no need for us to do this in the sequel since the D-closures below will be taken only in the whole semilattice (at the same time, for example, as N-closures are being taken in certain subsemilattices). The D-closure obviously satisfies the usual axioms for a closure operation. The D-closure of a set consisting of a single element x is denoted by $D(x)$. Clearly, $D(X) = \mathbf{U}_{x \in X} D(x)$. If $D(X) = X$ then X is said to be *D-closed*. We correspondingly obtain the notion of a D-closed subsemilattice.

From the corresponding definitions there immediately follows

Lemma 6.1. *Any D-closed subsemilattice is completely isolated.*

Let $S(x)$ denote the set of all of the elements y of a given semilattice such that $y \geq x$. Clearly, $S(x)$ is a D-closed subsemilattice, and $D(x) \subseteq S(x)$. $D(x)$ contains only the element x if and only if x is indecomposable.

Lemma 6.2. *If $D(x) \neq S(x)$ then $S(x)$ decomposes into the ordinal sum of the subsemilattices $D(x)$ and $S(x) \backslash D(x)$.*

Proof. By virtue of Lemma 6.1 the set $S(x) \backslash D(x)$ is a subsemilattice, and therefore we need only show that any element of $S(x) \backslash D(x)$ is greater than each element of $D(x)$ (it will then automatically follow that $D(x)$ is a subsemilattice). We assume the contrary: there exist elements $s \in S(x) \backslash D(x)$ such that s is not greater than y. Two cases are consequently possible.

1) $s < y$. Then $y \in D_i(x)$ for some $i \geq 1$ (see the definition at the beginning of this section). Consequently there exists an element $z \in D_i(x)$ such that $y \bar{\sigma} z$ and $y \wedge z = y_1 \in D_{i-1}(x)$. We will assume without loss of generality that i is the least of the numbers with the property that s is less than some of the elements of $D_i(x)$. Then $s < y_1$ is impossible, which implies, in particular, that $s \bar{\sigma} z$. If $s > y_1$ then $s \wedge z = (s \wedge y) \wedge z = s \wedge (y \wedge z) = s \wedge y_1 = y_1$, and hence, by definition, $s \in D_i(x) \subseteq D(x)$, which leads to a contradiction. Therefore $s \bar{\sigma} y_1$. We put $s \wedge y_1 = s_1$ and consider the element $s_1 \in S(x)$. First of all, $s_1 \notin D(x)$, since otherwise we

would get by definition that $s \in D(x)$. Since $s_1 < y_1 \in D_{i-1}(x)$, the least of the numbers j with the property that s_1 is less than some of the elements of $D_j(x)$ is strictly less than i. Repeating arguments analogous to those just made for s, we arrive at an element $s_2 \in S(x) \backslash D(x)$ for which the least of the numbers j with the property that s_2 is less than some of the elements of $D_j(x)$ is strictly less than the corresponding number for s_1. Thus, by continuing this process, we obtain an infinite decreasing sequence of natural numbers, which is impossible. Consequently case 1) is impossible.

2) $s \, \bar{\sigma} \, y$. Then $s \wedge y \notin D(x)$, since otherwise we would have $s \in D(x)$, in view of the fact that s is a proper divisor of $s \wedge y$. But $s \wedge y < y$, and this, as was just shown, leads to a contradiction. Consequently case 2) is also impossible.

Thus any element of $S(x) \backslash D(x)$ is greater than each element of $D(x)$, which proves the lemma.

It is easily seen that the subsemilattice $D(x)$ is indecomposable into an ordinal sum of semilattices. From this observation and Lemma 6.2 we obtain the following assertion.

Lemma 6.3. $D(x)$ *is the (first) component of the irreducible decomposition of* $S(x)$ *into an ordinal sum of semilattices.*

Lemma 6.4. *If* $N(x)$ *consists of more than one element (i.e. the corresponding semilattice contains elements that are incomparable with* x*), then* $D(x) \subseteq N(x)$.

Otherwise $N(x) \subseteq D(x)$, *and if* x *is decomposable then* $D(x) = \bar{N}(y)$ *for any other element* $y \in D(x)$.

Proof. Suppose $N(x)$ consists of more than one element. If $D(x)$ consists of only the element x then the inclusion $D(x) \subseteq N(x)$ is trivial. Suppose $D(x)$ contains elements other than x, and y is an arbitrary one of these elements. Recalling the definition of $D(x)$, we conclude then that there exist elements $z, z_1, \cdots, z_m$ such that $y \, \bar{\sigma} \, z$, $y \wedge z = y_1 \, \bar{\sigma} \, z_1, \cdots$ $\cdots, y_{m-1} \wedge z_{m-1} = y_m \, \bar{\sigma} \, z_m$ and $y_m \wedge z_m = x$. Clearly $z, z_1, \cdots, z_m \in D(x)$. Suppose $y \notin N(x)$. Then, by virtue of the fact that $N(x)$ is N-closed, y is comparable with any element of $N(x)$, and hence $z \notin N(x)$. Since $y > x$ and $z > x$ it follows by virtue of Lemma 5.1 that y and z are greater than any element of $N(x)$, whence $y_1 \geq u$ for any $u \in N(x)$. Inasmuch as $N(x)$ contains more than one element, $y_1 \notin N(x)$, since otherwise y_1 would be the greatest element in $N(x)$ and $N(x)$ would decompose into an ordinal sum of partially ordered sets, which contradicts Lemma 5.1. Repeating these arguments now, we get $z_1 \notin N(x)$ and $y_2 = y_1 \wedge z_1 \notin N(x)$, and, continuing by induction, we get $y_m \notin N(x)$, $z_m \notin N(x)$ and $y_m \wedge z_m = x \notin N(x)$. But the latter is impossible. This contradiction shows that $y \in N(x)$, which proves the first assertion of the lemma.

Let us prove the second assertion. Suppose $N(x)$ consists of only the element x, i.e. x is comparable with any element of the corresponding semilattice Γ. Then $N(x) \subseteq D(x)$. Suppose x is at the same time decomposable, i.e. $x \in C(\Gamma)$. Recalling the equivalent property of the elments of $C(\Gamma)$, we have $x \in \overline{N}(z) \setminus N(z)$ for some $z \in \Gamma$, with $x < y$ for any $y \in N(z)$, so that $\overline{N}(z) \subseteq S(x)$. Since $z \notin C(\Gamma)$ it follows by virtue of Lemma 5.6 that $\overline{N}(z)$ will be a component of the irreducible decomposition of Γ, and hence also of $S(x)$, into an ordinal sum of semilattices. This result together with the condition $\overline{N}(z) \cap D(x) \neq \emptyset$ and Lemma 6.3 implies that $D(x) = \overline{N}(z)$. Since $\overline{N}(z) = \overline{N}(y)$ for any $y \in N(z)$, we get $D(x) = \overline{N}(y)$. The lemma is proved.

Corollary. *For any element x of a given semilattice the set $\overline{N}(x) \cup D(x)$ is a subsemilattice that is equal to either $\overline{N}(x)$ or $D(x)$.*

We now formulate the basic result of the present section.

Proposition 6.1. *For any element x of a given semilattice the subsemilattice $\overline{N}(x) \cup D(x)$ is completely isolated, N-closed and indecomposable into an ordinal sum of semilattices.*

Proof. The fact that $\overline{N}(x) \cup D(x)$ is completely isolated is obtained by successively applying the corollary of Lemma 6.4 as well as Lemmas 5.6, 1.3 and 6.1. The fact that it is N-closed is obtained by applying the corollary of Lemma 6.4 and Lemma 6.4. Finally, its indecomposability follows from the corollary of Lemma 6.4 as well as Lemmas 5.6 and 6.3.

Proposition 6.1 implies by virtue of Lemma 5.2 the following

Corollary. *For any element x of a semilattice Γ the subsemilattice $\overline{N}(x) \cup D(x)$ is a component of the irreducible decomposition of Γ into an ordinal sum of semilattices.*

§ 7. Irregular pairs under weak isomorphisms

Let ϕ be a weak isomorphism of a semilattice Γ. A pair of elements $a, b \in \Gamma$ is said to be *regular* relative to ϕ if $\phi(a \wedge b) = \phi(a) \wedge \phi(b)$; otherwise it is said to be *irregular* relative to ϕ. It follows from the definition of a weak isomorphism that if the elements a and b form an irregular pair relative to ϕ then they are comparable and the following alternative holds: either 1) $a < b$ but $\phi(a) > \phi(b)$, or 2) $a > b$ but $\phi(a) < \phi(b)$.

This should be borne in mind in the sequel. An element $a \in \Gamma$ is said to be *regular* relative to ϕ if it forms a regular pair with any element of Γ; otherwise it is said to be *irregular* relative to ϕ. Clearly, ϕ will be an isomorphism if and only if all of the elements of Γ are regular relative to ϕ.

Everywhere below we will be considering a fixed weak isomorphism ϕ of a semilattice Γ, and when speaking of the regularity or irregularity of a given pair we will drop the words "relative to ϕ." The letters x, y, z and u in the lemmas given below denote elements of the semilattice Γ.

Lemma 7.1. *Suppose* $x \,\bar{\sigma}\, y$, $x > z$ *and* $y > z$. *The pair* (x, z) *will be regular if and only if the pair* (y, z) *is regular.*

Proof. Suppose one of the pairs (x, z) or (y, z) is regular while the other is irregular; for the sake of definiteness suppose (x, z) is regular while (y, z) is irregular. Then $\phi(x) > \phi(z)$ while $\phi(y) < \phi(z)$, and hence $\phi(x) > \phi(y)$. But the latter inequality implies that $x \,\sigma\, y$, which contradicts the hypothesis of the lemma. Q.E.D.

A completely analogous proof exists for

Lemma 7.2. *Suppose* $x \,\bar{\sigma}\, y$, $x < z$ *and* $y < z$. *The pair* (x, z) *will be regular if and only if the pair* (y, z) *is regular.*

Lemma 7.3. *Suppose* $z \notin N(x)$. *If* x *and* z *form a regular (irregular) pair then* z *forms a regular (irregular) pair with any element* $y \in N(x)$.

Proof. Lemma 5.1 implies that z is either less than each element of $N(x)$ or greater than each element of $N(x)$. Recalling the definition of $N(x)$, we directly obtain both assertions of the lemma: the first assertion by successively applying Lemma 7.1 and the second by successively applying Lemma 7.2.

Lemma 7.4. *If* $y \,\bar{\sigma}\, z$ *and* $x = y \wedge z$, *then the pairs* (x, y) *and* (x, z) *are regular.*

Proof. Since $y \,\bar{\sigma}\, z$, the elements y and z form a regular pair, i.e. $\phi(y \wedge z) = \phi(y) \wedge \phi(z)$. Hence $\phi(x \wedge y) = \phi(y \wedge z \wedge y) = \phi(y \wedge z) = \phi(y) \wedge \phi(z) = \phi(y) \wedge \phi(z) \wedge \phi(y) = \phi(y \wedge z) \wedge \phi(y) = \phi(x) \wedge \phi(z)$. Analogously, $\phi(x \wedge z) = \phi(x) \wedge \phi(z)$.

Lemma 7.5. *If* $y \,\bar{\sigma}\, z$, $x = y \wedge z$, $u \,\sigma\, y$ *and* $u \,\sigma\, z$, *then the pair* (u, x) *will be regular if and only if the pairs* (u, y) *and* (u, z) *are regular.*

Proof. By virtue of the preceding lemma the case $u = x$ is trivial. Suppose $u \neq x$. Since $y \,\bar{\sigma}\, z$, either 1) $u > y$ and $u > z$ or 2) $u < y$, $u < z$ and hence $u < x$. Therefore the elements u, y, z and x form a subsemilattice whose irreducible decomposition into an ordinal sum of semilattices consists of two components, one of which contains u and the other of which contains the remaining three elements. It follows on the basis of Proposition 5.1 that $\phi(u)$ is either less than each of the elements $\phi(y)$, $\phi(z)$, $\phi(x)$ or greater than each of these elements. This result implies the assertion of the lemma.

Recalling now the definition of $D(x)$ (see the preceding section), using Lemma 7.4 and 7.5 and proceeding by induction, we obtain the following assertion.

Lemma 7.6. *An element x forms a regular pair with any $y \in D(x)$.*

Lemma 7.7. *Suppose $z \notin N(x)$. If x and z form a regular (irregular) pair, than z forms a regular (irregular) pair with any $y \in \overline{N}(x)$.*

Proof. We take advantage of Lemma 5.5. If $\overline{N}(x) = N(x)$ then the required assertion simply follows from Lemma 7.3. Suppose the other case holds, i.e. $\overline{N}(x) \backslash N(x)$ consists of an element u such that $u = y_1 \wedge y_2$, where y_1, $y_2 \in N(x)$. The element z is comparable with any element of $\overline{N}(x)$ inasmuch as $z \notin \overline{N}(x)$. Since, in addition, $y_1 \, \overline{\sigma} \, y_2$, the conditions of Lemma 7.5 are satisfied. Consequently the pair (z, u) is regular if and only if the pairs (z, y_1) and (z, y_2) are regular. But by Lemma 7.3 the regularity (irregularity) of the pair (z, y_1) implies the regularity (irregularity) of the pair (z, y) for all $y \in N(x)$. From this result we immediately obtain both assertions of the lemma.

Lemma 7.8. *Suppose $z \notin D(x)$. If x and z form a regular (irregular) pair, then z forms a regular (irregular) pair with any $y \in D(x)$.*

Proof. Two cases are possible.

1) $z \, \sigma \, x$. Here either $z < x$, in which case z is less than any element of $D(x)$, or $z > x$, in which case z is greater than any element of $D(x)$ by Lemma 6.2. Therefore, by virtue of Lemma 6.3, the irreducible decomposition of the subsemilattice $\{z, D(x)\}$ into an ordinal sum of semilattices consists of two components: $\{z\}$ and $D(x)$. In this case, according to Proposition 5.1 we see that $\phi(z)$ is either less than any element of $\phi(D(x))$ or greater than any element of $\phi(D(x))$. From this result we obtain both assertions of the lemma.

2) $z \, \overline{\sigma} \, x$. Here x and z automatically form a regular pair. Let y an arbitrary element of $D(x)$. We show that y and z form a regular pair. If $z \, \overline{\sigma} \, y$ this immediately follows from the definition. If $z \, \sigma \, y$ then $z < y$ by virtue of the condition $z \, \overline{\sigma} \, x$, and we obtain the required assertion by successively applying Lemmas 7.6 and 7.2.

The lemma is proved.

The main goal of the present section is a proof of the following important result.

Proposition 7.1. *If a pair of elements* a *and* b *of a semilattice* Γ *is irregular relative to a weak isomorphism* ϕ *of this semilattice, then the isolated homogeneous subsemilattice generated by* a *and* b *is decomposable into an ordinal sum of semilattices, its irreducible decomposition consisting of two components containing* a *and* b *respectively.*

Proof. For the sake of definiteness we assume that $a < b$ (in the case $a > b$ the roles of a and b are interchanged). Lemma 7.6 implies that $b \in S(a) \backslash D(a)$ (see § 6), and therefore, by virtue of Lemma 6.3, b is greater than each element of $D(a)$. We consider the subsemilattice $\overline{N}(b, S(a))$ (see § 5) and put

$$\overline{N}(a, M) \cup D(a) = A.$$

Since $D(b) \subset S(a)$ it follows according to the corollary of Lemma 6.4 that B is a subsemilattice. By Lemma 6.3 any element of B is greater than each element of $D(a)$. Let M denote the set of elements of Γ that are less than each element of B. M is a subsemilattice and, as was just shown, $D(a) \subseteq M$. Put

$$\overline{N}(b, S(a)) \cup D(b) = B.$$

Since $D(a) \subseteq M$ it follows according to the corollary of Lemma 6.4 that A is a subsemilattice.

The subsemilattice $\{A, B\}$ generated by A and B is obviously their ordinal sum, so that, in particular, $\{A, B\} = A \cup B$. By the corollary of Proposition 6.1, A and B are components of the irreducible decomposition of $A \cup B$ into an ordinal sum. Let C denote the completely isolated homogeneous subsemilattice generated by a and b. We will show that $C = A \cup B$, which will complete the proof of the proposition.

Since C is homogeneous and completely isolated, the inclusions $D(a) \subset C$ and $D(b) \subset C$ are obvious. Any element of $N_1(b, S(a))$ is by definition incomparable with b and greater than a, which implies by virtue of the homogeneity of C that $N_1(b, S(a)) \subset C$ (if $N_1(b, S(a)) = \emptyset$, we at once have $N(b, S(a)) = \{b\} \subset C$). Continuing by induction, we get that $N(b, S(a)) \subset C$ and hence that $\overline{N}(b, S(a)) \subset C$. It is analogously shown that $\overline{N}(a, M) \subset C$. Comparing what has been said, we deduce the inclusion $A \cup B \subseteq C$.

For a proof of the converse inclusion it obviously suffices to show that $A \cup B$ is a completely isolated homogeneous subsemilattice.

We proceed to prove the homogeneity of $A \cup B$. We will show that if an arbitrary element of $\Gamma \setminus (A \cup B)$ is comparable with a then it is comparable with any element of $A \cup B$, and if it is incomparable with a then it is incomparable with any element of $A \cup B$. Clearly, we can assume that $A \cup B \neq \Gamma$, since in the case $A \cup B = \Gamma$ there is nothing to prove. Thus, let z be an arbitrary element of $\Gamma \setminus (A \cup B)$. There are three possibilities, each of which we consider separately.

1) $z > a$, i.e. $z \in S(a)$. According to Proposition 6.1 the subsemilattice B is N-closed in $S(a)$, and therefore z is comparable with any element of B. Since B is an indecomposable subsemilattice, z is either less than each element of B or greater than each element of B. In the first case $z \in M$, which implies, by virtue of the fact that A is N-closed in M (see Proposition 6.1), that z is comparable with each element of A; in the second case z is greater than each element of A.

2) $z < a$. In this case z is less than each element of B, i.e. $z \in M$. But then, as was just shown, z is comparable with any element of A and, in addition, z is less than any element of A, since A is an indecomposable subsemilattice (see Proposition 6.1).

3) $z \; \bar{\sigma} \; a$. This case is the most complicated of the three. We first note the following assertion, which is made use of below.

If u and v are arbitrary elements of A and B respectively, then the pair (u, v) is irregular.

Since $a \in A$, $b \in B$ and the pair (a, b) is irregular, this assertion follows immediately from Lemmas 7.7 and 7.8 by virtue of the corollary of Lemma 6.4.

Suppose z is comparable with an element $b_1 \in B$. Since $z \; \bar{\sigma} \; a$, we have $z < b_1$. But z cannot be less than every element of B (otherwise z would belong to $N(a, M) \subseteq A$), and hence there exists an element $b_2 \in B$ such that $z \; \bar{\sigma} \; b_2$. We consider the element $z \wedge b_2 = w$. The pair (b_2, w) is regular according to Lemma 7.4. Therefore, if $w \; \bar{\sigma} \; a_1$ for some $a_1 \in A$, the pair (b_2, a_1) is also regular by virtue of Lemma 7.2, which contradicts the above assertion. Consequently w is comparable with any element of A, and, in particular, $w < a$. Put $b_1 \wedge b_2 = b_3$. Since $w < b_1$, we have $z \wedge b_3 = z \wedge b_1 \wedge b_2 = w \wedge b_1 = w$, which implies, in particular, that $z \; \bar{\sigma} \; b_3$ and $b_3 \neq b_1$ (but it is possible that $b_3 = b_2$). Since by the above assertion the pair (b_1, a) is irregular it follows by virtue of Lemma 7.2, the the pair (b_1, z) is irregular, and hence, again by virtue of Lemma 7.2 the pair (b_1, b_3) is irregular. From this result and Lemma 7.5 (applied to the elements b_1, b_3, z and w) we deduce that the pair (b_1, w) is irregular. Summarizing what has been said, we get that $\phi(b_1) < \phi(w)$ and $\phi(w) < \phi(b_2)$. From $w < a$ it follows that w is less than any element of B; therefore $\phi(w)$ is comparable with any element of $\phi(B)$. But $\phi(w) \notin \phi(B)$, which implies by virtue of the relation $\phi(b_1) < \phi(w) < \phi(b_2)$ that $\phi(B)$ is decomposable into an ordinal sum of semilattices, and this contradicts Proposition 5.1 inasmuch as B is an indecomposable subsemilattice. Consequently z is incomparable with any element of B.

Suppose now z is comparable with an element $a_2 \in A$. Then, clearly, $z > a_2$. We consider the element $z \wedge a = p$. We will show that $p \in A$. We assume the contrary: $p \notin A$. Then it follows from $p < a$, as was shown in case 2), that p is less than any element of A, and, in particular, that $p < a_2 \wedge a$. But, since $z > a_2$, we have $p = z \wedge a \geq a_2 \wedge a$. We have obtained a contradiction. Thus $p \in A$, and hence, as was noted in the above assertion, the pair (p, b) is irregular. From this result and Lemma 7.1 we conclude that the pair (p, z) is irregular. But the latter contradicts Lemma 7.4. Consequently z is incomparable with any element of A, which completes the proof of the homogeneity of $A \cup B$.

We proceed to show, finally, that $A \cup B$ is completely isolated. Let x and y be arbitrary elements of $\Gamma \setminus (A \cup B)$. We will show that $x \wedge y \in \Gamma \setminus (A \cup B)$. We assume the contrary: $x \wedge y \in A \cup B$. If $x \wedge y \in B$ then $x > a$ and $y > a$, i.e. $x, y \in S(a)$. But then, in view of the fact that B is completely isolated in $S(a)$ according to Proposition 6.1, we have $x \wedge y \in S(a) \setminus B$. Therefore $x \wedge y \in A$, and hence, by virtue of the homogeneity of $A \cup B$, the elements x and y are comparable with any element of $A \cup B$. Since $x \, \bar{\sigma} \, y$ and B is an indecomposable subsemilattice, x and y are either greater than each element of B or less than each element of B, with the first case, as was just shown, being impossible. Consequently the second case holds, i.e. $x, y \in M$. But then $x \wedge y \in M \setminus A$, since A is completely isolated in M according to Proposition 6.1. Thus $x \wedge y \notin A \cup B$. Q.E.D.

Proposition 7.1 is completely proved.

§ 8. Main theorem

The results of the preceding sections permit us to prove

Proposition 8.1. *Every weak isomorphism is a canonical weak isomorphism to within an isomorphism. More precisely, every weak isomorphism is the product (i.e. composition) of a canonical weak isomorphism and an isomorphism.*

Proof. Suppose a semilattice Γ is weakly isomorphic to a semilattice Γ'. We must prove that Γ' is isomorphic to a semilattice obtained from Γ by permuting the components of the irreducible decompositions of some its saturated subsemilattices (see § 4). Let ϕ be a weak isomorphism of Γ onto Γ'. If ϕ is an isomorphism then there is nothing to prove. Suppose ϕ is not an isomorphism. Then Γ contains elements that are irregular relative to ϕ (see § 7). Suppose a is one of them, i.e. there exists an element b such that the pair (a, b) is irregular.[*] Then according to Proposition 7.1 the completely isolated homogeneous subsemilattice H generated by the elements a and b is decomposable, its irreducible decomposition consisting of two components containing a and b respectively. We will use the notation $Q(H) = Q(a, b)$ (see § 3). The subsemilattice $Q(a, b)$ is also decomposable by virtue of Corollary 2 of Lemma 3.4. We consider the subsemilattices $Q(a, x)$ for all of the elements $x \in \Gamma$ which form an irregular pair with a (it is clear that, in general, the equality $Q(a, x) = Q(a, y)$ is possible when $x \neq y$). Each of these subsemilattices is decomposable and saturated. The set of all distinct subsemilattices $Q(a, x)$ is denoted by $\mathfrak{N}(a)$.

We proceed analogously for all of the irregular elements of Γ: we obtain the set $\mathfrak{N}(y)$ for each such element y. The union of all of the sets $\mathfrak{N}(y)$ is denoted by $\mathfrak{N}$. The elements of $\mathfrak{N}$ are numbered by indices composing a set A of the same cardinality as $\mathfrak{N}$. Let $Q_\alpha (\alpha \in A)$ be an arbitrary subsemilattice belonging to $\mathfrak{N}$. The images of the components of the irreducible decomposition of Q_α under a weak isomorphism ϕ are the components of the irreducible decomposition of $\phi(Q_\alpha)$ (Proposition 5.1). Let ψ_α denote the permutation of the components of the irreducible decomposition of Q_α which arranges these components in the same order as their images in $\phi(Q_\alpha)$. Proceeding in this way for each $\alpha \in A$, we obtain a system $[\psi_\alpha] (\alpha \in A)$ of permutations.

We now show that the semilattice obtained from Γ by the permutations $[\psi_\alpha]$ is isomorphic to Γ'.

---

[*] Wherever it does not cause confusion, the words "relative to ϕ" will be dropped.

Let $A(u)$ denote the set of all indices of A corresponding to elements of $\Re(u)$. We first consider the semilattice Γ_u obtained from Γ by all of the permutations $[\psi_\alpha]$, where $\alpha \in A(u)$. The canonical weak isomorphism of Γ onto Γ_u is denoted by χ_u. The mapping $\phi\chi_u^{-1}$ will obviously be a weak isomorphism of Γ_u onto Γ'. We prove the following two assertions.

1. $\chi_u(u)$ *will be a regular element relative to the weak isomorphism* $\phi\chi_u^{-1}$.

On the basis of the definition of an irregular element (see §7) we need only consider the elements that are comparable with $\chi_u(u)$. Suppose $\chi_u(u) < \chi_u(v)$ (the case $\chi_u(u) > \chi_u(v)$ is analogous). We first assume that the pair (u, v) is irregular relative to ϕ. This implies that the elements u and v participated in a permutation (and hence in only one permutation; see §4) of $[\psi_\alpha]$, where $\alpha \in A(u)$, with $\chi_u(u)$ turning out to be in the same relation $(<, >)$ with $\chi_u(v)$ as $\phi(u)$ is with $\phi(v)$, i.e. in our case $\phi(u) < \phi(v)$. But $\phi(u) = \phi\chi_u^{-1}(\chi_u(u))$ and $\phi(v) = \phi\chi_u^{-1}(\chi_u(v))$, so that $\phi\chi_u^{-1}(\chi_u(u) \wedge \chi_u(v)) = \phi\chi_u^{-1}(\chi_u(u)) \wedge \phi\chi_u^{-1}(\chi_u(v))$.

The proof in the case when (u, v) is a regular pair relative to ϕ will result from the following assertion.

2. *If* $v, w \in \Gamma$ *form a regular pair relative to* ϕ, *then* $\chi_u(v), \chi_u(w) \in \Gamma_u$ *form a regular pair relative to* $\phi\chi_u^{-1}$.

The assertion is obvious if $v \,\bar\sigma\, w$. Suppose $v \,\sigma\, w$, assuming for the sake of definiteness that $v \leq w$ and hence $\phi(v) \leq \phi(w)$ (the roles of v and w are interchanged in the case $v > w$). Inasmuch as $\phi\chi_u^{-1}(\chi_u(v)) = \phi(v)$ and $\phi\chi_u^{-1}(\chi_u(w)) = \phi(w)$, we must prove that $\chi_u(v) \leq \chi_u(w)$. We put $\bigcup_{\alpha \in A(u)} Q_\alpha = Q$ and consider the two possible cases.

a) $v, w \in Q$. In the proof of Proposition 7.1 it was established that if a and b form an irregular pair then any elements belonging to different components of the irreducible decomposition of the completely isolated homogeneous subsemilattice generated by a and b also form an irregular pair. Hence v and w cannot belong to the components of the irreducible decomposition of a subsemilattice $Q_\alpha (\alpha \in A(u))$ the relative positions of which change under the permutations $[\psi_\alpha] (\alpha \in A(u))$, i.e. the images of v and w under the weak isomorphism χ_u are found to be in the same relation with each other as these elements themselves. But $v \leq w$, and consequently $\chi_u(v) \leq \chi_u(w)$. Q.E.D.

b) One of the elements v or w belongs to $\Gamma \setminus Q$. In this case v and w do not participate in a permutation of $[\psi_\alpha] (\alpha \in A(u))$ and therefore a fortiori $\chi_u(v) \leq \psi_u(w)$.

Assertion 2, and together with it assertion 1, is proved.

Let ψ be the canonical weak isomorphism of Γ onto the semilattice Γ_1 obtained from Γ by the permutations $[\psi_\alpha]\,(\alpha \in A)$. We show that $\psi(u)$ is regular relative to $\phi\psi^{-1}$ for any $u \in \Gamma$. If u is a regular element relative to ϕ, this follows from assertion 2. Suppose u is irregular relative to ϕ. Then $\psi = \chi'_u \chi_u$, where χ'_u is the canonical weak isomorphism of Γ_u onto Γ_1 corresponding to the permutations $[\psi_\alpha]\,(\alpha \in A \backslash A\,(u))$. In particular, it can happen that $\psi = \chi_u$. $\chi_u\,(u)$ is a regular element relative to $\phi \chi_u^{-1}$ by virtue of assertion 1, while $\chi'_u \chi_u\,(u) = \psi\,(u)$ is a regular element relative to $\phi \chi_u^{-1} \chi_u'^{-1} = \phi\psi^{-1}$ by virtue of assertion 2.

Thus all of the elements of Γ_1 are regular relative to the weak isomorphism $\phi\psi^{-1}$. Consequently $\phi\psi^{-1} = f$ is an isomorphism of Γ_1 onto Γ', and $\phi = f\psi$. The conjecture is proved.

We now formulate the main theorem of the article.

Theorem A. *In order for a semilattice Γ to project onto a semilattice Γ', it is necessary and sufficient that Γ' be isomorphic to a semilattice obtained from Γ by permutations of the components of the irreducible decompositions of some of its saturated subsemilattices into ordinal sums of semilattices.*

Proof. The necessity follows from Propositions 1.1 and 8.1, while the sufficiency follows from Propositions 4.1 and 1.1.

Theorem C. *In order for each projection of a given semilattice to be induced by an isomorphism of it, it is necessary and sufficient that all of the completely isolated homogeneous subsemilattices of this semilattice be indecomposable into an ordinal sum of semilattices.*

Proof. If a semilattice Γ contains a decomposable completely isolated homogeneous subsemilattice H, we can obtain a semilattice Γ' for which the canonical weak isomorphism of Γ onto Γ' is not an isomorphism by transposing any two different components of the irreducible decomposition of H (according to Corollary 2 of Lemma 3.4 this permutation can be considered as a permutation of the components in the corresponding saturated subsemilattice $Q\,(H)$; see § 3). By virtue of Proposition 1.1 this proves the necessity.

Conversely, suppose all of the completely isolated homogeneous subsemilattices of a semilattice Γ are indecomposable. By Corollary 2 of Lemma 3.4 they are then always saturated (and this, by virtue of Lemmas 1.3 and 2.6, is equivalent to the condition that all of the completely isolated homogeneous subsemilattices are indecomposable). The indecomposability of sub-semilattices immediately implies the following: any canonical weak isomor-phism of Γ is an isomorphism (which according to the remark following Proposition 4.1 essentially coincides with the identity automorphism). This result and Proposition 8.1 imply than an arbitrary weak isomorphism of Γ is an isomorphism. By virtue of Proposition 1.1 this proves the sufficiency.

BIBLIOGRAPHY

[1] R. Baer, *Linear algebra and projective geometry*, Academic Press, New York, 1952; Russian transl., IL, Moscow, 1955. MR 14, 675.

[2] M. Suzuki, *Structure of a group and the structure of its lattice of sub-groups*, Ergebnisse der Math., Heft 10, Springer-Verlag, Berlin, 1956; Russian transl., IL, Moscow, 1960. MR 18,715; 22 # 8057.

[3] P. G. Kontorovič, A. S. Pekelis and A. I. Starostin, *Lattice-theoretic problems of group theory*, Ural. Gos. Univ. Mat. Zap. 3 (1961), no. 1, 3–50. (Russian) MR 30 # 2082.

[4] L. N. Ševrin, *On lattice properties of semigroups*, Sibirsk. Mat. Ž. 3 (1962), 446–470. (Russian) MR 27 # 5847.

[5] L. A. Skornjakov, *Complemented modular lattices and regular rings*, Fiz-matgiz, Moscow, 1961; English transl., Oliver and Boyd, London, 1964. MR 29 # 3404; 30 #42.

[6] D. Sachs, *The lattice of subalgebras of a Boolean algebra*, Canad. J. Math. 14 (1962), 451–460. MR 25 # 1116.

[7] E. S. Ljapin, *Semigroups*, Fizmatgiz, Moscow, 1960; English transl., Transl. Math. Monographs, Vol. 3, Amer. Math. Soc., Providence, R. I., 1963. MR 22 # 11054; 29 # 4817.

[8] A. H. Clifford and G. B. Preston, *The algebraic theory of semigroups*. Vol. I, Math. Surveys, no. 7, Amer. Math. Soc., Providence, R. I., 1961. MR 24 # A 2627.

[9] M. I. Kargapolov, *Classification of ordered Abelian groups by their elementary properties*, Algebra i Logika Sem. 2 (1963), no. 2, 31–46. (Russian) MR 27 # 5823.

[10] L. A. Skornjakov, *Projectivities of modules*, Izv. Akad. Nauk SSSR Ser. Mat. 24 (1960), 511–520; English transl., Amer. Math. Soc. Transl. (2) 45 (1965), 233–244. MR 22 #5655.

[11] L. N. Ševrin, *Projectivities of semi-lattices*, Dokl. Akad. Nauk SSSR
154 (1964), 538–541 = Soviet Math. Dokl. 5 (1964), 142–146.
MR 30 #1958.

[12] V. V. Vagner, *Semigroups associated with a generalized group*, Mat.
Sb. 52 (94) (1960), 597–628; English tr ansl., Amer. Math. Soc. Transl. (2)
36 (1964), 351–381. MR 22 #12156.

[13] A. G. Kuroš, *Lectures in general algebra*, Fizmatgiz, Moscow, 1962;
English transl., Chelsea, New York, 1963; Internat. Series of Monographs
in Pure and Appl. Math., vol. 70, Pergamon Press, Oxford, 1965.
MR 25 #5097; 28 #1228; 31 #3448.

[14] G. Birkhoff, *Lattice theory*, 2nd rev. ed., Amer. Math. Soc. Colloq. Publ.,
vol. 25, Amer. Math. Soc., Providence, R. I., 1948; Russian transl., IL,
Moscow, 1952.   MR 10,673.

[15] L. N. Ševrin, *Semigroups with a modular lattice of subsemigroups*, Dokl.
Akad. Nauk SSSR 148 (1963), 292–295 = Soviet Math. Dokl. 4 (1963),
93–96.   MR 26 #6285.

[16] ————, *Semigroups with certain types of subsemigroup lattices*, Dokl.
Akad. Nauk SSSR 138 (196 1), 796–798 = Soviet Math. Dokl. 2 (1961),
737–739.   MR 26 #2527.

[17] N. Bourbaki, *Théorie des ensembles (fascicule de résultats)*, 2ième éd.,
Actualités Sci. Indust., no. 1141, Hermann, Paris, 1951; Russian transl.,
Fizmatgiz, Moscow, 1958.   MR 3,55; 5,102.

[18] J. Morgado, *Quasi-isomorphisms between complete lattices*, Portugal.
Math. 20 (1961), 17–31.   MR 24 #A 3093.

[19] ————, *Some remarks on quasi-isomorphisms between finite lattices*,
Portugal. Math. 20 (1961), 137–145.   MR 25 #2980.

Amer. Math. Soc. Transl.
(2) Vol. 96, 1970

PROJECTION OF LATTICES[*]

UDC 519.47

N. D. FILIPPOV

A *projection* of a lattice L on a lattice L' is an isomorphism of the lattice of sublattices of L onto that of L'.

In the theory of projections of algebraic systems one of the basic questions is the following, which we formulate directly for lattices:

1. To find necessary and sufficient conditions for the existence of a projection between two lattices.

It is easy to see that every isomorphism and dual isomorphism of a lattice induces a projection of it. Here there arises the following question:

2. To describe all lattices each projection of which is induced by an isomorphism or a dual isomorphism.

For semilattices both questions have been solved by L. N. Ševrin [1]. One might suppose that their solution for lattices would be in some sense analogous. And indeed, it has turned out to be so: here the basic role is played by the notion of a homogeneous completely isolated sublattice, which is the analog of the notion of a homogeneous completely isolated subsemilattice introduced in [1]. At the same time there arose the question of its specific character, conditioned by the fact that for lattices a dual isomorphism induces a projection whereas for semilattices this does not hold.

To avoid repetition we will not prove certain assertions if they are similar to the corresponding proofs for semilattices in [1]. In such cases the reader is referred to [1].

§2 below is devoted to answering question 2 (Theorem 2.1). §3 prepares the way for §4 where the main result is proved (Theorem 4.1) answering question 1. In §5 we consider projections of certain types of lattices.

We take this opportunity to express our deep gratitude to L. N. Ševrin for posing the problem and for unwavering guidance of this work, and also to V. T. Nagrebeckiĭ for several valuable remarks.

---

[*] Translation of Mat. Sb. 70 (112) (1966), 36–54.

§1. Basic definitions and preparatory lemmas

We will use the following notation. If x and y are elements of a lattice L, then, as usual, $x \vee y$ will denote their sup and $x \wedge y$ their inf. For brevity we will use the compound symbol $\underset{\wedge}{\vee}$, and $x \underset{\wedge}{\vee} y$ will mean either the sup or the inf, or both, depending on our need. It will always be clear from the context which is intended in a particular case. As in [1], if x and y are comparable in L we write $x\sigma y$, and, if not, $x\bar{\sigma}y$. If A and B are two subsets of L we will write $A\sigma B$ or $A\bar{\sigma}B$ or $A < B$ or $A > B$ if for any a in A and b in B we have $a\sigma b$ or $a\bar{\sigma}b$ or $a < b$ or $a > b$, respectively. $\Sigma(L)$ will denote the lattice of sublattices of L. The set-theoretic operations are denoted as usual: $\cup$, $\cap$, $\setminus$. The empty set is denoted $\varnothing$.

Definition 1. A one-to-one mapping ϕ of a lattice L onto another L' is called a *semi-isomorphism* if for any $x, y \in L$ we have

$$
(\alpha) \quad
\begin{aligned}
\varphi(x \vee y) &= \varphi(x) \vee \varphi(y), \\
\varphi(x \wedge y) &= \varphi(x) \wedge \varphi(y)
\end{aligned}
\qquad \text{or} \qquad
(\beta) \quad
\begin{aligned}
\varphi(x \vee y) &= \varphi(x) \wedge \varphi(y), \\
\varphi(x \wedge y) &= \varphi(x) \vee \varphi(y).
\end{aligned}
$$

If $x = y$ then we have both (α) and (β). If $x \neq y$ then clearly we have only one: either (α) or (β). Hence if (α) holds then none of (β) holds, and conversely. Note that in special cases semi-isomorphisms are isomorphisms if (α) holds for any pair in L, and are dual isomorphisms if (β) holds for any pair in L. It is easy to see that a semi-isomorphism carries comparable elements to comparable ones and incomparable ones to incomparable ones. In fact, if $x\sigma y$ then $\phi(x) = \phi(x \underset{\wedge}{\vee} y) = \phi(x) \underset{\wedge}{\vee} \phi(y)$, i.e. $\phi(x)\sigma\phi(y)$. If $\phi(x)\sigma\phi(y)$ then $\phi(x) = \phi(x) \underset{\wedge}{\vee} \phi(y) = \phi(x \underset{\wedge}{\vee} y)$, so that $x\sigma y$.

Definition 2. A pair x, y in L is called *regular relative to a given semi-isomorphism* ϕ if (α) holds for x and y, and is called *nonregular if* (β) *holds*.

Note that to check the regularity or nonregularity of a given pair it suffices to check at least one of equalities (α) or (β).

If $x\sigma y$ then clearly (x, y) is nonregular if and only if either $x \leq y$ and $\phi(x) \geq \phi(y)$ or $x \geq y$ and $\phi(x) \leq \phi(y)$.

If (x, y) is regular we write $x\rho y$, and $x\nu y$ if (x, y) is nonregular. It will always be clear relative to which semi-isomorphism the regularity is to be understood. For any x in L let P_x denote the set of elements of L which form a regular pair with x, and let N_x denote those forming a nonregular pair with x. Clearly $P_x \cap N_x = x$.

Let ϕ be a semi-isomorphism of L and let ψ be a semi-isomorphism of $\phi(L)$. Clearly the mapping $\psi \cdot \phi$ from L to $L' = \psi(\phi(L))$ is a semi-isomorphism of L onto L'. It is clear from the definition of semi-isomorphism that if ϕ is a semi-isomorphism from L onto L' then so is ϕ^{-1}, from L' onto L.

Definition 3. Let ϕ_1 be a semi-isomorphism of L_1 and ϕ_2 a semi-isomorphism of L_2 and let ϕ be a semi-isomorphism from L_1 onto L_2. Then ϕ_1 and ϕ_2 are called *adequate relative to* ϕ if for any x and y in L_1 the pair (x, y) is regular relative to ϕ_1 if and only if $(\phi(x), \phi(y))$ is regular relative to ϕ_2.

Clearly if ϕ_1 and ϕ_2 are adequate relative to ϕ then the same holds relative to ϕ^{-1}.

In the notation of Definition 3 we have

Lemma 1.1. *There exists a semi-isomorphism ϕ' from $\phi_1(L_1)$ onto $\phi_2(L_2)$ which is adequate to ϕ relative to ϕ_1.*

In fact, for ϕ' we can take $\phi_2 \cdot \phi \cdot \phi_1^{-1}$.

The value of the notion of semi-isomorphism becomes apparent in

Proposition 1.1. *The lattice L can be projected onto another L' if and only if L and L' are semi-isomorphic, where each projection of L is induced by some semi-isomorphism.*

Proof.[1] 1. Let L be projected onto L', i.e. there exists an isomorphism of $\Sigma(L)$ onto $\Sigma(L')$, say ψ. It assigns to each one-element sublattice $\{x\} \in \Sigma(L)$, which is an atom in $\Sigma(L)$, a unique one-element sublattice $\{x'\} \in \Sigma(L')$. Setting $\phi(x) = x'$ we establish a one-to-one correspondence between L and L'. Let x and y be in L, $x \neq y$. Let $\{x, y\}$ denote the sublattice generated by x and y. If $x \sigma y$ then $\{x, y\}$ is a two-element sublattice. To it there corresponds the two-element sublattice $\psi(\{x, y\}) = \{\phi(x), \phi(y)\}$, and we have $\phi(x) \sigma \phi(y)$. In this case clearly either (α) or (β) holds for x and y. Now let $x \bar{\sigma} y$. Then $\{x, y\}$ contains four distinct elements: x, y, $x \vee y$ and $x \wedge y$, and the sublattice $\psi(\{x, y\})$ contains also the distinct elements $\phi(x)$, $\phi(y)$, $\phi(x \vee y)$ and $\phi(x \wedge y)$. But then either (α) or (β) holds for x and y.

2. Let there be given a semi-isomorphism ϕ of L onto L'. Let $\phi(H)$ denote the image of H by ϕ for $H \subseteq L$. If H is a sublattice of L then $\phi(H)$

---

1) The proof of this proposition is a minor modification of that of Proposition 1.1 in [1].

is one in L', since $\phi(h_1) \overset{\vee}{\wedge} \phi(h_2) = \phi(h_1 \overset{\vee}{\wedge} h_2) \in \phi(H)$ for any h_1 and h_2 in H. If H' is a sublattice in L' then it is easy to show that $\phi^{-1}(H')$ is one in L. Thus ϕ induces a one-to-one correspondence from $\Sigma(L)$ onto $\Sigma(L')$. It is an isomorphism, since $H \subseteq F$ in L if and only if $\phi(H) \leq \phi(F)$ in L' for any $H, F \subseteq L$.

Assuming the notions of ordinal sum and cardinal product (see, for example, [2]), let us prove

Lemma 1.2. *A given lattice L decomposes into an ordinal sum of true sublattices if and only if $\Sigma(L)$ decomposes into the cardinal product of true sublattices.*[1]

Proof. Necessity. Each sublattice H in L decomposes uniquely into sublattices $H_\omega = H \cap A_\omega (\omega \in \Omega)$, where the A_ω are the components of the given decomposition of L into the ordinal sum. Clearly $H \subseteq F$, where H and F are sublattices, holds if and only if $H_\omega \subseteq F_\omega$ for all $\omega \in \Omega$. Since the set of sublattices of the form $H \cap A_\omega$ for a fixed ω and H ranging over all sublattices of L coincides with $\Sigma(A_\omega)$, it is easy to see that $\Sigma(L)$ decomposes into the cardinal product of the sublattices $\Sigma(A_\omega) (\omega \in \Omega)$. The necessity is proved. Note that these arguments are valid if as A_ω we take a disjoint class of ordinal summands.

Sufficiency. Let $\Sigma(L)$ be decomposed into the cardinal product of true sublattices. We express this fact: $\Sigma(L) = \cdots \times \Sigma_\omega \times \cdots (\omega \in \Omega)$. We denote the empty sublattice of L by $\theta = (\cdots, \theta_\omega, \cdots)$, where θ_ω is the zero of $\Sigma_\omega (\omega \in \Omega)$. Since there are no nonempty sublattices of L contained in a one-element sublattice, the one-element sublattice $\{x\} \in \Sigma(L)$ will have the form $\{x\} = (\cdots, \theta_\xi, \cdots, x_\eta, \cdots)$, where all the ξth components $(\xi \neq \eta)$ are equal to θ_ξ and x_η covers θ_η in Σ_η (see [2]). Let $\{x\} = (\cdots, \theta_\xi, \cdots, x_\eta, \cdots)$ and $\{y\} = (\cdots, \theta_\xi, \cdots, y_\zeta, \cdots) (\eta \neq \zeta)$ be two one-element sublattices in L. Then $(\cdots, \theta_\xi, \cdots, x_\eta, \cdots, y_\zeta, \cdots)$, where the ξth components $(\xi \neq \eta, \zeta)$ are equal to θ_ξ, is a sublattice in L (by the definition of the cardinal product), and hence $\{x, y\}$[2] $= (\cdots, \theta_\xi, \cdots, x_\eta, \cdots, y_\zeta, \cdots)$. Since in this case $\{x, y\}$ contains only three sublattices: $\theta, \{x\}$ and $\{y\}$, we have $x\,\sigma\,y$. Consider the set of elements of L to which there correspond one-element sublattices of the form $(\cdots, \theta_\xi, \cdots, x_\eta, \cdots)$ where η is fixed but arbitrary. Denote it by X_η. It is easy to see that X_η is a sublattice. Each element

---

1) A sublattice H of L is *true* if $H \neq L$. A sublattice F is *proper* if $F \neq L$ and F contains at least two elements.

2) $\{x, y\}$ is the sublattice of L generated by x and y in L.

$y \in L \backslash X_\eta$ decomposes X_η into two sublattices. One consists of the elements greater than y and the other of those less than y. (Of course, one of them may be empty.) Consider all possible such decompositions of X_η as y runs over $L \backslash X_\eta$. In the complete lattice of all decompositions of the lattice X_η (see [2]) take the infimum of all the constructed decompositions of X_η. This decomposition divides X_η into equivalence classes. Let A be such a class. A is a sublattice (since it is the intersection of sublattices). It is obvious too that $L \backslash A$ is a sublattice (since it is the union of an increasing sequence of sublattices). It is easy to see that $A \, \sigma (L \backslash A)$. Moreover, from the definition of our decomposition it follows that A is convex. This means that A is a component of some decomposition of L into an ordinal sum of true sublattices. The lemma is proved.

A decomposition of a lattice into an ordinal sum of sublattices is called *irreducible* if each component is indecomposable into an ordinal sum of true sublattices. Similarly, a decomposition of a lattice into the cardinal product of sublattices is called *irreducible* if each component of the decomposition is indecomposable into a cardinal product of true sublattices.

Clearly, any lattice L possesses a unique irreducible decomposition into an ordinal sum of sublattices. In the notation of Lemma 1.2 it is not hard to see that $\Sigma_\eta (\eta \in \Omega)$ is isomorphic to $\Sigma (X_\eta)$. But then from this lemma we have the

Corollary. *$\Sigma(L)$ possesses a unique irreducible decomposition into a cardinal product of sublattices, where between the set of factors of this product and the set of summands of the irreducible decomposition of L into an ordinal sum of sublattices there is a one-to-one correspondence, under which each cardinal factor is isomorphic to the lattices of the corresponding ordinal summand.*

Below throughout we will write ''decomposition of a given lattice L'' in place of ''decomposition of a given lattice L into an ordinal sum of true sublattices''.

If for components A and B of some decomposition of a given lattice L we have $A < B$ then we will say that A *precedes* B. Thus we have a *natural* order on the set of components.

If the set of components of a decomposition of a lattice L coincides up to isomorphism with the set of components of a decomposition of a lattice L', then we will say that *L' is obtained from L by permutations of the components of the given decomposition of L.*

If the natural order on the components of a decomposition of L coincides with that of a decomposition of L' where the corresponding components coincide up to isomorphism or dual isomorphism, then we will say that L' *is obtained from L by revolutions of the components of the given decomposition of L.*

If the set of components of a decomposition of L coincides up to isomorphism or dual isomorphism with the set of components of a decomposition of L', then we will say that L' *is obtained from L by pemutations and revolutions of the components of the given decomposition of L.*

If L is decomposable (i.e. there exists a decomposition of L), then it follows from the Corollary of Lemma 1.2 that revolution and permutation of a set of components of an irreducible decomposition of L leads to an isomorphism and to a permutation of the factors of an irreducible decomposition of $\Sigma(L)$ into a cardinal product. Hence we have

Lemma 1.3. *Let L' be obtained from L*

1) *by permutations of the components of an irreducible decomposition of L, or*

2) *by revolutions of the components of an irreducible decomposition of L, or*

3) *by revolutions and permutations of the components of an irreducible decomposition of L.*

Then L can be projected on L'.

Lemma 1.4. *A semi-isomorphism ϕ of L carries the components of an irreducible decomposition of L to the components of an irreducible decomposition of $\phi(L)$.*

The proof follows from Proposition 1.1 and the Corollary of Lemma 1.2.

At the end of this section we will give a series of definitions and lemmas which will be necessary later. These definitions and the formulations and proofs of lemmas agree essentially with the corresponding definitions and formulations and proofs of lemmas for semilattices given by L. N. Ševrin in [1], and hence the lemmas are given without proofs.

Definition 4. A sublattice H of L is called *completely isolated if* $L \backslash H$ is a sublattice in L.

Definition 5. A sublattice $H \subseteq L$ is called *homogeneous* if either $H = L$ or for any $x \in L \backslash H$ we have either 1) $x \, \sigma H$ or 2) $x \, \bar{\sigma} H$.

For a homogeneous sublattice $H \subseteq L$ we will let $L_\sigma(H)$ denote the set of all $x \in L \backslash H$ satisfying 1), and $L_{\bar\sigma}(H)$ all the $x \in L \backslash H$ satisfying 2). We have $L_\sigma(H) \cup L_{\bar\sigma}(H) = L \backslash H$ and $L_\sigma(H) \cap L_{\bar\sigma}(H) = \emptyset$.

Definition 6. A sublattice $H \subseteq L$ is called *saturated* if, firstly, it is a homogeneous completely isolated sublattice, and secondly, for any homogeneous completely isolated sublattice $F \subseteq L$ strictly containing H, we have the strong inclusion $L_{\bar\sigma}(F) \subset L_{\bar\sigma}(H)$.

Note that by this definition L itself is a saturated sublattice.

Lemma 1.5. *If H is a homogeneous sublattice of L and $x \in L_{\bar\sigma}(H)$, then $(x \wedge y) = (x \wedge z)$ and $(x \vee y) = (x \vee z)$ for any y, $z \in H$.*

Lemma 1.6. *If H is a saturated sublattice of L, then for any element $x \in L_\sigma(H)$ we have either $x > H$ or $x < H$.*

Lemma 1.7. *Under a semi-isomorphism of L the image of a homogeneous completely isolated sublattice is a homogeneous completely isolated sublattice, and the image of a saturated sublattice is a saturated sublattice.*

Lemma 1.8. *The union of an increasing sequence of homogeneous completely isolated sublattices is a homogeneous completely isolated sublattice.*

Lemma 1.9. *Let H be a homogeneous completely isolated sublattice in L. Among the homogeneous completely isolated sublattices $F \subseteq L$ for which $L_{\bar\sigma}(F) = L_{\bar\sigma}(H)$ there is a greatest. It is saturated.*

Lemma 1.10. *There exists a unique minimal saturated sublattice containing a given set M.*

The sublattice of Lemma 1.10 is called the saturated sublattice *generated by the given set M* and will be denoted by $H(M)$. If M lies in one component of an irreducible decomposition of $H(M)$ we will denote this component by $A(M)$.

Lemma 1.11. *If H is a saturated sublattice of L and F is a saturated sublattice of H, then F is a saturated sublattice of L.*

Proposition 1.2. *If H and F are any distinct saturated sublattices of a given lattice, then either H and F do not meet or one of them is strictly contained in some component of an irreducible decomposition of the other.*

§2. Lattices any projection of which is induced by an isomorphism or dual isomorphism

This section is devoted to the proof of

Theorem 2.1. *If in L there are no proper homogeneous completely isolated sublattices, then any semi-isomorphism ϕ of L is an isomorphism or a dual isomorphism.*

Let ϕ be a semi-isomorphism which is neither an isomorphism nor a dual isomorphism. We will show that L admits a proper homogeneous completely isolated sublattice. Our assumption means that in L there are regular and also nonregular pairs relative to ϕ. Then it is easy to see that there exist distinct elements a, b, c in L such that either 1) $a\rho b$, $a\rho c$, $b\nu c$, or 2) $a\nu b$, $a\nu c$, $b\rho c$.

In the second case we consider the semi-isomorphism $\phi_1 = \phi^* \cdot \phi$ of L, where ϕ^* is some dual isomorphism of $\phi(L)$. We have

$$\varphi_1(a \vee b) = \varphi^*[\varphi(a) \wedge \varphi(b)] = \varphi_1(a) \vee \varphi_1(b), \quad \varphi_1(a \wedge b) = \varphi_1(a) \wedge \varphi_1(b);$$
$$\varphi_1(a \vee c) = \varphi^*[\varphi(a) \wedge \varphi(c)] = \varphi_1(a) \vee \varphi_1(c), \quad \varphi_1(a \wedge c) = \varphi_1(a) \wedge \varphi_1(c);$$
$$\varphi_1(b \vee c) = \varphi^*[\varphi(b) \vee \varphi(c)] = \varphi_1(b) \wedge \varphi_1(c), \quad \varphi_1(b \wedge c) = \varphi_1(b) \vee \varphi_1(c);$$

Thus for the proof of Theorem 2.1 it suffices to consider the case where $a\rho b$, $a\rho c$, $b\nu c$ relative to the given semi-isomorphism ϕ, i.e.

$$(\alpha) \quad \varphi(a \vee b) = \varphi(a) \vee \varphi(b), \quad \varphi(a \wedge b) = \varphi(a) \wedge \varphi(b);$$
$$(\beta) \quad \varphi(a \vee c) = \varphi(a) \vee \varphi(c), \quad \varphi(a \wedge c) = \varphi(a) \wedge \varphi(c);$$
$$(\gamma) \quad \varphi(b \vee c) = \varphi(b) \wedge \varphi(c), \quad \varphi(b \wedge c) = \varphi(b) \vee \varphi(c).$$

We will say that an ordered triple of elements a, b, $c \in L$ is in the relation π_1 relative to ϕ, and write it $\pi_1(a, b, c)$, if for these elements we have conditions $(\alpha) - (\gamma)$. We will say that an ordered triple of elements x, y, $z \in L$ are in the relation π_2 relative to ϕ, and write it $\pi_2(x, y, z)$, if $x\nu y$, $x\nu z$, $y\rho z$ relative to ϕ. Throughout the sequel unless otherwise stipulated, we will mean the relations π_1 and π_2 relative to the given semi-isomorphism ϕ. It is clear that if we have $\pi_1(a, b, c)$ then we have $\pi_1(a, c, b)$, and if $\pi_2(x, y, z)$ then $\pi_2(x, z, y)$.

Lemma 2.1. *Let* $\pi_1(a, b, c)$. *Then*

1) *if* $a > b$, *then* $a > c$,

2) *if* $a < b$, *then* $a < c$,

3) *if* $a\,\bar{\sigma}\,b$ *then there are elements* $a_1 > b$ *and* $a_2 < b$ *such that* $\pi_1(a_1, b, c)$ *and* $\pi_1(a_2, b, c)$. *For* a_1 *and* a_2 *we can take* $a \vee b$ *and* $a \wedge b$, *respectively.*

Proof. 1) The element $d_0 = \phi(a) \vee \phi(c) = \phi(a \vee c) = \phi(a \vee b \vee c)$ is equal to one of $d_1 = \phi(a) \vee \phi(b) \vee \phi(c)$, $d_2 = [\phi(a) \vee \phi(b)] \wedge \phi(c)$ and to one of $d_3 = \phi(b) \wedge \phi(c) \wedge \phi(a)$, $d_4 = [\phi(b) \wedge \phi(c)] \vee \phi(a)$. The equality $d_0 = d_2$ is impossible, since $\phi(a) \vee \phi(c) \neq \phi(a) \wedge \phi(c) = d_2$ because $a \neq c$. Also, $d_1 \neq d_3$, since $a \neq b \neq c$.

There remains $d_1 = d_4$; but $\phi(b) \wedge \phi(c) \leq \phi(b) < \phi(a)$, and hence we get $\phi(a) \vee \phi(c) = \phi(a)$, i.e. $\phi(a) > \phi(c)$. But then $a > c$, as was required.

2) The proof of this is carried out dually.

3) First note that $a \bar{\sigma} c$, since if not we would have $a \sigma b$ from the preceding points, which is a contradiction. Let us show that $a_1 \rho b$ and $a_1 \rho c$, where $a_1 = a \vee b$. It suffices to check the first column of equalities (α) and (β), replacing a by a_1. Clearly $\phi(a_1 \vee b) = \phi(a \vee b \vee b) = \phi(a \vee b) = \phi(a) \vee \phi(b) = [\phi(a) \vee \phi(b)] \vee \phi(b) = [\phi(a \vee b)] \vee \phi(b) = \phi(a_1) \vee \phi(b)$, i.e. $a_1 \rho b$. For the proof of $a_1 \rho c$ suppose the contrary: $\phi(a_1 \vee c) = \phi(a \vee b \vee c) = \phi(a \vee b) \wedge \phi(c) \leq \phi(c)$. Clearly $\phi(a \vee b \vee c) = \phi(a \vee b \vee a \vee c) = \phi(a \vee b) \overset{\vee}{\wedge} \phi(a \vee c) = [\phi(a) \vee \phi(b)] \overset{\vee}{\wedge} [\phi(a) \vee \phi(c)] \geq \phi(a)$, and so $\phi(c) \geq \phi(a \vee b \vee c) \geq \phi(a)$, i.e. $\phi(c) \sigma \phi(a)$, which contradicts $c \bar{\sigma} a$. Thus $\phi(a_1 \vee c) = \phi(a \vee b) \vee \phi(c) = \phi(a_1) \vee \phi(c)$, i.e. $a_1 \rho c$. Dually one shows that $a_2 \rho b$, $a_2 \rho c$, where $a_2 = a \wedge b$. Since, clearly, $a_1 > b$ and $a_2 < b$, the lemma is proved.

Lemma 2.2. *Let* $\pi_2(x, y, z)$. *Then*

1) *if* $x > y$ *then* $x > z$,

2) *if* $x < y$ *then* $x < z$,

3) *if* $x \bar{\sigma} y$, *then there exist* $x_1 > y$ *and* $x_2 < y$ *such that* $\pi_2(x_1, y, z)$ *and* $\pi_2(x_2, y, z)$. *For* x_1 *and* x_2 *we can take* $x \vee y$ *and* $x \wedge y$, *respectively*.

The proof follows immediately from Lemma 2.1 if we apply to $\phi(L)$ a dual isomorphism and set $x = a$, $y = b$ and $z = c$.

Let a lattice L be given, and let $\pi_1(a, b, c)$. Also let $a < b$. In the case, from Lemma 1.1, setting $L_1 = L$ and $L_2 = \phi(L)$, and taking ϕ_1 and ϕ_2 as dual isomorphisms of L_1 and L_2, we get $\phi_1(a) \rho \phi_1(b)$, $\phi_1(a) \rho \phi_1(c)$, $\phi_1(b) \nu \phi_1(c)$ relative to some semi-isomorphism ϕ' of $\phi_1(L)$ which is adequate to ϕ relative ϕ_1. Moreover, clearly $\phi_1(a) > \phi_1(b)$.

Noting this and Lemmas 1.7 and 2.1, it suffices, to prove Theorem 2.1, to find a proper homogeneous completely isolated sublattice of L under the conditions $a > b$, $a > c$ and $\pi_1(a, b, c)$ relative to ϕ.

Now we turn to this problem.

Proposition 2.1. 1) N_u *is a sublattice in* L.

2) P_u *is a sublattice in* L, *where* u *is any element in* L.

Proof. 1) Let $x, y \in N_u$. The following cases are possible:

a) $x \sigma y$. In this case there is nothing to prove.

b) $x \bar{\sigma} y$, $x \sigma u$ and $y \sigma u$. Then either $x > u$ and $y > u$, or $x < u$ and

$y < u$. Then either $x > u$ and $y > u$, or $x < u$ and $y < u$. Let $x > u$ and $y > u$. Then $\phi(x) < \phi(u)$ and $\phi(y) < \phi(u)$. Hence $\phi(x) \overset{\vee}{\wedge} \phi(y) \le \phi(u)$, i.e. $\phi(x \overset{\vee}{\wedge} y) = \phi(x) \overset{\vee}{\wedge} \phi(y) \le \phi(u)$. And thus we have $(x \overset{\vee}{\wedge} y) \ge u$, $\phi(x \overset{\vee}{\wedge} y) \le \phi(u)$, i.e. $x \vee y$, $x \wedge y \in N_u$. The case $x < u$, $y < u$ is handled dually.

c) $x \bar{\sigma} y$ and $x \bar{\sigma} u$ $(y \bar{\sigma} u)$. Let us show that $x \vee y \in N_u$. Suppose not: $\phi(x \vee y \vee u) = \phi(x \vee y) \vee \phi(u)$. The following formula is obviously valid: $\phi(x \vee y \vee u) = \phi[(x \vee u) \vee (y \vee u)] = \phi(x \vee u) \overset{\vee}{\wedge} \phi(y \vee u) = [\phi(x) \wedge \phi(u)] \overset{\vee}{\wedge} [\phi(u) \wedge \phi(u)] \le \phi(u)$. This and the assumption give $x \vee y \vee u = u$, i.e. $x < u$ and $y < u$, which contradicts the hypothesis. And so $x \vee y \in N_u$. The proof that $x \wedge y \in N_u$ is handled dually.

2) If we apply to $\phi(L)$ the dual isomorphism ϕ^*, then for any $x \in P_u$, and only for $x \in P_u$, the pairs (x, u) are nonregular for the semi-isomorphism $\phi_1 = \phi^* \cdot \phi$ (see the proof in the beginning of §2). But then from what we have just proved it follows that P_u is a sublattice.

Proposition 2.2. $T = \mathbf{U}_{t \in N_b \cap P_a} N_t$, where a and b are the elements fixed above, is a sublattice of L.

Proof. Let $x, y \in T$, i.e. for some p and q let

$$x \nu p,\ p \nu b;\ y \nu q,\ q \nu b;\ p \rho a,\ q \rho a;\ b \rho a. \tag{1}$$

Let us show that $x \vee y \in T$. If $x \sigma y$ there is nothing to prove, and if $y \nu p$ then from Proposition 2.1 it follows that $(x \vee y) \nu p$, i.e. $x \vee y \in T$. And if $x \nu q$, then similarly $(x \vee y) \nu q$ and $x \vee y \in T$.

Thus let $x \bar{\sigma} y$, $y \rho p$ and $x \rho q$. If $x \nu y$ we have $\pi_2(x, y, p)$, and, applying Lemma 2.2 to x, y and p, we get $(x \vee y) \nu p$, i.e. $x \vee y \in T$.

Now let $x \rho y$. In this case $\pi_1(y, x, p)$, and, applying Lemma 2.1 (cases 1) and 3)) to y, x and p, we get $(x \vee y) \rho p$ and

$$(x \vee y) > p, \quad \varphi(x \vee y) > \varphi(p). \tag{2}$$

Similarly one proves that

$$(x \vee y) > q, \quad \varphi(x \vee y) > \varphi(q). \tag{3}$$

From (2) we have $\phi(x \vee y) > \phi(p) \ge \phi(p) \wedge \phi(b) = \phi(p \vee b)$, and hence $(x \vee y) \sigma (p \vee b)$. There are two possible cases.

1) $(x \vee y) < (p \vee b)$. In this case $(x \vee y) \nu (p \vee b)$ and, as in (1) and Proposition 2.1, it follows that $(p \vee b) \in N_b \cap P_a$, and then $(x \vee y) \in T$.

2) $(x \vee y) > (p \vee b)$. Then $(x \vee y) > b$. If $\phi(x \vee y) < \phi(b)$, then $(x \vee y)\nu b$ and $x \vee y \in T$. If $\phi(x \vee y) > \phi(b)$, we have $\phi(x \vee y) = \phi(x) \vee \phi(y) \geq \phi(x), \phi(y), \phi(p), \phi(q), \phi(b)$ and $\phi(x \vee y) = \phi(x \wedge p) \vee \phi(p \wedge b) \vee \phi(y \wedge q) \vee \phi(q \wedge b)$. Going back to the preimages, $x \vee y = [(x \wedge p) \overset{\vee}{\wedge} (p \wedge b)] \overset{\vee}{\wedge} [(y \wedge q) \overset{\vee}{\wedge} (q \wedge b)] \leq p \overset{\vee}{\wedge} q \leq p \vee q$. But from (2) and (3) we have $x \vee y \geq p \vee q$, and hence $x \vee y = p \vee q \in N_b \cap P_a$ (Proposition 2.1), i.e. $x \vee y \in T$. And thus, in every case, $x \vee y \in T$. Dually one shows that $x \wedge y \in T$. The proof is complete.

Consider the set $S = P_a \cap T = P_a \cap (\mathbf{U}_{t \in N_b \cap P_a} N_t)$, where a and b are the elements of L fixed above, i.e. $\pi_1(a, b, c)$ and $a > b$, $a > c$. From Propositions 2.1 and 2.2 it follows that S is a sublattice in L. Let us show that S is proper. It is easy to see that $b, c \in S$. Let us show that $a > S$. Let $x \in S$. Then for some $t \in N_b \cap P_a$ we have $x \nu t$. Since $\pi_1(a, b, t)$ and $a > b$, on applying Lemma 2.1 to a, b and t we have $a > t$. Further, since $\pi_1(a, t, x)$, on applying the lemma to a, t and x we have $a > x$. We have obtained $a > S$, i.e. $a \in L \backslash S$. Thus S is a proper sublattice. The fact that $a > S$ will be used below repeatedly.

Proposition 2.3. *The sublattice* $S = P_a \cap T$ *constructed above is homogeneous in* L.

Proof. Let $d \in L \backslash S$. Let us show that the comparability of d with one of the elements of S implies $d \sigma S$.

1) Consider first the case $d \bar{\sigma} a$. Let $d \sigma x$, $x \in S$, i.e. $x \nu t$ for some $t \in N_b \cap P_a$. Then $\phi(d) \sigma \phi(x)$. If $\phi(d) < \phi(x)$, then $\phi(d) < \phi(x) < \phi(a)$ and $d \sigma a$, which contradicts the hypothesis. Thus $\phi(d) > \phi(x)$ and $\phi(d) > \phi(x) \wedge \phi(t) = \phi(x \vee t)$, i.e. $d \sigma(x \vee t)$. It follows from Proposition 2.1 that $x \vee t \in S$. Further we have $d > (x \vee t)$, since $d < (x \vee t) < a$ contradicts the hypothesis. We have obtained $d > t$. After a sequence of applications of these arguments to the elements t, b, to b, t', and to t', y, where $t' \in N_b \cap P_a$ and $y \nu t'$, we get $d > b$, $d > t'$ and $d > y$, i.e. $d > S$.

2) Let $d \sigma a$ and $d \nu a$. If $d > a$ then $d > S$, since $a > S$. If $d < a$, then $\phi(d) > \phi(a) > \phi(S)$. We have obtained $d \sigma S$.

3) Now let $d \rho t$ for any $t \in N_b \cap P_a$; in particular, $d \rho b$.

3.1) Let $d \sigma x$, where $x \rho b$, $x \nu t$ and $t \in N_b \cap P_a$. Let us show that $d \sigma t$.

3.1.1) If $d \rho x$ then $\pi_1(d, x, t)$, and, applying Lemma 2.1 to the elements d, x, t, we get $d \sigma t$.

3.1.2) If $d \nu x$, we have

$$d \rho b, \ x \rho b, \ d \nu x, \ d \rho t, \ x \nu t, \ t \nu b; \ d \sigma x. \tag{1}$$

For definiteness assume, that

$$d > x. \tag{2}$$

(The case $(2')$ $d < x$ is considered dually.) We clearly have $\pi_2(x, d, t)$. Applying Lemma 2.2 to x, d and t, and using (2), we get

$$t > x, \tag{3}$$

$t > (x \wedge b)$. Applying Proposition 2.1 to t, x and b, we have $\phi(t) < \phi(x \wedge b) = \phi(x) \wedge \phi(b)$, i.e. $\phi(t) < \phi(b)$. But then

$$t > b. \tag{4}$$

Furthermore, we have $\phi(b) > \phi(t) \geq \phi(t) \wedge \phi(d) = \phi(t \wedge d)$, i.e. $b\sigma(t \wedge d)$. If $b < (t \wedge d)$ then $b < d$ and $\phi(t) < \phi(b) < \phi(d)$, i.e. $d\sigma t$, which is what we had to show. Now let $b > (t \wedge d)$. From (2) and (3) it follows that $(t \wedge d) \geq x$, and we have $b > (t \wedge d) \geq x$, i.e. $b > x$. But then $\phi(b) > \phi(x) > \phi(d)$, i.e. $b > d$, which with (4) gives $t > d$. And thus we have proved that if $d\sigma x$ then $d\sigma t$, where $x\nu t$, $t \in N_b \cap P_a$.

3.2) Let $d\sigma t$, where $t \in N_b \cap P_a$. Since we have $\pi_1(d, t, b)$, on applying Lemma 2.1 to d, t and b we get $d\sigma b$.

3.3) Let $d\sigma b$. Since $\pi_1(d, b, t')$, where t' is any element in $N_b \cap P_a$, on applying Lemma 2.1 to d, b and t' we get $d\sigma t'$.

3.4) Let $d\sigma t'$, where $t' \in N_b \cap P_a$. Let us show that $d\sigma y$, where $y\nu t'$ and $y\rho b$.

3.4.1) If $d\rho y$ then $\pi_1(d, t', y)$, and, applying Lemma 2.1 to d, t' and y, we have $d\sigma y$.

3.4.2) If $d\nu y$ we have

$$d\rho b, \ y\rho b, \ d\nu y, \ d\rho t', \ y\nu t', \ t'\nu b, \ b\sigma t'. \tag{5}$$

For definiteness assume that

$$d > t'. \tag{6}$$

(The case $(6')$ $d < t'$ is considered dually.) We have $\pi_1(d, t', b)$, and, by Lemma 2.1,

$$d > b. \tag{7}$$

From the hypothesis it follows that $\phi(t') < \phi(d) \leq \phi(d) \vee \phi(y) = \phi(d \wedge y)$, i.e. $t'\sigma(d \wedge y)$. If $t' < (d \wedge y)$, then $t' < y$ and $\phi(y) < \phi(t') < \phi(d)$, i.e. $d\sigma y$, as required. Now let $t' > (d \wedge y)$. Since $\pi_1(b, d, y)$, by Lemma 2.1, it follows from (7) that $y > b$. But then $(d \wedge y) \geq b$ and $t' > (d \wedge y) \geq b$, i.e. $t' > b$; $\phi(t') < \phi(b) < \phi(y)$ and $t' > y$, which together with (6) gives $d > y$.

Since, clearly, these three cases exhaust all the possibilities for d, on summing up the results 1), 2), 3.1), 3.2), 3.3) and 3.4), we have that the sublattice S is homogeneous. The proof is complete.

Proposition 2.4. *The sublattice* $S = P_a \cap T$, *where* $T = \mathbf{U}_{t \in N_b \cap P_a} N_t$, *is completely isolated in* L.

Proof. Let $x, y \in L \setminus S$. We have to show that $x \vee y, x \wedge y \in L \setminus S$. We try all cases.

1) $x \sigma y$. There is nothing to prove.

2) $x \bar{\sigma} z$ or $y \bar{\sigma} z$, where $z \in S$. Our assertion follows from the homogeneity of S (Proposition 2.3).

3) $x, y \in N_a$. From Proposition 2.1 we have $x \vee y, x \wedge y \in N_a$, i.e. $x \vee y, x \wedge y \in L \setminus S$.

4) $x, y \in P_a$. Since $x, y \in L \setminus S$, we have $x, y \in P_t$ for any $t \in N_b \cap P_a$. From Proposition 2.1 we get that $x \vee y, x \wedge y \in P_t$ for any $t \in N_b \cap P_a$; in particular, $x \vee y, x \wedge y \in P_b$. If $x \vee y \in N_t$ for some $t \in N_b \wedge P_a$, then clearly $x \vee y = t$. In such a case we have $x \vee y \in N_b$ and $x \vee y = b$, and similarly, $x \vee y = t'$ for any $t' \in N_b \cap P_a$, i.e. $N_b \cap P_a$ contains only one element. But this is not true, since $b, c \in N_b \cap P_a$. Thus $x \vee y \notin N_t$ for any $t \in N_b \cap P_a$, i.e. $x \vee y \in L \setminus S$. It is shown in exactly the same way that $x \wedge y \in L \setminus S$.

5) $x \bar{\sigma} y$, $x \sigma S$, $y \sigma S$, $x \rho a$ and $y \nu a$. At least one of x and y is not comparable with a since

    a) if $x > a$, $a > y$, then $x > y$;

    b) if $x < a$, $a < y$, then $x < y$;

    c) if $x > a$, $a < y$, then $\phi(x) > \phi(a) > \phi(y)$, i.e. $x \sigma y$;

    d) if $x < a$, $a > y$, then $\phi(x) < \phi(a) < \phi(y)$, i.e. $x \sigma y$.

Thus x (or y) $\bar{\sigma} a$. It follows that x (or y) $> S$ since, if not, then for some $z \in S$ we would have x (or y) $< z < a$ $(a > S)$, which contradicts what we have just shown. It follows in turn that y (or x) $> S$. Thus x and $y > S$. But then $x \vee y > S$, i.e. $x \vee y \in L \setminus S$. Now consider $x \wedge y \geq S$. Suppose that $x \wedge y \in S$. Then there is a $t \in S$ such that $(x \wedge y) \nu t$ and $t \in N_b \cap P_a$, and since $t \leq (x \wedge y)$ we have $\phi(t) \geq \phi(x \wedge y)$. Let us show that $\phi(t) > \phi(x), \phi(y)$. Indeed, clearly either $\phi(t) > \phi(x), \phi(y)$ or $\phi(t) < \phi(x), \phi(y)$. If the second is true then $\phi(x \vee y) = \phi(x) \vee \phi(y) \geq \phi(t)$; but we have $\phi(x \wedge y) < \phi(t)$, and thus $\phi(t) > \phi(x), \phi(y)$. We have $t < a$; hence $\phi(a) > \phi(t) > \phi(x), \phi(y)$, which contradicts x (or y) $\bar{\sigma} a$. Thus $x \wedge y \in L \setminus S$. The proposition is proved.

Hence under our assumptions there exists in L a proper homogeneous completely isolated sublattice S, which completes the proof of Theorem 2.1.

§3. Layers. The analysis of the action of a semi-isomorphism on a lattice L

Definition 7. The set of elements of a component A of an irreducible decomposition of a saturated sublattice H not lying in any proper saturated sublattice of L strictly contained in H is called the *layer* corresponding to the given component A of the saturated sublattice H.

We will denote the layer corresponding to the component A by $C(A)$. It is easy to see from the definition of a homogeneous completely isolated sub-lattice that a layer is a sublattice in L. Clearly, each element in L is in a unique layer, namely, in the layer corresponding to the component containing this element in an irreducible decomposition of the saturated sublattice generated by this element (see Lemma 1.10). This means that L is covered by disjoint layers.

Let x and y be distinct elements in L. Consider $H(x, y)$ (see the notation at the end of §1).

1. Let x and y be such that $A(x, y)$ exists. Let us show that $C(A(x, y))$ contains at least two elements.[1] If $H(x) = H(y) = H(x, y)$, then clearly $x, y \in C(A(x, y))$. Let, for example, $H(x) \subset H(x, y)$. Let G denote the union of all saturated sublattices containing x and not y. By Proposition 1.2 these sublattices form a chain under inclusion; hence by Lemma 1.8 G is a homogeneous completely isolated sublattice. Since $y \notin G$, $G \neq A(x, y)$.[2] In $A(x, y) \backslash G$ there is an element not comparable with x, since, if not, any element in $A(x, y) \backslash G$ would be comparable with G, and by the complete isolatedness of G $A(x, y)$ would be decomposable, which is impossible. Let $z \in A(x, y) \backslash G$ and $z \bar{\sigma} x$. Let us show that $x \vee z$ and $x \wedge z$ are not in any true saturated sublattice of L which is strictly contained in $H(x, y)$. Indeed if $x \vee z (x \wedge z) \in F$, where F is a saturated sublattice of L, then by the complete isolatedness of F it contains x or z. But then since $x \bar{\sigma} z$, by the homogeneity of F it contains also the second element. From the definition of G and Proposition 1.2 we have that $F \supseteq H(x, y)$. What we have

---

1) It is not difficult to construct an example where to a component A of a saturated sublattice H there corresponds the empty layer. This will always be the case if A is the union of an infinite increasing chain of saturated sublattices (see Lemma 1.8).

2) It is clear from this that $A(x, y)$ cannot be expressed in the form of an increasing chain of saturated sublattices.

just proved means that $x \vee z, x \wedge z \in C(A(x, y))$. And thus $C(A(x, y))$ has at least two elements.

Let us construct a mapping of the sublattice $A(x, y)$ into itself as follows. On $C(A(x, y)) \neq \emptyset$ let this mapping coincide with the identity isomorphism and let each maximal saturated sublattice of $A(x, y)$ be carried to some fixed element of this saturated sublattice. (Note that the maximal saturated sublattices in $A(x, y)$ do not intersect by Proposition 1.2.) From Lemmas 1.5 and 1.6 it follows that this mapping is an endomorphism. Let the image of $A(x, y)$ under this endomorphism be denoted $\overline{A}(x, y)$. Clearly, $\overline{x} \neq \overline{y}$ in $\overline{A}(x, y)$, where $\overline{x}$ is the image of x and $\overline{y}$ is the image of y. From Proposition 1.2 and Lemma 1.11 it follows that $\overline{A}(x, y)$ contains no proper saturated sublattices. $\overline{A}(x, y)$ does not even contain proper homogeneous completely isolated sublattices. In fact, let $\overline{F}$ be a true homogeneous completely isolated sublattice of $\overline{A}(x, y)$. Since $\overline{A}(x, y)$ contains no proper saturated sublattices, either the saturated sublattice in $\overline{A}(x, y)$ generated by $\overline{F}$ coincides with $\overline{A}(x, y)$ or it is a one-element sublattice. If the fomer is true then from Lemma 1.9, by the complete isolatedness of F, it follows that $\overline{A}(x, y)$ is decomposable. But then also $A(x, y)$ is decomposable (by the homogeneity of maximal saturated sublattices in $A(x, y)$), which cannot be. Thus the latter holds, i.e. $\overline{F}$ is not proper. And so $\overline{A}(x, y)$ has no proper homogeneous completely isolated sublattices.

From Lemmas 1.5 and 1.6 it is easy to see that the semi-isomorphism ϕ induces in $\overline{A}(x, y)$ a semi-isomorphism $\overline{\phi}$ which, by Theorem 2.1, is either the identity or a dual isomorphism.

From Lemmas 1.5, 1.6 and 1.7 it is easy to get that a pair (x, y) is regular relative to ϕ if and only if the pair $(\overline{x}, \overline{y})$ is regular relative to $\overline{\phi}$. Thus, since $\overline{\phi}$ coincides with ϕ on $C(A(x, y))$, and since $C(A(x, y)) \subseteq \overline{A}(x, y)$ and $C(A(x, y))$ contains not less than two elements, we conclude that the pair (x, y) is regular (nonregular) relative to ϕ if and only if ϕ induces on $C(A(x, y))$ the identity isomorphism (dual isomorphism).

2. Now let $x \in A$ and $y \in B$, where A and B are distinct components of an irreducible decomposition of $H(x, y)$. In this case, using Lemma 1.4, we find that the pair (x, y) is regular (nonregular) if and only if either $A < B$ and $\phi(A) < \phi(B)$ $(\phi(A) > \phi(B))$, or $A > B$ and $\phi(A) > \phi(B)$ $(\phi(A) < \phi(B))$.

§4. Criteria for the projectability of lattices

Let $\mathfrak{U}$ be any set of components of an irreducible decomposition of saturated sublattices of a lattice L. Let $C(\mathfrak{U})$ denote the set of layers corresponding to the components in $\mathfrak{U}$. We will now give the notation of the lattice obtained from the given lattice L by revolutions of the layers $C(\mathfrak{U})$. Let us introduce a binary relation μ on the set L. Set $x \mu y$ if and only if one of the following conditions holds:

a) x and y lie in some component A of $\mathfrak{U}$, where $A = A(x, y)$ and $x \geq y$;

b) there does not exist a component in $\mathfrak{U}$ with the above property and $x \leq y$.

Clearly the comparable elements in the lattice L and they alone are in the relation μ. Let us show that μ is a partial order on L. The reflexivity and antisymmetry are obvious. Let us show the transitivity. First let us state the following lemma, which follows from Proposition 1.2.

Lemma 4.1. 1) *Any three elements in L can be denoted x, y and z in such a way that $H(x, z) \subseteq H(x, y) = H(y, z) = H(x, y, z)$.*

2) *For any three elements x, y, $z \in L$ satisfying the condition $H(x, z) \subseteq H(x, y) = H(y, z) = H(x, y, z)$, if $A(x, y)$ exists, then $A(y, z)$ exists and $A(x, y) = A(y, z) = A(x, y, z)$.*

Let $x \mu y$ and $y \mu z$. Let us show that $x \mu z$. We consider all cases.

1. For x and y, and y and z, we have a). By Proposition 1.2 $A(x, y)$ and $A(y, z)$ are incident, and since $x > y > z$ it follows that $A(x, z)$ exists and is the greater of them, i.e. $A(x, z) \in \mathfrak{U}$. Thus in this case $x \mu z$.

2. For x and y, and y and z, we have b). If $A(x, z)$ does not exist, then it is clear that $x \mu z$. Suppose that $A(x, z)$ exists. Since $x < y < z$, then $y \in H(x, z)$ (Lemma 1.6). By Lemma 4.1.1), either $H(x, y) = H(x, y, z)$, or $H(y, z) = H(x, y, z)$. Suppose, for example, that $H(x, y) = H(x, y, z)$. Then $A(x, y)$ exists by the convexity of $A(x, z)$, and $A(x, z) = A(x, y) \notin \mathfrak{U}$, i.e. $x \mu z$. We have the same if $H(y, z) = H(x, y, z)$.

3. For x and y we have a), and for y and z we have b). Consider $A(x, y)$ and $H(y, z)$. It is easy to see that $A(x, y) \neq H(y, z)$. By Proposition 1.2 either (α) $A(x, y) \subset H(y, z)$ or (β) $A(x, y) \supset H(y, z)$.

Consider case (α). If $H(x, y) = H(y, z)$, then z and x lie in distinct components of the irreducible decomposition of $H(x, y)$, and since $y < z$, $x < z$. We have $x \mu z$ in this case. If $H(x, y) \subset H(y, z)$, then either x and z

lie in distinct components of the irreducible decomposition of $H(x, z)$, or by Lemma 4.1.2), $A(x, z) = A(y, z) \notin \mathfrak{A}$. Moreover, since $y < z$, it follows from Lemma 1.6 that $z < x$. Hence $x \mu z$.

Consider case (β). By Lemma 4.1.2), $A(x, z) = A(x, y) \in \mathfrak{A}$. Moveover, since $z > y$, Lemma 1.6 implies that $z > x$. Hence $x \mu z$.

4. For x and y we have b), for y and z we have a). This case is considered similarly to the preceding.

The transitivity is proven. Let L_1 denote the set L with the partial order we have just constructed. Let us show that L_1 is a lattice. We will denote the infimum of x and y by $\inf_\mu (x, y)$, and the supremum by $\sup_\mu (x, y)$. If $x \mu y$ then $\inf_\mu (x, y) = x$, and $\sup_\mu (x, y) = y$.

Let x and y be incomparable with respect to μ. Then from the definition of μ it follows that $x \bar{\sigma} y$. Let $z \mu x$ and $z \mu y$. For z and x and for z and y one has simultaneously either a) or b), since, if not, $x \sigma y$.

1. Let a) hold for z and x and for z and y. From Lemma 4.1.1), we have either $H(z, x) = H(x, y, z)$ and $A(z, x) = A(x, y, z)$, or $H(z, y) = H(x, y, z)$ and $A(z, y) = A(x, y, z)$. Clearly, $H(x \vee y, z) \subseteq H(x, y, z)$. On the other hand, since $x \bar{\sigma} y$, by the homogeneity and complete isolatedness of $H(x \vee y, z)$ we have $H(x, y, z) \subseteq H(x \vee y, z)$. Thus we have obtained $H(x \vee y, z) = H(x, y, z)$. It is also easy to see that $A(x \vee y, z)$ exists and $A(x \vee y, z) = A(x, y, z) \in \mathfrak{A}$. Similarly we get $H(x \wedge y, z) = H(x, y, z)$ and $A(x \wedge y, z) = A(x, y, z)$. Since, moreover, from the hypothesis it follows that $z > (x \wedge y)$ and $z \geq (x \vee y)$, we clearly have $z \mu (x \wedge y)$ and $z \mu (x \vee y)$. If $x \vee y$ and $x \wedge y$ satisfy a), then $(x \vee y) \mu (x, y) \mu (x \wedge y)$ and we have $\inf_\mu (x, y) = x \vee y$. If $x \vee y$ and $x \wedge y$ satisfy b) then $(x \wedge y) \mu (x, y) \mu (x \vee y)$, and we have $\inf_\mu (x, y) = x \wedge y$.

2. Let b) hold for z and x and for z and y. Just as above, it is shown that $H(x, y, z) = H(x \vee y, z) = H(x \wedge y, z)$ and either $H(z, x) = H(x, y, z)$ or $H(z, y) = H(x, y, z)$. For definiteness, let $H(z, x) = H(x, y, z)$. If $A(z, x)$ does not exist, then clearly $A(x \vee y, z)$ and $A(x \wedge y, z)$ do not exist, since $x, x \vee y$ and $x \wedge y$ cannot lie in distinct components of a decomposition of $H(x, y, z)$. If $A(z, x)$ exists, then by the same reasoning $A(x \vee y, z)$ and $A(x \wedge y, z)$ exist and $A(x \vee y, z) = A(x \wedge y, z) = A(x, z) \notin \mathfrak{A}$. Since moreover, $z \leq (x \wedge y)$, $z < (x \vee y)$, in both cases we have $z \mu (x \vee y)$ and $z \mu (x \wedge y)$. If $x \vee y$ and $x \wedge y$ satisfy a) then $(x \vee y) \mu (x, y) \mu (x \wedge y)$, and we have $\inf_\mu (x, y) = x \vee y$. If $x \vee y$ and $x \wedge y$ satisfy b) then $(x \wedge y) \mu (x, y) \mu (x \vee y)$, and we have $\inf_\mu (x, y) = x \wedge y$. We have proved that L_1 is a semilattice with respect to meets. Similarly one shows that L_1 is a

semilattice with respect to joins, where the sup also is either $x \vee y$ or $x \wedge y$. And thus, L_1 is a lattice. We call it *the lattice obtained from the given lattice L by revolutions of the layers $C(\mathfrak{U})$*.

Since for any $x, y \in L$ we have either $\inf_\mu (x, y) = x \vee y$ and $\sup_\mu (x, y) = x \wedge y$, or $\inf_\mu (x, y) = x \wedge y$ and $\sup_\mu (x, y) = x \vee y$, it follows that the mapping of L onto L' where each element is assigned to itself is a semi-isomorphism. We call it *the canonical semi-isomorphism of first degree corresponding to the layers $C(\mathfrak{U})$ and denote it by ϕ_1*.

Remark 4.1. A pair (x, y) will be nonregular relative to ϕ_1 if and only if $A(x, y)$ exists and $A(x, y) \in \mathfrak{U}$.

Lemma 4.2. *For any semi-isomorphism ϕ of the lattice L*

1) *there exists a semi-isomorphism ϕ_1' adequate to ϕ_1 relative to ϕ;*

2) *there exists a semi-isomorphism ϕ' adequate to ϕ relative to ϕ_1.*

Proof. 1) From Lemmas 1.7 and 1.4 it follows that under ϕ a component of an irreducible decomposition of a saturated sublattice of L goes onto a component of an irreducible decomposition of a saturated sublattice in $\phi(L)$ and a layer goes onto a layer. Thus it is clear that $A(x, y)$ exists if and only if $A(\phi(x), \phi(y))$ exists in $\phi(L)$. Noting Remark 4.1, it is easy to see that for ϕ_1' one can take the canonical semi-isomorphism of first degree of the lattice $\phi(L)$ corresponding to the layers $C(\phi(A_\xi))$, where A_ξ runs over $\mathfrak{U}$.

2) To prove this it is sufficient to set $\phi' = \phi_1' \cdot \phi \cdot \phi_1^{-1}$, where ϕ_1' is a semi-isomorphism adequate to ϕ_1 relative to ϕ.

Consider the set $L' = (L \backslash H) \bigcup H'$, where H is a saturated sublattice of L and H' is the lattice obtained from H by some permutation ψ of the components of its irreducible decomposition. From Lemmas 1.5 and 1.6 it is easy to see that L' is a lattice if we additionally set $x \vee y' = x \vee y$ and $x \wedge y' = x \wedge y$, for any $x \in L \backslash H$, $y \in H$ and $y' \in H'$. We will say that the lattice L' thus constructed is obtained from l by the given permutation ψ of the components of the irreducible decomposition of the saturated sublattice H.

Using this definition in a similar way to that in [1] for semilattices, we will now give the notion of the lattice obtained from a given lattice L by permutations of the components of the irreducible decompositions of some class of saturated sublattices. Let $[H_\omega] (\omega \in \Omega)$ be a class of saturated sublattices in L and let $[\psi_\omega] (\omega \in \Omega)$ be permutations of the components of the irreducible decompositions of these saturated sublattices. We will say that $x, y \in L$ participate in the permutation ψ_ω if x and y lie in distinct

components of the irreducible decomposition of H_ω. From Proposition 1.2
it follows that x and y cannot participate in more than one permutation.
Note that x and y participate in some permutation in $[\psi_\omega]$ $(\omega \in \Omega)$ if and
only if $H(x, t) \in [H_\omega]$ $(\omega \in \Omega)$ and $A(x, y)$ does not exist. Let us introduce
a binary relation λ on L. If x and y do not participate in any permutation
in $[\psi_\omega]$ set $x\lambda y$ if and only if $x \leq y$. If x and y do participate in some
permutation ψ_ω set $x\lambda y$ if and only if $x' \leq y'$, where x' and y' are the
images of x and y under some semi-isomorphism of H_ω corresponding to
ψ_ω. (Such a semi-isomorphism exists by Lemma 1.3 and Proposition 1.1.)
Similarly as was done in [1], it is shown that λ is a partial order relative to
which L is a lattice. Let us denote it L_2. Simultaneously one shows that
the identity mapping of L onto L_2 is a semi-isomorphism. Denote it ϕ_2.
We call L_2 *the lattice obtained from the given lattice L by the permutations*
$[\psi_\omega]$ $(\omega \in \Omega)$ *of the components of the irreducible decompositions of the*
saturated sublattices $[H_\omega]$ $(\omega \in \Omega)$. The semi-isomorphism ϕ_2 is called *the*
canonical semi-isomorphism of second degree corresponding to the permuta-
tions $[\psi_\omega]$ $(\omega \in \Omega)$ *of the irreducible decompositions of the saturated sub-*
lattices $[H_\omega]$ $(\omega \in \Omega)$.

Remark 4.2. Any pair (x, y) such that $A(x, y)$ exists will be regular
relative to ϕ_2.

Taking into account Lemma 4.2, for any canonical semi-isomorphisms of
first and second degree ϕ_1 and ϕ_2 of the lattice L, we can make

Definition 8. The lattice $L' = \phi_1' \cdot \phi_2(L)$, where ϕ_2 is the canonical
semi-isomorphism of second degree corresponding to the permutations
$[\psi_\omega]$ $(\omega \in \Omega)$ of the components of the irreducible decompositions of the
saturated sublattices $[H_\omega]$ $(\omega \in \Omega)$ and ϕ_1' is a semi-isomorphism adequate
relative to ϕ_2 to the canonical semi-isomorphism of first degree corresponding to the layers $C(\mathfrak{A})$, is called the *lattice obtained from the given lattice*
L by revolutions of the layers $C(\mathfrak{A})$ and permutations $[\psi_\omega]$ $(\omega \in \Omega)$ of the
components of the irreducible decompositions of the saturated sublattices
$[H_\omega]$ $(\omega \in \Omega)$.

Let ϕ be any semi-isomorphism of L. Let $\mathfrak{A}^0$ be the class of all components $A(x, y)$ for which the pair (x, y) is nonregular relative to ϕ, and
let $\mathfrak{H}$ be the class of all saturated sublattices in L. Let us construct the
canonical semi-isomorphism of first degree ϕ_1^0 of L corresponding to the
layers $C(\mathfrak{A}^0)$, and the canonical semi-isomorphism of second degree ϕ_2^0 of
the lattice L corresponding to those permutations of the components of the

irreducible decompositions of the saturated sublattices $\mathfrak{H}$ such that, for any two components A and B of any saturated sublattice of L, if $\phi(A) < \phi(B)$ then $\phi_2^0(A) < \phi_2^0(B)$. The existence of such a ϕ_2^0 follows from the definition of the lattice obtained from the given one by permutations of the components of the irreducible decompositions of the saturated sublattices.

Consider the semi-isomorphism $\epsilon = \phi'^0_1 \phi^0_2 \phi^{-1}$ of the lattice $\phi(L)$, where ϕ'^0_1 is the semi-isomorphism adequate to ϕ_1^0 relative to ϕ_2^0 constructed as in Lemma 4.2. As we proved in §3, the pair (x, y) is nonregular relative to ϕ only in two cases: 1) if $A(x, y)$ exists and $A(x, y) \in \mathfrak{U}^0$; 2) if $x \in A$ and $y \in B$, where A and B are distinct components of the irreducible decomposition of $H(x, y)$ and either $A > B$ but $\phi(A) < \phi(B)$, or $A < B$ but $\phi(A) > \phi(B)$. Taking this into account along with Remarks 4.1 and 4.2, it is easy to see that ϵ is an isomorphism. We have thus proved

Proposition 4.1. *Every semi-isomorphism ϕ can be written in the form of a product of a canonical semi-isomorphism of second degree, a canonical semi-isomorphism of first degree and an isomorphism.*

Let us now formulate the main result of the paper.

Theorem 4.1. *In order that a lattice L be projectable onto a lattice L' it is necessary and sufficient that L' be isomorphic to the lattice obtained from L by revolutions of the layers and permutations of the components of the irreducible decompositions of the saturated sublattices into an ordinal sum.*

The necessity follows from Propositions 4.1 and 1.1, and the sufficiency follows from Definition 8 and Proposition 1.1.

§5. Projections of lattices of certain types

Theorem 5.1. *A Dedekind lattice L can be projected onto a lattice L' if and only if L' is obtained from L by revolutions and permutations of the components of the irreducible decomposition of L into an ordinal sum.*

Proof. Let us show that L does not have proper saturated sublattices. Suppose not: let H be a proper saturated sublattice of L. Then there is an $e \in L_{\bar{\sigma}}(H)$ (see Lemma 1.9). Since H is proper it contains not less than two elements. Let $a, b, \in H$. Since H is a sublattice, we can assume $a \sigma b$. By Lemma 1.8, $e \vee a = e \vee b = c$ and $e \wedge a = e \wedge b = d$, and we have a five-element non-Dedekind sublattice $\{a, b, c, d, e\}$, contrary to the hypothesis. Thus L has no proper saturated sublattices. But then, as is easy to see, each nonempty layer coincides with some component of the irreducible decomposition, and the proof follows from Theorem 4.1.

Since, clearly, a dual isomorphism preserves the property of a lattice being Dedekind or distributive, we have the following corollaries.

Corollary 1. *The projective image of a Dedekind lattice is a Dedekind lattice.*

Corollary 2. *The projective image of a distributive lattice is a distrubutive lattice.*

Clearly, lattices with complements are indecomposable into an ordinal sum, and 0 and 1 lie in one layer. But then from Theorem 4.1 and Proposition 1.1 we have

Theorem 5.2. *The projective image of a lattice with complements is also a lattice with complements.*

Theorem 5.3. *Any projection of a lattice L with relative complements is induced by either an isomorphism or a dual isomorphism.*

Proof. Let us show that L has no proper homogeneous completely isolated sublattices. Suppose not: let F be a proper homogeneous completely isolated sublattice of L. Then L is not a two-element lattice and, as is easy to see, is indecomposable into an ordinal sum. This and Lemma 1.9 give that $L_{\bar{\sigma}}(H(F)) \neq \varnothing$. Let $c \in L_{\bar{\sigma}}(H(F))$. Since F is a proper sublattice, there exist $a, b \in H(F)$ such that $a > b$. By the homogeneity of $H(F)$ we have $c_1 = c \vee b \in L_{\sigma}(H(F))$. Since $b < c_1$, by Lemma 1.6 we have $a < c_1$. Consider the closed interval $[b, c_1]$. By hypothesis there is an a' such that $a \vee a' = c_1$ and $a \wedge a' = b$. Clearly, $a'\bar{\sigma}a$. Since, moreover, $a'\sigma b$, by the homogeneity of $H(F)$ we have $a' \in H(F)$. But this is impossible since then $a \vee a' = c_1 \in H(F)$, which contradicts what we proved earlier. Thus our supposition is false, i.e. L has no proper homogeneous completely isolated sublattices, and the theorem follows from Theorem 2.1.

Theorem 5.4. *Any projection of a lattice with unique complements is induced by either an isomorphism or a dual isomorphism.*

Proof. Let L be a lattice with unique complements. Let us show that L has proper homogeneous completely isolated sublattices. Suppose not: let F be a proper homogeneous completely isolated sublattice. Let us show that 0 and 1 do not lie in F. In fact, if $0 \in F$ or $1 \in F$, then, by the complete isolatedness of F, for any two complementary elements x and x' at least one of them is in F. Since an element is comparable with its complement only if it is 0 or 1, for any elements distinct from 0 and 1, say x and x', which are complementary, by the complete isolatedness and homogeneity of F we get x, $x' \in F$ and $0 = x \wedge x' \in F$, $1 = x \vee x' \in F$. We have $F = L$, which contradicts the assumption. Thus 0 and 1 do not lie in F. Since F is a proper sublattice there are a, b, $\in F$ such that $a \neq b$ and $a \sigma b$. Let a' be the complement of a. Then $a' \notin F$, since, if not, $1 = a \vee a' \in F$. We also have $a \overline{\sigma} a'$. Then $a' \overline{\sigma} F$. By Lemma 1.5 we have $b \vee a' = a \vee a' = 1$ and $b \wedge a' = a \wedge a' = 0$. We have for a' two complements a and b, which contradicts the hypothesis. Thus L has no proper homogeneous completely isolated sublattices, and our assertion follows from Theorem 2.1.

Corollary. *Any projection of a Boolean algebra is induced by either an isomorphism or a dual isomorphism.*

BIBLIOGRAPHY

[1] L. N. Ševrin, *Basic problems in the theory of projections of semilattices*, Mat. Sb. 66 (108) (1965), 568–597; English transl., Amer. Math. Soc. Transl. (2) 96 (1970), 1–35.   MR 35 #2799.

[2] G. Birkhoff, *Lattice theory*, Amer. Math. Soc. Colloq. Publ., vol. 25, rev. ed., Amer. Math. Soc., Providence, R. I., 1948; Russian transl., IL, Moscow, 1952.   MR 1, 325;   MR 10, 673.

Translated by:
J. Ellis

Amer. Math. Soc. Transl.
(2) Vol. 96, 1970

PARTIAL IDEMPOTENT OPERATIONS
ASSOCIATED WITH ORDERED SETS[*]

UDC 519.50

V. V. ROZEN

A set A with a given order relation ω is said to be an ordered set; it can be considered as a pair $(A; \omega)$. Associated with every order relation ω there is a binary relation $S(\omega)$ representable as the set of all pairs of elements (a_1, a_2) for which there exists a greatest lower bound with respect to ω. Since the greatest lower bound of a pair of elements, if it exists, is unique, on every ordered set $(A; \omega)$ we obtain in a natural way a partial binary operation by taking the greatest lower bound of pairs of elements, where $S(\omega)$ is the defining relation of this partial operation.

If an ordered set is a semilattice with respect to greatest lower bounds (i. e. every pair of elements has a greatest lower bound), then the corresponding partial operation becomes an operation and its properties have been thoroughly studied [3]. However, in applications we frequently are concerned with ordered sets which do not have this property, and therefore the study of this partial operation is of interest in the general case and is the goal of the present article.

A set A with a given partial binary operation o is called a partial binary operative; it can be considered as the pair $(A; o)$. To every partial binary operative $(A; o)$ there corresponds a global operative whose operation $\breve{o}$ is defined for subsets of the set A. We will set $\breve{o}(a_1, a_2) = a_1 a_2$. In the future we will make use of certain facts connected with the theory of relations and with the theory of partial binary operatives [2].

Let $(A; \omega)$ be an ordered set. We will use the notation $a_1 \leqslant a_2$ as being the equivalent of $(a_1, a_2) \in \omega$. If a pair of elements (a_1, a_2) has a greatest lower bound, we will denote it by $\inf (a_1, a_2)$, indicating, if necessary, the order relation with respect to which the greatest lower bound is being considered. We will say that a partial operative $(A; o)$ is associated with the ordered set $(A; \omega)$ if o coincides with the partial operation of taking the greatest lower bound of pairs of elements with respect to the relation ω. We note that the order relation ω can be recaptured with the aid of the

---

[*] Translation of Izv. Vysš. Učebn. Zaved. Matematika 1966, no. 4 (53), 96–103.

partial operation o. Thus an ordered set and a partial binary operative associated with it are completely concomitant with one another, and thus examining one of them is completely equivalent to examining the other. In this we see one of the distinctions between the partial operatives considered here and the lattice-oids introduced in [4].

Theorem 1. *In order that a partial binary operative* $(A; o)$ *be associated with some ordered set, it is necessary and sufficient that the following conditions be satisfied:*

$$\{a\}\{a\} = \{a\}, \tag{1}$$
$$\{a_1\}\{a_2\} = \{a_2\}\{a_1\}, \tag{2}$$
$$\{a_1\}\{a_2\} \neq \varnothing \wedge \{a_2\}\{a_3\} \neq \varnothing \rightarrow (\{a_1\}\{a_2\})\{a_3\} = \{a_1\}(\{a_2\}\{a_3\}), \tag{3}$$
$$\left(\bigvee_{\bar{a}} A\{a_1\} \cap A\{a_2\} = A\{\bar{a}\}\right) \rightarrow \{a_1\}\{a_2\} \neq \varnothing. \tag{4}$$

Proof. Necessity. Let $(A; o)$ be a partial binary operative associated with the ordered set $(A; \omega)$. Conditions (1) and (2) are then fulfilled in an obvious manner. Suppose premise (3) is fulfilled; that is, $\inf(a_1, a_2)$ and $\inf(a_2, a_3)$ exist. We suppose that $(\{a_1\}\{a_2\})\{a_3\} \neq \varnothing$; that is, $\inf(\inf(a_1, a_2), a_3)$ exists and is equal to $\bar{a}$. Then $\bar{a} \lesssim a_1$, $\bar{a} \lesssim a_2$ and $\bar{a} \lesssim a_3$. Since $\inf(a_2, a_3)$ exists by hypothesis, we obtain $\bar{a} \lesssim a_1$ and $\bar{a} \lesssim \inf(a_2, a_3)$. Suppose for some a we have $a \lesssim a_1$ and $a \lesssim \inf(a_2, a_3)$; then $a \lesssim a_1$, $a \lesssim a_2$ and $a \lesssim a_3$. Since $\inf(a_1, a_2)$ exists by assumption, we obtain $a \lesssim \inf(a_1, a_2)$ and $a \lesssim a_3$, and so $a \lesssim \bar{a}$. We have shown that $\bar{a} = \inf(a_1, \inf(a_2, a_3))$; that is, the conclusion of the implication (3) holds if $(\{a_1\}\{a_2\})\{a_3\} \neq \varnothing$. In an analogous manner we can show that the conclusion of (3) holds if $\{a_1\}(\{a_2\}\{a_3\}) \neq \varnothing$. This proves (3). Further, it is easy to verify that $A\{a\}$ is the set of all lower bounds of the element a, from which (4) follows immediately.

Sufficiency. We consider a partial binary operative $(A; o)$ for which the conditions (1)–(4) are fulfilled. We introduce a binary operation $\omega \subset A \times A$, defining it by the formula

$$(a_1, a_2) \in \omega \longleftrightarrow \{a_1\}\{a_2\} = \{a_1\}. \tag{5}$$

Using (1) and (2) we immediately obtain that ω is reflexive and antisymmetric. Let $(a_1, a_2) \in \omega$ and $(a_2, a_3) \in \omega$, i.e. $\{a_1\}\{a_2\} = \{a_1\}$ and $\{a_2\}\{a_3\} = \{a_2\}$. Then, in particular, $\{a_1\}\{a_2\} \neq \varnothing$ and $\{a_2\}\{a_3\} \neq \varnothing$. Therefore, in accordance with (3), we have $(\{a_1\}\{a_2\})\{a_3\} = \{a_1\}(\{a_2\}\{a_3\})$, from whence it follows that $\{a_1\}\{a_3\} = \{a_1\}$; that is, $(a_1, a_3) \in \omega$. We have shown that ω is an order relation. Let $(a_1, a_2) \in \delta$, where δ is the defining relation of the partial binary operative $(A; o)$. Then, using $\{a_2\}\{a_1\} \neq \varnothing$

and $\{a_1\}\{a_1\} \neq \emptyset$, we obtain $\{a_1 a_2\}\{a_1\} = (\{a_1\}\{a_2\})\{a_1\} = (\{a_2\}\{a_1\})\{a_1\} = \{a_2\}(\{a_1\}\{a_1\}) = \{a_2\}\{a_1\} = \{a_1\}\{a_2\} = \{a_1 a_2\}$, i.e. $(a_1 a_2, a_1) \in \omega$. Analogously $(a_1 a_2, a_2) \in \omega$. Suppose that for some a we have $(a, a_1) \in \omega$ and $(a, a_2) \in \omega$, i.e. $\{a\}\{a_1\} = \{a\}$ and $\{a\}\{a_2\} = \{a\}$. Then, taking into account $\{a_1\}\{a_2\} \neq \emptyset$ and $\{a\}\{a_1\} \neq \emptyset$, we obtain $\{a\}\{a_1 a_2\} = \{a\}(\{a_1\}\{a_2\}) = (\{a\}\{a_1\})\{a_2\} = \{a\}\{a_2\} = \{a\}$, i.e. $(a, a_1 a_2) \in \omega$. We have shown that if $(a_1, a_2) \in \delta$, then $a_1 a_2$ is the greatest lower bound of the pair of elements (a_1, a_2) with respect to the order relation defined by formula (5). We will show that the following equation is satisfied:

$$\overset{-1}{\omega}\langle a\rangle = A\{a\}. \tag{6}$$

We have $a' \in \omega^{-1}\langle a\rangle \leftrightarrow (a', a) \in \omega \leftrightarrow \{a'\}\{a\} = \{a'\} \rightarrow a' \in A\{a\}$. Conversely, if $a' \in A\{a\}$, then for some a'', $\{a'\} = \{a''\}\{a\}$; then $(a'', a) \in \delta$ and by what has been shown a' is the greatest lower bound of the pair (a'', a), from whence it follows, in particular, that $a' \in \omega^{-1}\langle a\rangle$. Suppose a pair (a_1, a_2) has a greatest lower bound with respect to ω, and write $\bar{a} = \inf(a_1, a_2)$. Then $\omega^{-1}\langle a_1\rangle \cap \omega^{-1}\langle a_2\rangle = \omega^{-1}\langle \bar{a}\rangle$. Using (6), we obtain $A\{a_1\} \cap A\{a_2\} = A\{\bar{a}\}$, and, by (4), $(a_1, a_2) \in \delta$. We have shown that the partial binary operative $(A; o)$ is associated with the ordered set $(A; \omega)$, where ω is defined by formula (5).

Obtained in the course of the proof is the following

Corollary. *If a partial binary operative $(A; o)$ satisfies conditions (1)–(3), then the binary relation ω defined by formula (5) is an order relation, where, moreover, if $(a_1, a_2) \in \delta$ (where δ is the relation of determination for $(A; o)$), then $a_1 a_2$ is the greatest lower bound of the pair (a_1, a_2) with respect to ω.*

It is clear that conditions (1)–(3) are elementary; condition (4) can also be expressed in an elementary form, and therefore the class of partial binary operatives associated with ordered sets are elementarily axiomatizable. We will next consider the question of axiomatizability of a given class with the aid of identities.

An algebraic system $(A; o_1, o_2)$, where o_1 and o_2 are partial binary operations given on A, is called a partial binary bioperative. We will say that a partial binary bioperative $(A; o_1, o_2)$ is associated with an ordered set $(A; \omega)$ if o_1 coincides with the partial operation of taking greatest lower bounds and o_2 coincides with the partial operation of taking least upper bounds of pairs of elements.

Theorem 2. *In order that a partial binary bioperative $(A; o_1, o_2)$ be* *associated with some ordered set, it is necessary and sufficient that for o_1 and o_2 conditions (1)–(4) be fulfilled and also that conditions*

$$\{a_1\} \, o_1 \, \{a_2\} \neq \varnothing \longrightarrow \{a_1\} \, o_2 \, (\{a_1\} \, o_1 \, \{a_2\}) = \{a_1\}, \tag{7}$$

$$\{a_1\} \, o_2 \, \{a_2\} \neq \varnothing \longrightarrow \{a_1\} \, o_1 \, (\{a_1\} \, o_2 \, \{a_2\}) = \{a_1\} \tag{8}$$

be fulfilled

Proof. The necessity of conditions (1)–(4) follows from Theorem 1 and the principle of duality; the necessity of conditions (7) and (8) can easily be verified. We will prove the sufficiency of these conditions. We introduce two binary relations $\omega_1 \subset A \times A$ and $\omega_2 \subset A \times A$, defined by the formulas

$$(a_1, a_2) \in \omega_1 \longleftrightarrow \{a_1\} \, o_1 \, \{a_2\} = \{a_1\}, \quad (a_1, a_2) \in \omega_2 \longleftrightarrow \{a_1\} \, o_2 \, \{a_2\} = \{a_1\}.$$

We will show that $\omega_1^{-1} = \omega_2$. If $(a_1, a_2) \in \omega_1^{-1}$, then $\{a_1\} o_1 \{a_2\} = \{a_2\}$, and, in particular, $\{a_1\} o_1 \{a_2\} \neq \varnothing$. From here, using (7), we obtain $\{a_1\} o_2 \{a_2\} = \{a_1\} o_2 (\{a_1\} o_1 \{a_2\}) = \{a_1\}$, and therefore $(a_1, a_2) \in \omega_2$. Analogously one proves the reverse inclusion with the aid of (8). By Theorem 1 o_1 coincides with the partial binary operation of taking greatest lower bounds of pairs of elements with respect to ω_1 while o_2 coincides with the partial operation of taking greatest lower bounds with respect to ω_2. Taking into account the fact that $\omega_2 = \omega_1^{-1}$, we find that $(A; o_1, o_2)$ is associated with the ordered set $(A; \omega_1)$.

Let $(A_i; o_i)$ $(i \in I)$ be an arbitrary family of partial binary operatives associated with the ordered sets $(A_1; \omega_i)$ $(i \in I)$. It is easy to verify that their direct product $(\times_i A_i; \square_i o_i)$ is again a partial binary operative and is associated with the ordered set $(\times_i A_i; \square_i \omega_i)$. Therefore the class of partial binary operatives under consideration is closed under direct products.[1]

Let $(A; \omega)$ be an ordered set and $H \subset A$. Then $\omega \cap H \times H$ is an order relation on the set H which is said to be the order relation on H induced by the relation ω.

Theorem 3. *In order that a stable subset H of a partial binary operative $(A; o)$, associated with an ordered set $(A; \omega)$, be a partial operative associated with some ordered set, it is necessary and sufficient that*

$$S \, (\omega \cap H \times H) \subset S \, (\omega). \tag{9}$$

Proof. Necessity. Suppose that a stable subset H is a partial binary operative associated with the ordered set $(H; \bar{\omega})$, and suppose $(a_1, a_2) \in \bar{\omega}$. Then $a_1, a_2 \in H$ and $\inf(a_1, a_2)$ with respect to $\bar{\omega}$ exists and is equal to a_1; this latter is equivalent to $a_1 a_2 = a_1$, from whence it follows that

---

[1] The direct product of a family of relations in the sense of [2].

$\inf(a_1, a_2)$ with respect to ω exists and is equal to a_1 and $(a_1, a_2) \in \omega$. We have shown that $\overline{\omega} \subset \omega \cap H \times H$. Conversely, if $(a_1, a_2) \in \omega \cap H \times H$, then $a_1 a_2$ is defined in H and is equal to a_1, so that $\inf(a_1, a_2)$ with respect to $\overline{\omega}$ is equal to a_1, whence $(a_1, a_2) \in \overline{\omega}$. Thus $\overline{\omega} = \omega \cap H \times H$. Let $(a_1, a_2) \in S(\omega \cap H \times H)$. Then $(a_1, a_2) \in S(\overline{\omega})$ and consequently $a_1 a_2$ is defined, and therefore $(a_1, a_2) \in S(\omega)$.

Sufficiency. Suppose for two elements $a_1, a_2 \in H$ their product is defined, say $a_1 a_2 = a_3$; then a_3 is $\inf(a_1, a_2)$ with respect to ω, and, by virtue of the stability of H, $a_3 \in H$. It is easy to verify that a_3 is $\inf(a_1, a_2)$ with respect to the order relation $\omega \cap H \times H$. Conversely, if $\inf(a_1, a_2)$ with respect to $\omega \cap H \times H$ exists, then in accordance with (9) $\inf(a_1, a_2)$ with respect to ω exists, and therefore $a_1 a_2$ is defined and, moreover, $a_1 a_2 \in H$.

Suppose $(A; o)$ is an arbitrary binary operative and $H \subset A$. It is clear that if H is stable, then the set consisting of the single element sets $\{a\}$, where $a \in H$, together with the empty set $\varnothing$, will be a suboperative of the elementary global operative connected with $(A; o)$. One can easily construct an example of a partial binary operative associated with an ordered set such that for a stable subset of it condition (9) does not hold. To every partial binary operative associated with an ordered set there corresponds an elementary global operative. Then we see that this class of operatives is not closed with respect to subalgebras, and consequently is not primitive. Therefore the class of partial binary operatives associated with ordered sets cannot be axiomatized with the aid of identities expressed in the language of elementary global operatives.

Lemma. *If f is a strong homomorphism[2] of a partial binary operative $(A; o_A)$ associated with the ordered set $(A; \omega)$ to the partial binary operative $(B; o_B)$, then $f\,\overset{\smile}{\Box}\,f(\omega)$ is an order relation on the set B, where, moreover,*

$$(b_1, b_2) \in f\,\overset{\smile}{\Box}\,f(\omega) \leftrightarrow \{b_1\}\{b_2\} = \{b_1\}. \tag{10}$$

Proof. The global mapping $\overset{\smile}{f}$ will be a homomorphism from the elementary global operative connected with $(A; o_A)$ to the elementary global operative connected with $(B; o_B)$. Using this, it is easy to show that conditions (1)–(3) of Theorem 1 are fulfilled for $(B; o_B)$, and by the Corollary to this theorem we obtain the fact that the binary relation $\omega_B \subset B \times B$ de-

---

[2] Strong homomorphism is understood in the sense of [2].

fined by the formula $(b_1, b_2) \in \omega_B \longleftrightarrow \{b_1\}\{b_2\} = \{b_1\}$ is an order relation. Suppose $(b_1, b_2) \in \omega_B$ so that, in particular, $\{b_1\}\{b_2\} \neq \emptyset$. Therefore one can find $a_1', a_2 \in A$ such that $\{a_1'\}\{a_2\} \neq \emptyset$ and $f(a_1') = b_1$, $f(a_2) = b_2$. Set $a_1 = a_1'a_2$. Then $(a_1, a_2) \in \omega$. We have $f(\{a_1\}) = f(\{a_1'\,a_2\}) = f(\{a_1'\}\{a_2\}) = f(\{a_1'\})f(\{a_2\}) = \{f(a_1')\}\{f(a_2)\} = \{b_1\}\{b_2\} = \{b_1\}$, and hence $f(a_1) = b_1$. Thus $(b_1, b_2) \in \overbrace{f\,\square\,f}(\omega)$.

Conversely, suppose $(b_1, b_2) \in \overbrace{f\,\square\,f}(\omega)$; that is, for some a_1 and a_2 we have $f(a_1) = b_1$, $f(a_2) = b_2$ and $(a_1, a_2) \in \omega$. Then $\{a_1\}\{a_2\} = \{a_1\}$, and using the fact that f is a homomorphism, we obtain $\{b_1\}\{b_2\} = \{b_1\}$, i.e. $(b_1, b_2) \in \omega_B$. We have shown that $\overbrace{f\,\square\,f}(\omega) = \omega_B$.

Theorem 4. *In order that the image of a partial binary operative* $(A; o_A)$, *associated with the ordered set* $(A; \omega)$, *under a strong homomorphism* f *be a partial binary operative associated with some ordered set, it is necessary and sufficient that*

$$S(\overbrace{f\,\square\,f}(\omega)) \subset \overbrace{f\,\square\,f}(S(\omega)). \tag{11}$$

Proof. Necessity. Suppose (B, o_B) is the image under a strong homomorphism f of a partial binary operative $(A; o_A)$ and $(B; o_B)$ is associated with the ordered set $(B, \bar{\omega})$. Then the condition $(b_1, b_2) \in \bar{\omega} \longleftrightarrow \{b_1\}\{b_2\} = \{b_1\}$ is fulfilled. Taking into account that f is a strong homomorphism and that $S(\omega)$ and $S(\bar{\omega})$ are the relations of determination for $(A; o_A)$ and $(B; o_B)$, we get $\overbrace{f^{-1}\,\square\,f^{-1}}(S(\bar{\omega})) \subset S(\omega)$, whence $S(\bar{\omega}) \subset \overbrace{f\,\square\,f}(S(\omega))$. Since, in accordance with the lemma above, $\bar{\omega}$ coincides with $\overbrace{f\,\square\,f}(\omega)$, we obtain the required inclusion.

Sufficiency. Suppose f is a strong homomorphism and (11) is fulfilled. Then conditions (1)–(3) hold for $(B; o_B)$. By the Corollary to Theorem 1 we get, using the lemma, the fact that if $(b_1, b_2) \in \delta_B$ (where δ_B is the relation of determination for $(B; o_B)$), then $b_1 b_2$ is the greatest lower bound for the pair (b_1, b_2) with respect to $\overbrace{f\,\square\,f}(\omega)$. Since $S(\omega)$ is the relation of determination for $(A; o_A)$, then $\overbrace{f\,\square\,f}(S(\omega)) \subset \delta_B$; from this, in accordance with (11), we get $S(\overbrace{f\,\square\,f}(\omega)) \subset \delta_B$. Thus we have shown that o_B coincides with the partial operation of taking greatest lower bounds of pairs with respect to the order relation $\overbrace{f\,\square\,f}(\omega)$.

It is possible to cite an example to show that condition (11) is not always satisfied. Accordingly, we assign to every partial binary operative associated with an ordered set an elementary global operative; consequently we see that this class of operatives is not closed with respect to homomorphisms.

Theorem 5. *In order that in a partial binary operative $(A; o)$ associated with the ordered set $(A; \omega)$, the operation o be globally associative, it is necessary and sufficient that whenever a pair of elements has a lower bound in $(A; \omega)$, it has a greatest lower bound.*

Proof. Necessity. Suppose a pair (a_1, a_2) has a lower bound $\bar{a}$. Then $\{\bar{a}\}\{a_1\} = \{\bar{a}\}$ and $\{\bar{a}\}\{a_2\} = \{\bar{a}\}$. As a consequence of associativity of the global operation we get $\{\bar{a}\}(\{a_1\}\{a_2\}) = (\{\bar{a}\}\{a_1\})\{a_2\} = \{\bar{a}\}\{a_2\} = \{\bar{a}\}$. From this it follows that $\{a_1\}\{a_2\} \neq \emptyset$, i.e. the pair (a_1, a_2) has a greatest lower bound.

Sufficiency. We consider $(\{a_1\}\{a_2\})\{a_3\}$ and suppose that $(\{a_1\}\{a_2\})\{a_3\} \neq \emptyset$. Then set $\bar{a} = \inf(\inf(a_1, a_2), a_3)$. It is clear that $\bar{a} \lessgtr a_1$, $\bar{a} \lessgtr a_2$ and $\bar{a} \lessgtr a_3$. Since $\bar{a}$ is a lower bound of the pair (a_2, a_3), $\inf(a_2, a_3)$ exists and, moreover, $\bar{a} \lessgtr \inf(a_2, a_3)$. Then $\bar{a}$ is a lower bound for $(a_1, \inf(a_2, a_3))$, and therefore $\bar{\bar{a}} = \inf(a_1, \inf(a_2, a_3))$ exists and $\bar{a} \lessgtr \bar{\bar{a}}$. In an analogous manner we show that $\bar{\bar{a}} \lessgtr \bar{a}$. Thus if $(\{a_1\}\{a_2\})\{a_3\} \neq \emptyset$, then

$$(\{a_1\}\{a_2\})\{a_3\} = \{a_1\}(\{a_2\}\{a_3\}). \tag{12}$$

is fulfilled. In exactly the same way we obtain from $\{a_1\}(\{a_2\}\{a_3\}) \neq \emptyset$ that (12) holds. Thus (12) is always valid.

Corollary. *Conditions (1), (2) and (12) completely characterize those partial binary operatives associated with ordered sets having the property that every pair of elements admitting a lower bound admits a greatest lower bound.*

Proof. Since (3) follows from (12), according to the Corollary to Theorem 1, if $(a_1, a_2) \in \delta$ (δ is the relation of determination of the partial binary operative), then $a_1 a_2$ is the greatest lower bound of the pair (a_1, a_2) with respect to the order relation defined by the formula $(a_1, a_2) \in \omega \longleftrightarrow \{a_1\}\{a_2\} = \{a_1\}$. We suppose that (a_1, a_2) has a greatest lower bound a; then, in particular, $\{a\}\{a_1\} = \{a\}$ and $\{a\}\{a_2\} = \{a\}$, and as a consequence of global associativity $\{a\}(\{a_1\}\{a_2\}) = (\{a\}\{a_1\})\{a_2\} = \{a\}\{a_2\} = \{a\}$. From this it follows that $(a_1, a_2) \in \delta$. We have shown that the partial operation coincides with the partial operation of taking greatest lower bounds of pairs of elements. Since condition (12) holds, by Theorem 5 every pair having a lower bound in the ordered set $(A; \omega)$ has a greatest lower bound.

From Theorem 5 it follows that in general the partial operation of taking greatest lower bounds of pairs of elements in an ordered set will not be globally associative; it satisfies a type of associativity, weaker than global associativity, expressed by formula (3).

Among the other properties of partial operations that we shall consider is the catenary property which plays an important role in understanding categories. A partial operation is said to be left catenary if the corresponding global operation satisfies the condition

$$\{a_1\}\{a_2\} \neq \varnothing \wedge \{a_2\}\{a_3\} \neq \varnothing \rightarrow \{a_1\}(\{a_2\}\{a_3\}) \neq \varnothing;$$

and right catenary if

$$\{a_1\}\{a_2\} \neq \varnothing \wedge \{a_2\}\{a_3\} \neq \varnothing \rightarrow (\{a_1\}\{a_2\})\{a_3\} \neq \varnothing;$$

it is said to be catenary if it is catenary from the left and the right. In the case of partial binary operatives associated with ordered sets, the notions of left catenary, right catenary and catenary coincide.

Theorem 6. *In order that in a partial binary operative $(A; o)$ associated with the ordered set $(A; \omega)$ the operation o be catenary, it is necessary and sufficient that for any element the set of its lower bounds under the induced order be a semilattice with respect to greatest lower bounds.*

Proof. Necessity. Let a_1 and a_2 be two lower bounds of a. Then $\{a_1\}\{a\} = \{a_1\}$ and $\{a\}\{a_2\} = \{a_2\}$, and, in particular, $\{a_1\}\{a\} \neq \varnothing$ and $\{a\}\{a_2\} \neq \varnothing$. From this it follows that $\{a_1\}\{a_2\} \neq \varnothing$, i.e. $\inf(a_1, a_2)$ exists; it is clearly a lower bound for a and therefore is the greatest lower bound for (a_1, a_2) under the induced order.

Sufficiency. We suppose that $\{a_1\}\{a_2\} \neq \varnothing$ and $\{a_2\}\{a_3\} \neq \varnothing$. Since $\inf(a_1, a_2)$ and $\inf(a_2, a_3)$ are lower bounds of a_2, there exists a greatest lower bound $\bar{a}$ for the pair $(\inf(a_1, a_2), \inf(a_2, a_3))$ under the order induced by ω on the subset $\omega^{-1}\langle a_2 \rangle$. It is clear that $\bar{a} \lesssim a_1$, $\bar{a} \lesssim a_2$ and $\bar{a} \lesssim a_3$, whence $\bar{a} \lesssim a_1$ and $\bar{a} \lesssim \inf(a_2, a_3)$. Now suppose for some a we have $a \lesssim a_1$, $a \lesssim \inf(a_2, a_3)$. Then $a \lesssim a_1$, $a \lesssim a_2$ and $a \lesssim a_3$, whence $a \lesssim \inf(a_1, a_2)$ and $a \lesssim \inf(a_2, a_3)$. Using the fact that $a \in \omega^{-1}\langle a_2 \rangle$, we find that $a \lesssim \bar{a}$. We have shown that $\bar{a} = \inf(a_1, \inf(a_2, a_3))$, so that $\{a_1\}(\{a_2\}\{a_3\}) \neq \varnothing$, i.e. the partial binary operation o is left catenary and thus catenary.

In studying an ordered set it is convenient to decompose it into connected components which coincide with the classes of the equivalence relation γ_ω, the equivalence closure of the binary relation ω. γ_ω is called the connectivity relation of the ordered set. One can show that

$$\gamma_\omega = \bigcup_{n=1}^{\infty} \overbrace{\omega^{-1} \circ \omega}^{n} = \bigcup_{n=1}^{\infty} \overbrace{\omega \circ \omega^{-1}}^{n} = \bigcup_{n=1}^{\infty} \overbrace{\omega \bigcup \omega^{-1}}^{n}.$$

The least ternary equivalence relation containing o is called the connectivity relation for a partial binary operative $(A; o)$, [2]. Let $(A; \omega)$ be an ordered set. One can show that the decomposition of A into connected components corresponding to the connectivity relation of the ordered set $(A; \omega)$ and the decomposition corresponding to the connectivity relation of the partial binary operative associated with the ordered set coincide.

We consider an arbitrary family $(A_i; \omega_i)$ $(i \in I)$ of greatest-lower-bound semilattices (we assume that the A_i are pairwise disjoint). Their cardinal sum is an ordered set.

Theorem 7. *In order that an ordered set $(A; \omega)$ be the cardinal sum of a family of greatest-lower-bound semilattices, it is necessary and sufficient that the relation of determination of the partial binary operative associated with the ordered set $(A; \omega)$ be transitive.*

Proof. Necessity. Suppose the ordered set $(A; \omega)$ is the cardinal sum of some family $(A_i; \omega_i)$ $(i \in I)$ of greatest lower bound semilattices. If $(a_1, a_2) \in \delta$ and $(a_2, a_3) \in \delta$, then for some i $a_1, a_2, a_3 \in A_i$, and since $(A_i; \omega_i)$ is a greatest-lower-bound semilattice, the pair (a_1, a_3) has a greatest lower bound, i.e. $(a_1, a_3) \in \delta$.

Sufficiency. Since the relation of determination of any partial binary operative associated with an ordered set is reflexive and symmetric, if δ is transitive it will be an equivalence relation. Thus, for any two elements a_1 and a_2 belonging to the same equivalence class, $(a_1, a_2) \in \delta$ holds, i.e. $\inf(a_1, a_2)$ exists. Since $\inf(a_1, a_2)$ belongs to this same equivalence class, $\inf(a_1, a_2)$ will be the greatest lower bound of the pair (a_1, a_2) relative to the order induced on this equivalence class. Consequently every equivalence class of δ is a greatest-lower-bound semilattice. It is not difficult to verify that if the relation of determination of a partial binary operative associated with an ordered set is transitive, then it coincides with the connectivity relation of that ordered set, and therefore every equivalence class of δ is a connected component of the ordered set $(A; \omega)$, while every ordered set is the cardinal sum of its connected components. From Theorem 7 we get

Corollary. *In order that an ordered set $(A; \omega)$ be the cardinal sum of*

lattices, it is necessary and sufficient that the relations of determination of the partial operations o_1 and o_2 of the partial binary bioperative $(A; o_1, o_2)$ associated with $(A; \omega)$ coincide and be transitive.

In conclusion I wish to express my deep gratitude to V. V. Vagner for valuable advice during the writing of this article.

BIBLIOGRAPHY

[1] G. Birkhoff, *Lattice theory*, Amer. Math. Soc. Colloq. Publ., vol. 25, rev. ed., Amer. Math. Soc., Providence, R. I., 1948; Russian transl., IL, Moscow, 1952.　　MR 10, 673.

[2] V. V. Vagner, *The theory of relations and the algebra of partial mappings*, Theory of Semigroups and Appl. I, Izdat. Saratov. Univ., Saratov, 1965, pp. 3–178. (Russian)　　MR 35 #109.

[3] E. S. Ljapin, *Semigroups*, Fizmatgiz, Moscow, 1960; English transl., Transl. Math. Monographs, vol. 3, Amer. Math. Soc., Providence, R. I., 1963.　　MR 22 #11054; 29 #4817.

[4] Ju. I. Sorkin, *On the imbedding of latticoids in lattices*, Dokl. Akad. Nauk SSSR **95** (1954), 931–934. (Russian)　　MR 15, 926.

Translated by:
Gretchen Wagner

Amer. Math. Soc. Transl.
(2) Vol. 96, 1970

ORDERED ALGEBRAIC SYSTEMS[*]

A. A. VINOGRADOV

The first basic studies of partially ordered algebraic systems appeared at the beginning of this century. One can acquaint oneself with the history of the development of the theory of such systems in the books of G. Birkhoff [1] and L. Fuchs [60]. In the surveys of L. A. Skornjakov [54] and L. M. Gluskin [12] appear new results on ordered rings and semigroups obtained in 1960–62. In the series "Itogi Nauki" there is still no report devoted to ordered groups. We will give a short survey of the works in ordered groups, rings, and semigroups which have been reviewed in the Referativnyĭ Žurnal "Matematika" from 1963 to 1965, and as well as earlier results in the theory of ordered groups which are omitted in [60].

§ 1. Linearly ordered groups

1. **Archimedean groups.** Any archimedean group satisfies the following axiom (A) (expressed in terms of mathematical logic):

$$\exists x H(x) \wedge \exists x \sim H(x) \wedge \forall x,\ y\,(H(x) \wedge y < x) \to H(y)$$
$$\to \forall c\,[c > 0 \to \exists d\,(H(d) \wedge \sim H(d + c))].$$

The symbol $H(x)$ means that the element x belongs to the subset H. Rautenberg [209] noted that there exist non-archimedean groups for which (A) holds, and that linearly ordered groups satisfying (A) are abelian.

The notion of a linearly ordered abelian group can be formulated in the restricted predicate calculus by three predicates and a set of axioms based on them. Let $E(x, y)$ denote the two-place predicate: "the element x is equivalent to the element y", let $Q(x, y)$ be the predicate: "the element x is less than the element y or equivalent to it", and let $S(x, y, z)$ be the predicate: "z is the sum of the elements x and y". A correctly formed formula of the restricted predicate calculus is called an elementary formula of the theory of groups if it contains no predicates other than E, Q and S.

---

* Translation of *Algebra. Topology. Geometry.* 1965, Akad. Nauk SSSR Inst. Naučn. Informacii, Moscow, 1967, pp. 83–131.

Two linearly ordered groups are called elementarily equivalent if they have the same elementary properties.

We will say that two elements x and y of an abelian group G are congruent modulo a positive integer n if there exists a $z \in G$ such that $x = y + nz$. Let $[n]G$ denote the maximal number of pairwise congruent modulo n elements of G. A linearly ordered group G is called regularly discrete if it is discrete and such that $[p]G = p$, for each prime p. G is regularly dense if for every positive integer n and any $a, b \in G$ $(a < b)$ there exists an $x \in G$ $(a < x < b)$ such that $x = nz$ for some $z \in G$. G is called regular if it is regularly discrete or regularly dense. This definition of a regular group, introduced by Robinson and Zakon [216], turned out to be equivalent to the following (Zakon [262]): A linearly ordered group G is regular if and only if for each convex subset $S \subset G$ and each positive integer n there exists a $g \in G$ such that $ng \in S$. Each complete linearly ordered abelian group is regular.

Conrad [114] introduced the notion of an l-regular group as a linearly ordered group G such that for each infinite interval (a, b) in G and each positive integer n there exists a $g \in G$ such that $ng \in (a, b)$. If G is regular then it is l-regular. There exist linearly ordered abelian groups which are l-regular but not regular. Regular and l-regular groups were studied without the assumption of commutativity. Each linearly ordered abelian regular group is elementarily equivalent to some archimedean group (Robinson and Zakon [215]), and all archimedean groups are regular (Zakon [262]), and hence l-regular. Thus, abelian regular and l-regular groups can be considered as generalized archimedean groups.

Suppose that $p_1, p_2, \cdots, p_n$ is the increasing sequence of all primes, and that $m_1, m_2, \cdots, m_n$ is any sequence, where each m_n is either a non-negative integer or ∞. There always exists a dense archimedean group G such that $[p_n]G = P_n^{m_n}$ (Zakon [262]). All abelian regular discrete groups are elementarily equivalent. Two abelian regular dense groups G_1 and G_2 are elementarily equivalent if and only if $[p]G_1 = [p]G_2$ for every prime p.

M. I. Zaĭceva [16] and Trevisan [245] found, independently, all the ways of linearly ordering a free abelian group with a finite number of generators. Also, Zaĭceva gave ways of archimedeanly ordering such a group. It turned out that the power of the set of all archimedean orderings (for $n \geq 2$) is equal to that of the continuum. The cardinality of the set of all orderings of a free abelian group with a finite number $(n \geq 2)$ of generators is also that of the continuum. The same conclusion about these cardinalities was reached by Matsushita [196]. Ja. V. Hion [63] described all ways of archimedeanly ordering an archimedean orderable group, and in doing so generalized the result of Zaĭceva and Trevisan on the archimedean orderability of a group with a finite number of generators. He also showed in which case two isomorphisms of an archimedean group G with a subgroup of the additive group of reals R arise from the same ordering of G.

Torsion free abelian groups of rank 1, and only they, admit an archimedean linear ordering but they do not admit any non-archimedean linear ordering (Podderjugin [52]).

Krull [181] studied the order preserving or reversing endomorphisms of an archimedean group. Their totality forms a commutative semigroup.

2. η_α-groups. Let T be a linear set with subsets H and K. We write $H < K$ if $h < k$ for any $h \in H$ and $k \in K$. Suppose that α is an ordinal. T is called an η_α-set if for any subsets H and K in T such that $|H| + |K| < \aleph_\alpha$ and $H < K$, there exists a $t \in T$ such that $H < \{t\} < K$ ($|H|$ denotes the power of H). A linearly ordered group G is called an η_α-group if as a linear set it is an η_α-set. Abelian η_α-groups were studied by Alling [74-76] and Fleischer [129]. They gave a characterization of η_α-groups. Alling, in particular, showed that for $\alpha > 0$, any two complete abelian η_α-groups of power $\aleph_\alpha$ are isomorphic, but that there exist nonisomorphic abelian η_α-groups of power $\aleph_\alpha$.

3. **Symmetrically linearly ordered groups.** The linearly ordered additive group of integers possesses two characteristic properties: 1) as an ordered set, it is symmetric in the sense that each initial interval of it is anti-isomorphic to the complementary final interval, and 2) it is hereditarily discrete, i.e. each subgroup of it is also discrete. Sankaran and Venkataraman [218] carried out a generalization of the ordered additive group of integers in two directions, namely linearly ordered abelian groups which are hereditarily discrete and symmetric abelian groups. As a result of their studies it turned out that these classes coincide.

4. **Solvable linearly ordered groups.** Note first three theorems formulated by Ree [210,211]: 1) A linearly ordered solvable group with a finite number of generators is nilpotent if and only if it satisfies the maximality condition for subgroups. 2) If a linearly ordered group satisfies the maximality condition for subgroups then it is nilpotent in the generalized sense, i.e. it is a ZD-group. 3) If G is a linearly ordered nilpotent group which is torsion free and finitely generated and if A is the group of monotone automorphisms of G, then A is a finitely generated nilpotent torsion free group.

A. A. Vinogradov [7] gave an example of a non-nilpotent linearly ordered solvable group with three generators and with the maximality condition for subgroups. With this, it is proved that Theorem 1) of Ree, cited above, is false. He noted also that the proof of Theorem 2) is not convincing. The truth of Theorem 2) is open. One still has to verify Theorem 3) since its proof is based on Theorems 1) and 2).

A. I. Kokorin [23] found a series of necessary conditions for a group to be uniquely linearly orderable, and these conditions are sufficient in the class of nilpotent groups. He also gave an example of a non-abelian group which can be uniquely linearly ordered.

We call a subgroup H of a group G strictly isolated if $xg_1^{-1}xg_1 \cdots$ $\cdots g_n^{-1}xg_n \in H$ implies that $x, g_1^{-1}xg_1, \cdots, g_n^{-1}xg_n \in H$. A 2-solvable group is linearly orderable if and only if its unit is strictly isolated (A. I. Kokorin [26]).

D. M. Smirnov [55] proved that in a locally nilpotent ordered group every convex subgroup is normal, and B. I. Plotkin [50] brought attention to linearly ordered groups all of whose convex subgroups are normal. He established that the isolator of the commutant of such a group contains a central system and the group is an extension of a group with a central system by an abelian torsion free group.

An isomorphism ϕ mapping the lattice $S(G)$ of semigroups of a group G onto the lattice $S(G^*)$ of semigroups of a group G^* is called a semigroup lattice isomorphism (SL-isomorphism). K. M. Kutyev [37] proved that if G^* is the image of a linearly ordered locally nilpotent group G with semigroup of positive elements P under an SL-isomorphism ϕ, then G^* is locally nilpotent and it can be linearly ordered so that $\phi(P)$ becomes the semigroup of positive elements of G^*.

5. Representations (realizations, embeddings) of linearly ordered groups. Suppose that we have a well-ordered sequence of linearly ordered sets $M_0, M_1, \cdots, M_\alpha, \cdots$ with order types $\xi_0, \xi_1, \cdots, \xi_\alpha, \cdots$. The product of these types $\xi_0\xi_1 \cdots \xi_\alpha \cdots$ or their ω-product is the order type of the set of finite systems $(m_{a_1}, \cdots, m_{a_k})$, $m_i \in M_i$, compared lexicographically by the distinct elements. In case $\xi_0 = \xi_1 = \cdots = \xi$, set $\xi_0\xi_1 \cdots \xi_\alpha \cdots = \xi^\beta$, where β is the order type of the sequence. Let ζ denote the order type of the integers, η that of the rationals and ξ that of the reals. A. I. Mal'cev [44] showed that for a linearly ordered abelian group G to be an ω-direct sum of λ cyclic groups (λ an ordinal) it is necessary and sufficient that the order type of G have the form ζ^λ. Similarly, a linearly

ordered abelian group G is an ω-direct sum of λ additive groups of real numbers if it has order type ξ^λ. The order type of any countable linearly ordered group has the form $\zeta^\lambda \eta^\epsilon$, where λ is an ordinal and $\epsilon = 0, 1$. For each countable linearly ordered group G there exists an abelian group having the same order type as G. It is unknown whether this is true for uncountable groups.

Note that the theorem of Michiura [199] on representing a linearly ordered group in the form of a lexicographic sum of copies of the additive groups of reals turned out to be mistaken (Fleischer [127]).

V. M. Kopytov [34] showed that for any linearly ordered group G with positive cone P there exists a linearly ordered group G^* such that: 1) $G \subseteq G^*$, 2) the center of G^* is complete, 3) $P \subseteq P^*$, and so on.

Let Γ denote the set of all pairs of convex subgroups G_γ and G^γ of a linearly ordered group G such that $G_\gamma \subset G^\gamma$ and there are no other convex subgroups of G between them. Γ is linearly ordered by set-theoretic inclusion of subgroups. If Γ is well ordered, then we call G a group with well-ordered rank. Conrad [107,109] showed that if, for a group G with well-ordered rank, each factor group G^γ/G_γ is order isomorphic to a subgroup D_γ of the additive group R of reals and the o-automorphisms of the group D_γ are only multiplications by some positive rationals, then the group of o-automorphisms of G can be linearly ordered. In the subsequent work [110], Conrad showed that for each group Q of o-automorphisms of a linearly ordered group G which is orderable, one can construct a new linearly ordered group H containing Q and G.

We note in passing that the o-automorphisms of linearly ordered groups were studied by Harvey [142].

For linearly ordered groups G and H, let $G \subseteq H$. We call H: an a-extension of G if for each $e < h \in H$ there exists a $g \in G$ and a positive integer n such that $h \leq g \leq h^n$; a b-extension of G if for each $e < h \in H$ there exists a $g \in G$ and a positive integer n such that $g \leq h \leq g^n$; a c-extension of G if for each $e < h \in H$ there exists a $g \in G$ such that $(hg^{-1})^n < h$ for every integer n; a t-extension of G if for each $h, h',$ $h'' \in H$, $h < h' < h''$, there exists a $g \in G$ such that $h < g < h''$. If $x \in \{a, b, c, t\}$, then G is called x-closed if G has no proper x-extension. Holland [148] showed that a t-extension is a b-extension, each c-extension is a b-extension, and each b-extension is an a-extension. He also clarified other properties of x-extensions, among which are conditions for x-closure.

We call an isomorphism of the lattice of subsemigroups of two groups an

SL-isomorphism. K. M. Kutyev [38-40] showed that SL-isomorphism is a consequence of the isomorphism or anti-isomorphism of two groups when one of them is orderable.

We will not touch on the theorem of Hahn [141] on embedding abelian groups. A detailed exposition can be found in the book of L. Fuchs [60]. Note only one result of Wang Shih-chiang [4] related to this theorem: If a linearly ordered set of archimedean equivalent classes of an ordered abelian group consists of a finite number $(n + 1)$ of elements: $0 \ll \chi_1 \ll \chi_2 \ll \cdots \ll \chi_n$, then the group is isomorphic to a subgroup of the ordered additive group of vectors in n-dimensional real space.

6. **Conditions for the orderability of a group (o-groups).** In the book of Fuchs [60] there are considered in detail results obtained by many authors relative to conditions under which a group can be linearly ordered (briefly, an o-group). There, in particular, are noted the basic theorems of A. I. Mal'cev [44,45] on linearly ordered groups. We will restrict ourselves only to supplementary ones.

L. N. Ševrin [68] gave a characterization of an o-group by means of the lattice of all semigroups of the group. This solved the question of B. I. Plotkin [51] on the existence of a lattice subsemigroup characterization of orderable groups.

A. I. Kokorin [26] showed that the factor group G/Z_1 of an o-group G by a complete subgroup Z_1 of the center Z is orderable, and V. M. Kopytov [34] noted that the factor group of an o-group by its center is also an o-group.

Tully [246] showed that for a group G the following are equivalent: 1) G is an o-group; 2) for every $g \in G$, $g \neq e$, there exists a subset $S \subseteq G$ such that $g \in S$, $e \notin S$ and for each g', $h' \in G$ either $g'Sh' \subseteq S$ or $S \subseteq g'Sh'$; and 3) there exists a collection Σ of subsets of G such that $e = \cap S$ and for any $g, h \in G$ and $S \in \Sigma$ either $gSh \subseteq S$ or $S \subseteq gSh$.

Shepperd [220] found necessary and sufficient conditions for a torsion free group containing a normal subgroup of index 2 to be an o-group.

Set $[x, y] = x^{-1}y^{-1}xy$ and $[x, y, z] = [[x, y], z]$ for elements of a group. Several relations for elements of an o-group were established by B. H. Neumann [204]: thus, for example, he showed that if a and b are elements of an o-group and $[a^m, b] = e$ for each integer $m \neq 0$, then $[a, b] = e$, or if $[a^m, b, a] = e$ for each integer $m \neq 0$, then $[a, b, a] = e$. Chehata [98] constructed an example of an o-group in which the following relations hold:

$[a, b^m, a, a] = e$, $m > 1$; $[a, b, a, a] \neq e$.

For any group G set $G_1 = G$, $G_{i+1} = [G_i, G]$. For any sequence of natural numbers $n_1, n_2, \cdots, n_i, n_{i+1}, \cdots$ $(n_i > 1)$, introduce the chain of normal subgroups

$$G \supseteq G_{n_1} \supseteq G_{n_1, n_2} \supseteq \ldots \supseteq G_{n_1, n_2, \ldots, n_i} \supseteq G_{n_1, n_2, \ldots, n_i, n_{i+1}} \supseteq \cdots,$$

where $G_{n_1, \cdots, n_i, n_{i+1}} = (G_{n_1, n_2, \cdots, n_i})_{n_{i+1}}$. G is called seminilpotent if $G_{n_1, \cdots, n_s} = E$. All nilpotent and solvable groups are seminilpotent. If F is a free group then the factor group $F/F_{n_1, \cdots, n_k}$ is a free seminilpotent group. A. L. Šmel'kin [70,71] showed that every free seminilpotent group is an o-group.

7. **The elementary theory of linearly ordered abelian groups.** A formula of the restricted predicate calculus not containing other nonlogical symbols than $=, +, <$, is called a V-formula if in its normal form it does not contain an existential quantifier. Two linearly ordered abelian groups G_1 and G_2 are called V-equivalent if every V-formula true in G_1 is true in G_2, and conversely. Ju. S. Gurevič and A. I. Kokorin [15] showed that linearly ordered abelian groups are V-equivalent. Ju. S. Gurevič [13,14] established the solvability of the elementary theory of linearly ordered abelian groups. Elementary properties of linearly ordered groups are those which can be expressed in the language of the restricted predicate calculus applied to the basic operations and relations of the groups. The report of M. I. Kargapolov [21] was devoted to the classification of linearly ordered abelian groups by their elementary properties.

8. **Other questions in the theory of linearly ordered groups.** Krantz [179] considered various properties of linearly ordered groups whose elements are equivalence classes $(\overline{A, P})$ defined on the direct product $A \times P$ of sets A and P.

§ 2. Lattice ordered groups

1. **Definition of an l-group.** The study of l-groups began after the appearance of the work of L. V. Kantorovič [18] on semiordered spaces. The systematic study of non-abelian l-groups was first looked into by Birkhoff [85]. He noted that an l-group can be defined as a group with a unary operation $a \longrightarrow a^*$ which satisfies the following conditions: 1) $0^* = 0$, 2) $c = c^* \cdot ((c^{-1})^*)^{-1}$, 3) the operation $(ab^{-1})^* \cdot b$ is associative, i.e.

$\{[(ab^{-1})^* \cdot b] \cdot c^{-1}\}^* \cdot c = \{a \cdot [(bc^{-1})^* \cdot c]^{-1}\}^* \cdot [(bc^{-1})^* \cdot c]$. Michiura [197] noted that this definition is not complete: to axioms 1)–3) one must add axiom 4): $ax^*a^{-1} = (axa^{-1})^*$. Besides the set of postulates 1)–4), Michiura [199] gave four other systems, each of which is equivalent to 1)–4). In [80] Balmaña gave yet another set of postulates. A set of postulates defining an l-group was also given by Vaida [249].

We call an element x of an l-group G a peak (German: Spitze) of the element $a > e$ if x is a minimal element in G with the property $a \geq x \geq e$ and $ax^{-1} \wedge x = e$. An $x \in G$ is called a peak in G if it is a peak for some element of G. Šik [223] showed that an l-group is a linearly ordered group if and only if each of its element greater than zero is a peak.

Teh [241] showed than an l-group is linearly ordered if it satisfies at least one of the following conditions: 1) each pair $a, b > e$ has a lower bound $c > e$, 2) $a, b > e$ implies $a \wedge b \neq e$, 3) $a \wedge b = e$ implies $a = e$ or $b = e$, 4) $a, b > e$ implies $ab > a \vee b$, 5) there exists an element $b > e$ such that for each $a > e$ there is an integer n such that $a^n \geq b$, 6) for any two elements $a, b > e$ there exists an integer n such that $a^n \geq b$.

Vaida [248] showed that for an l-group G to be linearly ordered it is necessary and sufficient that its lattice of l-ideals be a linearly ordered set and that for any $x, t \in G$, $x \vee (tx^{-1}t^{-1}) \geq e$.

2. **Archimedean l-groups.** There are several known proofs of the theorem of the commutativity of an archimedean l-group. Yet another proof of this theorem was given by Goffman [139].

Jakubík [72] showed that for a given archimedean l-group G there exists an l-group of real functions isomorphic to G.

We will say than an l-group G has property (f) if the fact that $\{x_i\} \subseteq G^+$ and any two distinct elements of this set are disjoint implies $\vee x_i$ exists in G. A. G. Pinsker [48] showed that any archimedean l-group can be embedded in a complete l-group G' having property (f). Jakubík [72] reproved this result in another way.

In reviewing the reports of Choe [99,100], Conrad noted that many theorems were in error. The following theorem of Choe is true with a correction by Conrad: If G is an archimedean l-group in which there exist distinct elements $x_1, \cdots, x_n$ such that each x_i exceeds zero and such that $y > 0$ implies $y \geq x_i$ for some i, then G is the cardinal product of n infinite cyclic groups.

3. **Some other classes of l-groups.** Elements $a_1, \cdots, a_n$ of an l-group

G are called disjoint if they are strictly positive and $a_i \wedge a_j = 0$ for all $i \neq j$. Conrad and Clifford [117] gave the structure of an l-group having exactly two disjoint elements. Such an l-group can be obtained as follows: one takes two linearly ordered groups A and B, forms their cardinal sum $A + B$ and then an l-group is obtained as one in which $A + B$ is an l-ideal, and each positive element not in $A + B$ exceeds $A + B$. Conrad [112] gave the structure of l-groups with a finite number of disjoint elements. These groups are obtained by means of a construction similar to that for groups with two disjoint elements. He also gave the structure of an l-group in which each element $a > 0$ exceeds some finite number of disjoint elements.

For each $g \neq 0$ in an l-group G let R_g denote the subgroup of G generated by the set of all l-ideals in G not containing g. R_g is an l-ideal in G. The radical $R(G)$ of G is defined by the equation $R(G) = \bigcap R_g$ $(e \neq g \in G)$. Conrad [115] studied l-groups with radical 0.

An l-group with two generators was considered by Jakubík [173]. Let x be an element of an l-group G. If $0 < x$ and there is no z such that $0 < z < x$, then we write $0 \rightarrow x$; if $nx < y$ for each integer n, then $x \ll y$. Let $A \dotplus B$ denote the direct sum of l-groups A and B. Let C be the additive group of integers with the natural order. In [173] Jakubík constructed examples of l-groups with two generators, denoted G_m ($m = 0$, $1, 2, \cdots$). He showed that if G has two generators x and y and $0 \rightarrow x \ll y$, then G is order isomorphic to one of the G_m. If $0 \rightarrow x$, then G is isomorphic to one of the following l-groups: C, $C \dotplus C$, $C \dotplus C \dotplus C$, G_m, $G_m \dotplus C$ ($m = 0, 1, 2, \cdots$). Jakubík studied also l-groups with set of generators $B = \{y\} \cup A$ where for each $a \in A$ one has $0 \rightarrow a \ll y$.

For any set $A \subset G^+$ of an l-group G we define the set $K'(A)$ to be all $y \in G^+$ such that $x \in A \Rightarrow x \wedge y = 0$. Set $K'(K'(A)) = K(A)$. We say that G satisfies property (P) if for $A \subset G^+$, for any $x \in G^+$ there are elements $y \in K(A)$, $z \in K'(A)$ such that $x \leq y \wedge z$. l-groups with property (P) were studied by Jakubík [164]. Šik [224] showed that a group with property (P) is a subdirect sum of linearly ordered groups. We say that an l-group G satisfies property (Q) if $A \subset G^+$, $A \neq \{0\}$, implies that $K(A)$ is not bounded above. Jakubík [167] studied groups with property (Q). Each l-group satisfying property (P) also satisfies property (Q) but the converse does not hold. If an l-group G with property (Q) satisfies the descending chain condition in the lattice $[G]$ of its l-ideals, then it can be represented in the form of a direct cardinal sum of a finite number of linearly ordered groups. If an l-group G with property (Q) satisfies the ascending chain condition in

$[G]$, then either it is decomposable as a direct sum or it has a maximal l-ideal.

An l-group G is called compactly generated if for any $A \subset G$ such that $\bigvee A$ exists in G, there exists a finite subset $B \subset A$ such that $\bigvee B = \bigvee A$. Šik [230] showed that a nontrivial compactly generated complete l-group is isomorphic to a direct sum of copies of the linearly ordered group of integers, and, conversely, a direct sum of copies of the linearly ordered additive group of integers is a compactly generated l-group.

Let X be an abelian l-group with lattice operations $\cap$ and $\cup$. Suppose that X_0 is a subgroup of X which is an l-group relative to the order in X with lattice operations $\wedge$ and $\vee$. For $x \in X_0$, set

$$|x|_X = (x \cup 0) + ((-x) \cup 0), \quad |x|_{X_0} = (x \vee 0) + ((-x) \vee 0).$$

In the work of Kantorovič, Vulih and Pinsker [19], the following problem is raised: find complete l-groups X and X_0 such that for some $x \in X_0$ the inequality $|x|_{X_0} \supset |x|_X$ holds. Jakubík [169] solved this problem in the affirmative.

Let α be an infinite cardinal, let S be a lattice and $A \subset S$. Also suppose that A is a lattice in the order induced from S. We call A an α-sublattice of the lattice S if the following holds: if $B \subset A$, $|B| \leq \alpha$, and $s = \sup_S B$ exists, then $s = \sup_A B$. Jakubík [169] showed that if α is a regular cardinal, then there exists a complete l-group X and a subset $X_0 \subset X$ which is an l-group such that X_0 is an $\mathfrak{f}$-sublattice but not an α-sublattice in X for every cardinal $\mathfrak{f} < \alpha$.

An l-group G is called autometrizable, or an l-space, if it admits a mapping $d\colon G \times G \to G$ such that $d(a, b) \geq 0$ (where equality holds only for equal pairs), $d(a, b) = d(b, a)$ and $d(a, c) \leq d(a, b) + d(b, c)$. Swamy [238] showed that an abelian l-group G is autometrizable under the distance $d(a, b) = |a - b|$. Not every non-abelian l-group is autometrizable. He established other properties of l-spaces.

Every l-group is a distributive lattice, but there are l-groups which satisfy stronger distributivity conditions. An l-group is called completely distributive if for all $g_{ij} \in G$ ($i \in I$, $j \in J$) we have the equality

$$\bigwedge_{i \in I} \bigvee_{j \in J} g_{ij} = \bigvee_{\rho} \bigwedge_{i} g_{i\rho(i)},$$

where $\rho \in J^l$ and it is assumed that all the indicated sup's and inf's exist. Completely distributive l-groups were studied by Weinberg [254,255] and Conrad [115].

4. **Convex subgroups of an l-group.** Interest in the convex subgroups of an l-group is due to the fact that l-groups may enjoy various properties depending on the existence of certain individual convex subgroups, or on the character of certain subsets of convex subgroups. For example, whether an l-group G is a complete distributive lattice depends on the lattice L of all l-ideals of G. For proving many theorems on l-groups, one uses convex subgroups. Thus Holland [149], in proving that an l-group can be presented as a group of order preserving automorphisms of a linearly ordered set, constructed this set from conjugate classes of convex l-subgroups of G.

Yet again, Birkhoff [85] noted that L is a complete distributive lattice. Jakubíková [73] and Lorentz [189] showed that the set Γ of all convex l-subgroups is also a distributive lattice. The study of the lattice of all convex l-subgroups of an l-group is the subject of the works of Conrad [116] and Šik [235]. They established, in particular, that the lattice Γ is complete. The lattice K of all convex subgroups of an l-group was studied by Jakubíková [73]. She proved that if an l-group is not linearly ordered, then K is not a Dedekind lattice and does not satisfy the descending chain condition. Vaida [248] showed that if L satisfies the ascending chain condition, then the representation of G in the form of a direct product of indecomposable factors is unique. If the lattice L of an l-group G satisfies the ascending chain condition and $a \wedge x^{-1} ax = e$ implies $a = e$ in G, then either G is the direct cardinal sum of two of its l-ideals, or G contains a maximal l-ideal containing all the other proper l-ideals of G (Lorentz [189]). This generalizes a result of Birkhoff [1] on abelian groups.

It is known that an abelian l-group has two proper l-ideals if it is not linearly ordered. Jakubík [167] showed that there is a proper l-ideal in an l-group which is not linearly ordered and satisfies condition (Q) (see subsection 3). Holland [151] showed when an l-group G is a simple l-group, i.e. an l-group with no proper l-ideals. Birkhoff [85] showed that if each chain consisting of elements of an abelian l-group has length not exceeding a fixed number r, then each of its l-ideals is principal. This result was generalized by Jakubík [165] to non-abelian l-groups. He also negatively answered Birkhoff's problem 99 [1]. A theorem on l-ideals of an l-group useful in the theory of normed spaces was proved by Nakano [203].

Šik introduced what has become a very useful notion, namely that of a component (or polar) of an l-group G as a set $A' = \{x \in G: |x| \wedge |a| = 0, a \in A\}$,

$\emptyset \neq A \subset G$. He also [222] showed that the components of an l-group, under set inclusion, form a complete Boolean algebra Γ_0; the components which are also l-ideals form a complete subalgebra Γ_1 of Γ_0; and the l-ideals which are direct factors of G form a (not necessarily complete) subalgebra Γ_2 of Γ_1. The paper [234] of Šik is devoted to the lattice of components of an l-group.

Among all l-ideals, one singles out the l-ideals I in an l-group G such that the naturally ordered quotient group G/I is linearly ordered and for $a \in G/I$ one has $b > a$ for every $b \in I$ or $a > b$ for every $b \in I$. Such l-ideals are called strong. Šik [224] showed than an l-ideal of an l-group is strong if and only if it contains all proper components.

5. **The lattice of filets of an l-group.** Jaffard [159] found necessary and sufficient conditions for an l-group to have a finite number of filets. He also [155] raised the question if the lattice of filets Y of an abelian l-group is a relatively complemented lattice if it satisfies the descending chain condition. Jakubík [170] gave a positive answer for archimedean l-groups and gave an example showing that the answer is negative for general abelian l-groups.

6. **Free l-groups.** The free l-group with one generator was considered by Birkhoff [85]. He showed that such a group is l-isomorphic to the cardinal sum of two copies of the additive group of integers. All ways of lattice ordering a free abelian group with a finite number of generators were found by A. I. Kokorin [25], and in doing so he completed the solution of Birkhoff's 102nd problem in [1].

Let $[G, P]$ denote a partially ordered group G with positive cone P. An abelian l-group $[G(P), G(P)^+]$ is called a free l-group over $[G, P]$ if: 1) there exists an isomorphism ψ of the partially ordered group $[G, P]$ into $[G(P), G(P)^+]$; 2) $\psi([G, P])$ generates $[G(P), G(P)^+]$; 3) if $\phi: [G, P] \to [L, L^+]$ is an isomorphism of the partially ordered group into an l-group, then there exists an l-homomorphism $\phi': [G(P), G(P)^+] \to [L, L^+]$ such that $\phi = \phi' \psi$. Weinberg [256] showed that there is a free l-group over the partially ordered group $[G, P]$ if and only if the latter is semiclosed (i.e. $np \in P$ implies $p \in P$). If G is a partially ordered semiclosed archimedean abelian group of finite rank, then the free l-group over $[G, P]$ is archimedean. He considered the lattice of filets of the free l-group over $[G, \{0\}]$, where G is an abelian group of rank > 1. Let I^α denote the free group with set of free generators of power α. Then $A_\alpha = [I^\alpha(\{0\}), I^\alpha(\{0\})^+]$ is the free l-group with set of free generators of power α. He established that each free l-group A_α is archimedean. In [255] it is shown that for $\alpha > 1$, A_α has no nontrivial direct factor, and that A_α is a subdirect sum of a family of copies of the linearly ordered group of integers. In a special way one defines the free abelian l-group over a distributive lattice.

7. **Embedding of l-groups.** Vaida [248] showed that an l-group G can be represented as a subdirect product of linearly ordered groups if and only if for any of its elements x and t we have $x \vee (tx^{-1}t^{-1}) \geq e$. Conditions of embeddability of an l-group into a direct product of linearly ordered groups were studied by Banaschewski [82]. Jaffard [155] found necessary and sufficient conditions expressed in terms of filets. Lorentzen [190] showed that every abelian l-group is l-isomorphic to a proper subgroup of the full direct product of linearly ordered groups. Jaffard [155] made this more exact by showing that an abelian l-group admits an irreducible representation if and only if the set of minimal filets $\leq \bar{x}$ is not empty for every filet $\bar{x} \neq \bar{0}$. The theorem on the uniqueness of the irreducible representation of an l-group was proved topologically by Jaffard [162]. In [158] he considers l-representations of complete l-groups.

We call a representation of an l-group G complete if $G \supset \bar{G}_\nu$ for all ν, where $\bar{G}_\nu$ is the set of all elements $x(\cdot) \in \Pi G_\nu$ (in the full cardinal product of the linearly ordered groups G_ν) for which $x(\mu) = 0$ for $\mu \neq \nu$. A representation is called an α- or a β-representation if $G \cap \bar{G}_\nu \neq e$ or $G \cap \bar{G}_\nu = e$, respectively. A representation $(G_\nu \colon x\nu \in I)$ of G is called reduced if for any $\beta, \gamma \in I$, $\beta \neq \gamma$, there exists an $x(\cdot) \in \Pi_{\nu \in I} G_\nu$ such that $x(\beta) > e$ and $x(\gamma) < e$. Šik [223,224] found conditions equivalent to the existence in an l-group of the various kinds of representations, i.e. α-representation, β-representation, reduced representation, reduced β-representation, etc. To characterize the various types of representations one uses the system of components of the l-group and the set of relatively complemented l-ideals. Works [227,228,235 and 82] are devoted to the problem of representation.

Ribenboim [212] showed that every abelian l-group has a representation which is Hausdorff, proper, and completely regular. The studies of Ribenboim are expounded systematically in the book [213].

Holland [149] devoted his work to the problem of embedding an l-group into an l-group of automorphisms of a linearly ordered set. Consider the group $A(S)$ of all order isomorphisms of the linearly ordered set S, and partially order it by setting $f \leq g$ if for all $x \in S$ one has $xf \leq xg$. Under this order, $A(S)$ is an l-group. Each l-group is l-isomorphic to an l-subgroup of an l-group $A(S)$, for some linearly ordered set S. Suppose that S and T are linearly ordered sets and that a given l-group G is l-isomorphic to a transitive l-subgroup of $A(S)$ and also to a transitive l-subgroup of $A(T)$. In [150] it is cleared up how the sets S and T can be related in this circumstance.

Holland [149] showed that every l-group can be embedded in a divisible l-group (i.e. where the equation $nx = g$ is solvable). Jakubík [72] and Weinberg [256] proved the possibility of embedding an archimedean l-group in a divisible archimedean l-group.

Birkhoff [85] raised the problem of extending an l-group G_1 by another one G_2 in the following form: To find all l-groups G containing G_1 as an l-ideal so that the factor group G/G_1 is l-isomorphic to G_2. He also showed that this problem has at least one solution: it suffices to form the cardinal product of the l-groups G_1 and G_2; and if G_1 is linearly ordered then there is another possibility: the lexicographic product of G_1 and G_2. In [160] Jaffard solved the problem of extending l-groups as stated above.

Papangelou [206] introduced various definitions of order convergence of directed nets in an abelian l-group, i.e. of families $(x_i)_{i \in I}$ of elements of a given l-group, the domain I of indices of which is a directed partially ordered set. He introduced the notion of fundamental sequence, by means of which he constructed an l-group containing the given one as an l-subgroup.

Let Γ be a partially ordered set and H_γ a partially ordered group for each $\gamma \in \Gamma$. Let $V = V(\Gamma, H_\gamma)$ be the following subset of the full direct sum of the groups H_γ: an element $v = (\cdots, v_\gamma, \cdots)$ belongs to V if and only if $S_v = \{\gamma \in \Gamma: v_\gamma \neq 0\}$ contains no infinite ascending sequence. V forms a subgroup. Define $\Gamma^v = \{\gamma \in \Gamma: v_\gamma \neq 0$ and $v_\alpha = 0$, all $\alpha > \gamma\}$. The components v_γ, $\gamma \in \Gamma^v$, are called the maximal components of v. A non-zero element $v \in V$ is, by definition, positive if each maximal component v_γ of v is positive in the order of H_γ. V thus becomes a partially ordered group. If Γ is a linearly ordered set and, for each $\gamma \in \Gamma$, $0 \neq H_\gamma$ is a subgroup of the additive group of reals with the natural order, then V is linearly ordered. Hahn [141] showed that every abelian linearly ordered group can be embedded in such a group V. Conrad, Harvey, and Holland [118] generalized the theorem of Hahn for abelian lattice ordered groups.

8. **Topology in an l-group.** If G is an l-group, then the interval topology of G is the topology obtained when as basis of closed sets one takes the sets of the form $\{f \in G: f \geq f_0\}$, $\{g \in G: g \leq g_0\}$ and the set G. A linearly ordered group is a topological group and a topological lattice in its interval

topology. Birkhoff [1] raised the question (problem 104) if every l-group is a topological group and a topological lattice in its interval topology. Northam [205] showed that there exists an l-group which is not a Hausdorff space in its interval topology and hence is not a topological group. Jakubík [168] showed that if in an l-group G there exist archimedean elements a and c, $a \wedge c = 0$, then G is not a topological l-group in its interval topology. Here an $a > 0$ in G is called archimedean if for each $b \in G$ there exists a natural number n such that $a^n \leq b$. Wider classes of l-groups which are not topological in their interval topologies were found by Wolk [260]. He also indicated classes of l-groups such that if an l-group of these classes is a Hausdorff space in its interval topology, then it is linearly ordered (see Choe [100], Wolk [260], Conrad [113], and Jakubík [171]). Holland [152] gave an example of an l-group which is not linearly ordered but is a topological group and a topological lattice in its interval topology. Topologies in l-groups were also studied by Šik [222,235] and Zamansky [263].

§ 3. Partially ordered groups

1. **Directed groups.** Jakubík [166] showed that if C is a maximal and convex chain in a directed group G containing e, then for C to be a direct factor of G it is necessary and sufficient that the following condition (f) hold: if $r \in C^+$ and $z \in G^+$, then $r \wedge z$ exists in G. There is known to be an example showing that condition (f) does not hold in all directed groups. This theorem is a generalization of another theorem of Jakubík given on Russian page 120 of the book of Fuchs [60].

Suppose that G is a directed abelian group with semigroup of positive elements G^+ satisfying the following conditions: 1) each nonempty lower directed subset in G^+ has an infimum in G^+; 2) (the decomposition property) $x \leq a + b$ for $x, a, b \in G^+$ implies the existence of elements $a', b' \in G^+$ such that $x = a' + b'$, $a' \leq a$, $b' \leq b$. Bauer [83] showed that such a group is a complete l-group.

Non-abelian directed groups with the decomposition property were studied by Cristescu. Here an important role is played by the notion of component. Let G be a directed group and let Q be a subgroup of G. The set Q is called a component in G if for any $x \in G^+$ there exist $x' \in G^+$ and $x'' \in (Q^+)^*$ such that $x = x' \cdot x''$ (where $(G^+)^*$ denotes the set of all $y \in G^+$ for which $\inf(x, y) = e$ for any $x \in Q^+$). In [119] Cristescu studied the components of a directed group and in [120] he studied the components of a linearly ordered group.

Šik [222] gave necessary and sufficient conditions for a directed (more generally, a partially ordered) group to be an l-group.

A partially ordered group G is called γ-simple if in G there are no non-unit elements x and y such that $x \wedge y = e$. Šik [232] found necessary and sufficient conditions for a directed group to be isomorphic to a direct product of γ-simple partially ordered groups.

Aubert [79] showed that a directed abelian group is isomorphic to a subgroup of the additive group of reals if and only if its S_v-ideals (in the sense of Lorentzen) form a group relative to S_v-multiplication: $A_v O_v B_v = (AB)_v$ (where O_v is the symbol for the multiplication).

2. o^*-groups. These are the partially ordered groups each partial order of which can be extended to a linear order. Detailed results on the theory of o^*-groups are expounded in [60]. Here we give only some further ones.

A free n-solvable group with k generators for $n \geq 3$ and $k \geq 2$ is not an o^*-group (M. I. Kargapolov [22]).

We call a subgroup H of a group G strictly isolated if $x g_1^{-1} g x_1 \cdots \cdots g_n^{-1} x g_n \in H$ implies $x, g_1^{-1} x g_1, \cdots, g_n^{-1} x g_n \in H$. A. I. Kokorin [26] showed that the factor group G/H of an o^*-group G is an o^*-group if and only if H is a normal strictly isolated subgroup.

We call a subgroup H of a group G relatively convex if there exists a linear order on G such that H is a convex subgroup. H is called infra-invariant if $\forall g \in G[(g^{-1} H g \subseteq H) \vee (g^{-1} H g \supseteq H)]$. An infra-invariant strictly isolated subgroup of an o^*-group is relatively convex (A. I. Kokorin [27]).

A 2-solvable o-group is an o^*-group (A. I. Kokorin [27]). However, as indicated above, a 3-solvable o-group is not always an o^*-group (M. I. Kargapolov [22]). A 2-solvable group with strictly isolated unit is an o^*-group.

An example of an o^*-group which is not locally nilpotent was given by Ja. B. Livčak [41]. A. I. Mal'cev [45] showed that every locally nilpotent group not containing elements of finite order is an o^*-group. He also showed that if every subgroup with a finite number of generators in a partially ordered group is an o^*-group, then the group itself is also an o^*-group. In [47] Mal'cev showed that in an o^*-group, a partial order with cone P is the intersection of linear orders if and only if one has the following condition: $P' \cap S(x) \neq \emptyset$ implies $x \in P'$ for any $x \in G$, where $S(x)$ is the smallest normal semigroup containing x and $P' = P \cup \{1\}$.

3. V-groups. A group is called a V-group (respectively, a V^*-group) if

each linear (respectively, partial) order of any of its subgroups can be extended to a linear order of the whole group. A group is called a VN-group (respectively, a VA-group) if each linear order of any (respectively, any abelian) of its normal subgroups can be extended to a linear order of the whole group.

A. A. Terehov [58] showed that a nilpotent V-group is necessarily abelian, and a locally solvable one is solvable of length 2. He also [59] found necessary and sufficient conditions that a locally solvable group be a V-group. Next, this result was generalized by M. I. Kargapolov [20], who proved that any torsion free group G is a V-group if and only if there exists in G a normal abelian subgroup A such that the factor group G/A is abelian and for $a \in A$ and $b \in G \backslash A$ the element $b^{-1}ab$ is equal to a^{α} for some positive rational $\alpha \neq 1$. Thus it turned out that every V-group is 2-solvable. As a consequence of the main theorem on the structure of V-groups, it is the case that if a V-group has a central system with torsion free factors, then it is abelian.

A. I. Kokorin [28] showed that a solvable VN-group is a V^{*}-group. If the class of V-groups, VN-groups and VA-groups are denoted, respectively, by (V), (VN) and (VA), then we have the following inclusions: $(V) \subset (VN) \subset (VA)$. The direct product of V-groups is an o^{*}-group. If every subgroup with a finite number of generators in G is a VN-group, then G is a VN-group. A nilpotent VA-group is abelian. A nilpotent normal subgroup of a VN-group is abelian.

A. I. Kokorin and V. M. Kopytov [30] showed that the classes of solvable V^{*}-groups, V-groups and VA-groups coincide, and that they all are o^{*}-groups. A group G is in these classes if and only if: 1) G contains an abelian normal subgroup A which is torsion free; 2) for any $g \notin A$ there exist distinct integers m and n of the same sign such that $g^{-1}a^{m}g = a^{n}$ for all $a \in A$. A free group is a VA-group but not a V-group. There is given an example of a 2-solvable nilpotent V-group which is not a VA-group.

Every V-group can be embedded in a complete V-group (V. M. Kopytov [35]).

4. **Solvable and nilpotent groups.** A. I. Mal'cev [46] proved the following local theorems: If every subgroup H with a finite number of generators of a partially ordered group G has a (solvable) central system of (normal) convex subgroups, then so does G. He also established the validity of the analogous local theorems for the existence of (solvable) central systems all factors of

which are torsion free. He also proved in [47] that the assertion of A. A. Vinogradov [6] on conditions for embeddability of a partially ordered nilpotent group into a cardinal product of linearly ordered groups, in general, is not true, but remains true for metabelian groups.

A. A. Vinogradov [8] generalized the results of Šik [225] for abelian partially ordered groups to metabelian partially ordered groups.

A semigroup σ in a group G is called isolated if for any $g \in G$ and any natural number n $g^n \in \sigma$ implies $g \in \sigma$. K. M. Kutyev [37] showed that if a locally nilpotent partially ordered group G with isolated semigroup of positive elements is SL-isomorphic to a group G^*, then G^* is locally nilpotent and can be partially ordered by means of an isolated semigroup of positive elements.

5. **Certain other classes of partially ordered groups.** By an ω-series in a partially ordered group G we mean a countable sequence of elements $g_1 < g_2 < \cdots$ such that for any $g \in G$ there exists a natural number n such that $g < g_n$. A partially ordered group G is called ω-partially ordered if it has an ω-series. ω-partially ordered groups were studied by K. S. Sibirskiĭ and A. M. Stahi [53]. An ω-partially ordered group is directed.

Partially ordered groups without proper convex subgroups were studied by Michiura [201], E. Gabovič [9], A. A. Vinogradov [5], K. S. Sibirskiĭ and A. M. Stahi [53], and others. A nontrivial partially ordered group with no proper convex subgroups which is semiclosed ($x^n \geq e$ implies $x \geq e$) is isomorphic to a subgroup of the additive group of reals [201]. Later it was seen [9] that the latter conclusion holds for abelian partially ordered groups without the hypothesis of semiclosedness. It was also established in [53] that a partially ordered group with no nontrivial convex subgroups is ω-partially ordered, and hence directed.

Let G be a partially ordered group, $M(x, y) = \{z \in G: z \geq x, z \geq y\}$, and $N(x, y)$ the set of all minimal elements in $M(x, y)$. G is called a J-group if $N(x, y) \neq \emptyset$ for any two $x, y \in G$. J-groups were studied by Vaida [250]. If $M(x, y) \neq \emptyset$ for each pair $x, y \in G$, and for each $z \in M(x, y)$ there exists a $z_1 \in N(x, y)$ such that $z_1 \leq z$, then G is called a multilattice ordered group. Vaida raised the question: do there exist J-groups which are not multilattice ordered? Jakubík [172] gave a positive answer to this question by showing that each l-group can be embedded in a partially ordered J-group which is not a multilattice.

A partially ordered group of real functions defined on an arbitrary set was studied by Krull [180] and Jaffard [163].

Michiura [200] considered strongly archimedean groups. A partially ordered group G is called strongly archimedean if for each $a \neq 0$ in G and each $x \in G$ we have $na > x$ for some integer n.

Let G be a partially ordered group. If a and b in G have a sup and an inf, write $a \sim b$. Let $a \sim b$. The intersection of all intervals containing a and b is called a D-interval and is written $\langle a, b \rangle = \langle b, a \rangle$. G is called D-distributive if $x \sim a$, $y \sim a$ and $\langle x, a \rangle = \langle y, a \rangle$ implies $x = y$. D-distributive groups were studied by Burgess [95]. He showed, in particular, that each partially ordered archimedean group is D-distributive.

Write $x \| y$ if x and y are elements in an abelian partially ordered group G and are not comparable in the given order. We call the order in G stable if the relation $\|$ is transitive. In a group G with a stable order define a topology whose neighborhoods of zero consist of the $t \in G$ such that $-x < t < x$ where $x > 0$. Relative to this topology the group G becomes a topological group. Topological abelian groups with this topology were studied by Lavis [185]. Topological partially ordered abelian groups were also studied by Bauer [83].

Archimedean partially ordered Ω-groups were considered by E. Gabovič [10]. The work [56] of D. M. Smirnov was devoted to the study of Ω-groups.

6. Embedding of partially ordered groups. We call a partially ordered group archimedean if for $x > 0$ and each y there exists an integer n such that $nx \geq y$. It is called para-archimedean if for $x \nleq 0$ and y there exists an n such that $nx \nleq y$. Jaffard [154,155] showed that a nontrivially ordered archimedean group in which $n > 0$ and $nx \geq 0$ implies $x \geq 0$ can be represented as a group of real functions if and only if it is para-archimedean. There are given necessary and sufficient conditions that a partially ordered group be isomorphic to a subgroup of the additive group of reals.

Jaffard [156] studied the question of embedding a given partially ordered abelian group into a lattice ordered group. He proved that for such an embedding it is necessary and sufficient that in the given group one have the condition of semiclosedness ($nx \geq 0$ implies $x \geq 0$).

A partially ordered abelian group G is called a partially ordered extension of a partially ordered group Q by a partially ordered group H if 1) G is an ordinary group-theoretic extension of Q by H, 2) the partial order in G induces the original partial order on H, 3) H is a convex subgroup of G, and 4) the natural partial order defined on the quotient group G/H coincides with the original partial order on Q. The group G can be defined as the set

$H \times Q$ with the group operation $(a, x) + (b, y) = (a + b + f(x, y), x + y)$, where f is some mapping of $Q \times Q$ into H. The partially ordered extension G is called lexicographic if $(a, x) \geq 0$ if and only if $x > 0$ or $x = 0$ and $a \geq 0$. The problem of constructing a partially ordered extension of a partially ordered abelian group is the subject of work [157] by Jaffard. In particular, he shows that a lexicographic extension G of a partially ordered group Q by a partially ordered group H is an l-group if and only if one has one of the following two situations: 1) H is an l-group and Q is a linearly ordered group; 2) $H = \{0\}$ and Q is an l-group. Jaffard's work [161] is also devoted to the problem of embedding a partially ordered group in an l-group. In [160] there are proved the following theorems: Let G be a partially ordered extension of a partially ordered group H by a partially ordered group Q. In order that G be directed it is necessary and sufficient that Q be directed. A partially ordered extension G is linearly ordered if and only if it is lexicographic and the groups H and Q are linearly ordered.

Teller [242] found necessary and sufficient conditions that a partially ordered extension G of a partially ordered abelian group Q by a partially ordered abelian group H be an l-extension, i.e. that G be an l-group. He also studied l-extensions of an abelian l-group Q with finite basis. Here an element g in an l-group Q is called basic if $0 < g$ and the set $\{x \in Q: 0 < x \leq g\}$ is a linearly ordered set. A subset S in Q is called a basis if it is a maximal set of disjoint elements and each element $g \in S$ is basic. The analog of the Schreier extension problem for partially ordered groups was studied by Fuchs [131]. The extension theory he constructed was applied by Loonstra [188] to the case when both the extendor and extendee group are isomorphic to the linearly ordered additive group of integers.

Jakubík [72] and Weinberg [256] showed how a partially ordered abelian torsion free group can be embedded in a divisible partially ordered abelian group.

An abelian partially ordered group G with strictly positive semigroup cone R will be denoted (G, R). A partial order L (respectively, an extension L of the partial order R) is called quasilinear if $L \cup -L \supset G_\infty$, where G_∞ is the set of all elements of infinite order of G. Let $(\widetilde{G, R})$ denote the set of all additive and isotone functionals defined on the partially ordered group (G, R), i.e. the set of all homomorphic and isotonic mappings of the partially ordered group (G, R) into the linearly ordered additive group of reals. Šik [226] gave two constructions of quasilinear extensions of the partial order R

of the group G by means of additive and isotonic functionals on (G, R). His report [233] is devoted to questions connected with extending functionals.

Cohen and Goffman [105] introduced a notion of a system of neighborhoods of zero in a nondiscrete linearly ordered abelian group relative to which the group becomes a topological group. Using the definition of convergence of a countable sequence, Everett [124] obtained an extension of an l-group to one where each fundamental sequence converges. Here one applies a method similar to the Cantor method of getting the reals, and transfinite sequences are not used. There is shown there the relation between the Cantor and the Dedekind extensions of an abelian l-group. Everett and Ulam [125] extended some of Everett's results relative to countable convergent sequences to non-abelian l-groups. The report [81] of Banaschewski generalized this still further. In [81] one considers arbitrary partially ordered groups, and topological notions are used in a form appropriate to contemporary set-theoretic topology. Similarly to the way that the reals can be obtained from the rationals in two ways (by cuts, à la Dedekind, or by the topological method of Cantor) he studies the relationship between the Dedekind and the topological extensions of partially ordered groups.

Ky Fan [126] characterized the set of all continuous real functions defined on a compact space as a partially ordered abelian group. Clifford [104] gave another characterization. Fleischer [128] solved the same problem, but by application of lattice notions the solution is obtained more concisely and simply, together with the fact that it gives a certain generalization of the results of Ky Fan.

7. **Some properties of partially ordered groups.** Lavis [186] found which normal subgroups H of a partially ordered group G are such that the naturally ordered quotient group G/H is linearly ordered. To this end, one defines the subgroup R_0 of G which is generated by all the $x \in G$ which are not comparable with e, and considers the smallest convex subgroup K_0 containing R_0. The property mentioned above then is seen to be possessed only by the subgroups H which are normal convex subgroups of G containing K_0.

For $x, y \in G^+$ in a partially ordered group G we set $x \epsilon y$ if $x \wedge y = e$. For two elements $x, y \in G$ let $x \delta y$ mean that there exist $a, b \in G^+$ such that $a \geq x \geq a^{-1}$, $b \geq y \geq b^{-1}$ and $a \wedge b = e$. The binary relations ϵ and δ are called respectively ϵ-disjunctivity and δ-disjunctivity. Let A be a nonempty subset of G. Let A^ϵ denote the set $\{x \in G : x \epsilon A\}$. Similarly we define the set A^δ. A set A is called an ϵ-component (δ-component) if $A = A^{\epsilon\epsilon}$ ($A = A^{\delta\delta}$). A subset $\mathfrak{A}$ of a partially ordered group G is called a γ-subgroup if there exists an ϵ-component A in G such that $\mathfrak{A} = AA^{-1}$. Šik [231] studied γ-subgroups and $\epsilon(\delta)$-components in a partially ordered group. In particular, he showed that the set of all γ-subgroups in a partially ordered group G forms a complete Boolean lattice (under set-theoretic inclusion).

We will say that a partially ordered set Z is the cardinal product of its subsets X and Y if 1) each $z \in Z$ is uniquely expressable in the form $z = a \vee b$ with $a \in X$, $b \in Y$ and 2) for $z_1 = a_1 \vee b_1$, $z_2 = a_2 \vee b_2$, $a_i \in X$, $b_i \in Y$ $(i = 1, 2)$ the relation $z_1 \geq z_2$ holds if and only if $a_1 \geq a_2$ and $b_1 \geq b_2$. Moreover, it is assumed, for any pair a and b with $a \in X$ and $b \in Y$, that there exists a $z \in Z$ such that $z = a \vee b$. The subsets X and Y are called cardinal factors of Z. Just as above, G^+ is the semigroup of positive elements of G. E. P. Šimbireva [69] showed that if X is a cardinal factor of the partially ordered set G^+, then X is a direct factor of the partially ordered semigroup G^+. With this, one generalizes Jakubík's result for l-groups. Jakubík showed that every maximal convex chain containing e in an l-group is a direct factor, and Šimbireva gave an example showing that this assertion is not generally true for directed groups.

V. I. Frenkel′ [61], [62] took up the solution of algorithmic problems in partially ordered groups. For partially ordered groups given by a finite number of generators and defining inequalities (defining relations) the word problem and the conjugacy problem were solved.

§ 4. Generalizations of partially ordered groups

1. **Partially ordered groups with preservation of ideals.** We will consider a partially ordered set P. For a nonempty subset F of P let F^* (respectively F_*) denote the set of all $x \in P$ such that $x \geq f$ (respectively, $x \leq f$) for all $f \in F$. A nonempty set $I \subset P$ is called an ideal (respectively, a dual ideal) if for each finite nonempty subset $F \subset I$ the set F^* (respectively, F_*) is nonempty and $(F^*)_* \subset I$ (respectively, $(F_*)^* \subset I$). A group G is said to be partially ordered with preservation of ideals if G is a partially ordered set and the set $x \cdot I \cdot y$ is an ideal or dual ideal for any ideal or dual ideal $I \subset G$. Andrus and Butson [78] showed that every partially ordered group with preservation of ideals is an extension of a partially ordered group by means of a partially ordered group with preservation of ideals, the length of each chain in which is not greater than two. Thus the class of partially ordered groups with preservation of ideals contains the class of all partially ordered groups. Elements a, $b \in P$ are called connected if there exists a finite sequence of elements $a = x_1, x_2, \cdots, x_n = b$ in P such that each pair of

these elements x_i, x_{i+1} satisfies one of the relations $x_i \lessgtr x_{i+1}$. A group G is called connected if each pair of its elements is connected. In the same report [78] are found necessary and sufficient conditions for a group to be made into a connected partially ordered group with preservation of ideals.

2. **One-sided partially ordered groups.** If in the definition of a partially ordered group we weaken the monotonicity law and express it in the form: "$a \leq b$ implies $ac \leq bc$ for each element c in the group", then we get a definition of a right partially ordered group. Similarly we define a left partially ordered group. Right partially ordered groups were first studied by M. I. Zaĭceva [17]. She established sufficiency criteria for a right linear ordering of a group. She proved, in particular, that every group of type RN with torsion free factors can be right linearly ordered, and gave an example of a solvable right linearly orderable group which is not linearly orderable in the usual sense. By means of her theory, she found the order type of a countable homogeneous set, i.e. of a linearly ordered set with a transitive group of automorphisms.

We call a subset R of a group G stable if it is closed under the group operation. Let $H^{\#}$ denote the smallest stable subset in G containing the subset H. We write $[H] = \{T: \text{ there exists } K \subseteq H \text{ such that } T = K \cup \overline{H - K}\}$, where $H - K$ denotes set-theoretic difference and $\overline{H - K}$ is the set of all elements of G inverse to the elements of the set $H - K$. Let $H^{*} = \{a: a \in T^{\#} \text{ for each } T \in [H]\}$. A subset H of G is called a subset of finite order if H^{*} contains the unit e of G. Giampa [103] found necessary and sufficient conditions for the one-sided linear orderability of a group, expressed in terms of stability and order of sets, and also gave conditions that a one-sided partial ordering can be extended to a one-sided linear ordering.

One-sided partially ordered groups were studied by Conrad [111]. He gave necessary and sufficient conditions for the existence of a one-sided partial ordering, and a sufficiency criterion for extending a one-sided partial ordering to a linear one. He carried over some theorems on ordered groups to one-sided ordered ones.

Necessary and sufficient conditions for the one-sided linear ordering of a group were given by Tully [246].

D. M. Smirnov studied the group of automorphisms of group rings of right orderable groups.

3. **Semihomogeneous partially ordered groups.** The monotonicity law in the definition of a partially ordered group can be weakened in another way. In place of it, we will assume that each element x in the group possesses

exactly one of the properties: (A) $a < b$ implies $xa < xb$, $ax < bx$; (B) $a < b$ implies $xa > xb$, $ax > bx$. Partially ordered groups with this weakened definition were studied by P. G. Kontorovič and A. I. Kokorin [32] and were called by them semihomogeneous partially ordered (lattice ordered, linearly ordered) groups. They found necessary and sufficient conditions for a group G to be semihomogeneous partially ordered (lattice ordered, linearly ordered). They indicated the structure of a semihomogeneous lattice ordered group with one generator, and also considered some properties of such groups with a finite number of generators.

4. **Almost partially ordered groups.** Again we weaken the monotonicity law, now in the following direction: if $a < b$, then $ga < gb$ for a "majority" of elements g in the group G and $ah < bh$ for a "majority" of elements h in G. Here the phrase "for a majority of elements in G" must be taken to mean that a set S with "a majority of elements of G" satisfies the following conditions: 1) if S_1 and S_2 contain a "majority" of elements of G, then $S_1 \cap S_2$ is not empty; 2) if S contains a "majority" of elements of G, then so do the sets gS and Sg for each $g \in G$; and 3) if S_1 contains a "majority" of elements in G and $S_1 \subset S_2$, then S_2 contains a "majority" of elements in G. Almost partially ordered groups were studied by Britton and Shepperd [94].

5. **Quasi partially ordered groups.** We arrive at a generalization of partially ordered groups by excluding the law of antisymmetry from the list of defining postulates. Groups generalized thus are called quasi partially ordered groups, or alternately pre-ordered groups (see, for example, Bourbaki [3]). They were studied by Lorentzen [191], who gave conditions for embedding them in l-groups, and Choe [101] who carried over certain properties of partially ordered groups to quasi partially ordered ones.

6. **Generalized partially ordered groups.** Let G be a non-abelian group and let G^+ be a nonempty semigroup in G. Note the following axioms: 1) $e \in G^+$, 2) $(x \in G^+)$ implies $(x^{-1} \notin G^+)$, 3) $(x \neq e)$ implies $(x \in G^+$ or $x^{-1} \in G^+)$, 4) $x^{-1}G^+x \subset G^+$ for each $x \in G$. Consider a group $G(1, 2)$ satisfying axioms 1) and 2). We introduce in it two partial orders: I) $(x^{-1}y \in G^+)$ iff $(x <_s y)$ and II) $(yx^{-1} \in G^+)$ iff $(x <_d y)$. Set $(a <_s b, a <_d b)$ iff $(a < b)$. If for $y > e$ and every integer n we have $x^n <_s y$ or $x^n <_d y$, then x is called infinitely small on the left (on the right) relative to y; notation: $x \ll_s y$, $x \ll_d y$. Axiom 5): the collection of incomparable elements of a group $G(1, 2)$ is a proper normal subgroup; 6) $(\epsilon \ll_s a)$ if and only if $(\epsilon \ll_d a)$. Busulini [97] gave the simplest properties of the groups $G(1, 2)$, $G(1, 2, 3)$,

$G(1, 2, 5)$ and $G(1, 2, 6)$. He noted that in geometric applications one encounters groups $G(1, 2, 5)$.

7. **Betweenness groups.** The notion of a group with such an appelation was introduced by Shepperd [220]. A betweenness group is a group in which there is defined a ternary relation (a, b, c) satisfying the axioms: 1) for any a, b, c, either (a, b, c), or (b, c, a), or (c, a, b); 2) (a, b, c) and (a, c, b) hold simultaneously if and only if $b = c$; 3) (a, b, c) implies (c, b, a); 4) (a, b, c) and (a, c, d) imply (b, c, d); 5) (a, b, c) implies (gah, gbh, gch) for any g and h in the group. According to this definition a betweenness group is a generalization of a linearly ordered group: The relation "between" is induced by a linear order if we say (a, b, c) iff $a \leq b \leq c$ or $c \leq b \leq a$. Betweenness groups can be finite as well as infinite. The former class is exhausted by groups of order four. There are found conditions so that an infinite group can be made into a betweenness one, and the connection is made between linearly ordered groups and betweenness ones.

Greve [140] studied partial betweenness groups. A set $G = \{a, b, c \cdots\}$ is called partially between if there is defined on it a ternary relation (a, b, c) satisfying the following axioms: 1) (a, b, c) implies (c, b, a); 2) $(a, b, c) \wedge (a, c, b)$ if and only if $b = c$; 3) $(a, b, c) \wedge (a, d, b)$ implies (d, b, c); 4) $(a, b, c) \wedge (b, c, d) \wedge b \neq c$ implies (a, b, d). The symbol $\wedge$ denotes logical conjunction. A partial betweenness group G is a partially between set G relative to a ternary relation (a, b, c) which is a group under a composition so that: α) $(-a, 0, a)$ for all $a \in G$; β) (a, b, c) implies $(x + a + y, x + b + y, x + c + y)$ for all $x, y \in G$; γ) $(-a, x, a)$ implies $(-a, x, 0)$ or $(0, x, a)$. Note that partial betweenness vector spaces were studied by Behrend [84].

Very close to the notion of betweenness group is that of cyclic group introduced by Rieger [214]. Cyclically ordered groups were considered by Swierczkowski [240]. In the book of Fuchs [60] there is given a detailed description of the properties of such groups.

8. **Separation groups.** The notion of a separation group was also introduced by Shepperd [221]. By such a group is meant one in which there is defined a quaternary relation (a, b, c, d) satisfying the axioms: 1) for any a, b, c, d, either (a, b, c, d) or (a, c, b, d) or (a, b, d, c); 2) (a, b, c, d) and (a, b, d, c) hold simultaneously if and only if $a = b$ or $c = d$; 3) (a, b, c, d) implies (d, c, b, a); 4) (a, b, c, d) implies (b, c, d, a); 5) (a, b, c, d), (a, e, b, c) and $b \neq c$ implies (a, e, b, d); 6) (a, b, c, d) implies (gah, gbh, gch, gdh) for any g and h in the group. The relation

(a, b, c, d) is obtained from a linear order if we set (a, b, c, d) if and only if one of the relations $a \leq b \leq c \leq d$ or $d \leq c \leq b \leq a$ holds, or one of the relations obtained from these by a cyclic permutation of the elements. Separation groups can have elements of finite order. In [221] there are given necessary and sufficient conditions that a group can be made into a separation group, and there is found the connection between linearly ordered groups and separation groups.

9. **Locally partially ordered groups.** Locally partially ordered groups are those which satisfy all the postulates defining a partially ordered group except the postulate of transitivity, which is replaced by the following weakened postulate: if $a \leq b$, $b \leq c$, $c \leq d$, and $a \leq d$ then $a \leq c$. Locally partially ordered groups were studied by A. G. Pinsker [49], and he stressed their value in the theory of K-spaces.

10. **Abelian Γ-groups.** Let Γ be a partially ordered set. A commutative operator R-group G is called a Γ-group if to each $\gamma \in \Gamma$ there corresponds a pair of admissible subgroups G_γ, G^γ of G satisfying the conditions: 1) $G_\gamma \subseteq G^\gamma$ for all $\gamma \in \Gamma$; 2) $\alpha < \beta$ implies $G^\alpha \subseteq G_\beta$; 3) for each $a \in G$ there exists at least one γ such that $a \in G^\gamma$ and $a \notin G_\gamma$; 4) if $a \notin G^\gamma$, then there exists a $\beta > \gamma$ such that $a \in G^\beta$ and $a \notin G_\beta$. Every abelian partially ordered group with the property that $na > 0$ implies $a > 0$ can be viewed as a Γ-group. The theory of Γ-groups was developed by Conrad [106]. In the case when Γ is a linearly ordered set studies on Γ-groups can be found in his work [108].

§ 5. Ordered rings and algebras

1. **Partially ordered rings.** If for an element a in a ring A there exist elements $a_1, \cdots, a_n \in A$ such that a equals the product of $2n$ of the elements $a_1, \cdots, a_n$, taken in some order, then we write $a = a(a_1, \cdots, a_n)$. An ideal M of the ring A is called an F-ideal if the relation

$$a(a_1, a_2, \ldots, a_m) + b(b_1, b_2, \ldots, b_n) + \cdots$$
$$\cdots + d(d_1, d_2, \ldots, d_r) \in M$$

implies

$$a_1 a_2 \ldots a_m \in M, \; b_1 b_2 \ldots b_n \in M, \; \ldots, \; d_1 d_2 \ldots d_r \in M.$$

If the ideal $\{0\}$ is an F-ideal, then the ring A is called an F-ring. Such rings were studied by Thierrin [244]. He showed, in particular, that a partially ordered ring $A \neq \{0\}$ is isomorphic to a subdirect sum of linearly ordered rings with no zero divisors if and only if A is an F-ring.

The paper [144] of Hayes is devoted to the representation theory of partially ordered rings of a certain class.

A study was carried out by Strodt [237] on the algebraic closure of certain fields with a specific determination of the partial order on them.

The work [219] of P. Scherk has some bearing on the theory of ordered sfields.

The method by which the order on the integers is extended to the rationals is used to extend the order of a ring to a ring of quotients. Let S be a ring containing a ring R. Suppose that for each $a \in S \backslash R$ there exist a, $b \in R$ such that $a(b)$ is not a left (right) zero divisor in S and $a\alpha, ab \in R$. Such a ring S is called a ring of quotients of the ring R. Fuchs [132] showed that every linear order of an associative ring R can be extended to a linear order of a ring S of quotients of R uniquely.

2. **Lattice ordered rings. f-rings.** A lattice ordered ring is called an f-ring if $a \wedge b = 0$ and $c \geq 0$ imply $ca \wedge b = 0 = ac \wedge b$. Hayes [144] showed that a partially ordered ring R without nilpotent elements is an f-ring iff for each $a \in R$ there are $b, c \in R$ such that $b \geq 0$, $c \geq 0$, $a = b - c$ and $bc = cb = 0$.

Johnson [175] and Kist [177] proved a theorem on embedding archimedean f-rings with no nonzero nilpotent elements into rings of real continuous functions defined on topological spaces.

Brainerd [93] studied the problem of embedding an f-ring R with no nonzero left and right annihilators into a special f-ring consisting of endomorphisms of the additive group of the ring R.

Anderson [77] gave necessary and sufficient conditions for an f-ring to be isomorphic to a subdirect sum of a finite number of linearly ordered rings.

We call an f-ring unitable if it is embeddable in an f-ring with a multiplicative unit. Henriksen and Isbell [147] showed that the class of unitable f-rings is primitive. Each linearly ordered ring is decomposable into a lexicographic direct sum of two summands, of which one contains all the ring idempotents and the other is a null ring. They affirmatively solved problem 107 of Birkhoff [1].

A lattice ordered ring is called a d-ring if $a(x \vee y) = ax \vee ay$ and $(x \vee y)a = xa \vee ya$ for every $a \geq 0$ and any x and y. d-rings were studied by Kudláček [183]. He proved, in particular, that every f-ring is a d-ring, but not conversely.

Brainerd [92] gave a characterization of the ring $C(X)$ of real continuous functions on completely regular spaces X as an archimedean lattice ordered algebra over the reals.

3. **Norm (valuation) theory.** Linearly ordered fields with a norm were considered by Gemignani [137]. Ion [153] showed that every linearly ordered abelian group is a norm group for some sfield. Lattice ordered groups were used for seminorming sfields.

4. **Betweenness and separation rings.** A. I. Čeremisin [65], [66] introduced the notions of betweenness and separation rings, and described them. A ring R is called a betweenness ring if its additive group is a betweenness group relative to a relation (a, b, c) (see §4.7) and for any $a, b, c, x \in R$, (a, b, c) implies (ax, bx, cx) and (xa, xb, xc). There is given a complete table of all finite betweenness rings and it is shown that the class of all betweenness rings is, in the main, exhausted by rings which are linearly ordered and in which the relation (a, b, c) of betweenness is determined by the order relation.

A ring R is called a separation ring if its additive group is a separation group (see §4.8) relative to a relation (a, b, c, d) and for any $a, b, c, d, x \in R$, (a, b, c, d) implies (ax, bx, cx, dx) and (xa, xb, xc, xd). There are given cases in which infinite rings can be made into separation ones, and the connection between separation rings and their convex maximal linearly ordered ideals is outlined.

5. **Ordered semirings.** A linearly ordered semiring is a set P with two algebraic operations (addition and multiplication) relative to each of which P is a semigroup; these operations are related by the (two-sided) distributive laws; moreover, there is defined on P an order relation making P a linearly ordered set so that: 1) $a < b$ implies $a + c < b + c$ and $c + a < c + b$ for all $c \in P$; 2) $a < b$ implies $ac < bc$ and $ca < cb$ for positive $c \in P$, and $ac > bc$ and $ca > cb$ for negative $c \in P$. Here by positive (negative) we mean a $c \in P$ such that $c < c + c$ ($c + c < c$). Lugowski [192] studied linearly ordered semirings such that for $a \in P$ and $b \in P$ with $a < b$ there exists $x \in P$ with $a + x = b$. He considered the problem of completing such semirings.

The paper [87] of Bleicher and Bourne is devoted to the problem of embedding partially ordered semirings into partially ordered rings.

By a semi-sfield we mean a semiring whose nonzero elements form a group under multiplication. Weinert [257] showed that if at least one element of a semi-sfield is idempotent (i.e. $a + a = a$), then the same holds for all its elements, and that any semi-sfield with no zero element and with a commutative and idempotent addition can be viewed as a lattice ordered group, and conversely.

Archimedean semirings were studied by Endler [122].

6. **Linearly ordered modules over linearly ordered rings.** Let A be a linearly ordered ring, and let E be a unitary left module over A. We will call E linearly ordered if E is a linearly ordered group and $x \geq 0$, $x \in E$, $\alpha \geq 0$ and $\alpha \in A$ imply $\alpha x \geq 0$. A subset P is called a cone if: 1) $0 \in P$; 2) $P + P \subseteq P$; and 3) $\alpha P \subseteq P$ for all $\alpha \geq 0$. Bleicher and Schneider [88] devoted their effort to the problem of decomposing cones in a module E.

7. **Lattice ordered algebras.** We call a vector space V with real scalars a partially ordered vector space if it is a partially ordered group under addition and for any positive scalar λ the correspondence $x \to \lambda$ is an automorphism. If a partially ordered vector space V is a lattice, it is called a vector lattice. If a lattice ordered ring A is a vector lattice it is called a lattice ordered algebra. If an f-ring B is also a vector lattice B is called an f-algebra. A lattice ordered ring A is called archimedean if for each $a \in A$ different from 0 the set $\{na: n = \pm 1, \pm 2, \cdots\}$ has no lower bound in A. An f-algebra which is an archimedean ring is called an archimedean f-algebra. An archimedean f-algebra with unit 1 is called a Φ-algebra. The works of Henriksen and Johnson [147] and Henriksen, Isbell and Johnson [146] are devoted to the study of Φ-algebras.

Linearly ordered normed algebras were studied by Urbanik [247].

§ 6. Ordered semigroups

1. **Linearly ordered semigroups.** If elements of a semigroup S are connected by the relation $aba = a$ and $bab = b$, then they are called regularly conjugate. Let S be a linearly ordered semigroup. The set E of all idempotents of S, if it is nonempty, is a subsemigroup of S. Saito [217] classified the subsemigroups of a semigroup S generated by pairs of regularly conjugate nonidempotents, and in [216] he considered arbitrary linearly ordered semigroups S of idempotents.

B. M. Bredihin [2] clarified the law of distribution of generating elements in a linearly ordered semigroup.

Doneddu [121] gave necessary and sufficient conditions that a linearly ordered semigroup be order isomorphic to the additive semigroup of positive reals.

Order endomorphisms of linearly ordered semigroups were considered by
E. Gabovič [11]. In particular, he found all order endomorphisms of the
additive semigroup of positive reals.

The two reports [193,194] of Lugowski are devoted to linearly ordered semigroups. There one considers semigroups S such that for $a \in S$, $b \in S$ and
$a < b$ there exist $x \in S$ and $y \in S$ such that $a + x = y + a = b$.

The analog of the ring of quotients is the notion of a semigroup of
quotients. Let S be a semigroup. An element $a \in S$ is called left (right)
cancellable if $ax = ay$ $(xa = ya)$ implies $x = y$. Suppose that T is a semigroup containing the semigroup S and such that for each $\alpha \in T \backslash S$ there exist
elements a and b in S such that a (b) is left (right) cancellable in T and
$a\alpha$, $\alpha b \in S$. In this case T is called a semigroup of quotients of S. Fuchs
[132] showed that a linear order of a semigroup S can be uniquely extended to
a linear order of a semigroup of quotients T.

A commutative semigroup P is called archimedean if 1) $s + r = t + r$
always implies $s = t$ $(r, s, t \in P)$; 2) $r > s$ if and only if $r = s + t$ for some
$t \in P$; 3) for each pair $r, s, r \neq s$ implies $r > s$ or $s > r$; and 4) for each
pair $r, s, \in P$ there exists a natural number n such that $nr > s$. Archimedean
semigroups were studied by Endler [122]. He gave conditions for embedding
such semigroups in semigroups which he called eudoxic. A eudoxic semigroup can be generally described as one in which for each triple of elements
there exists a "fourth proportional element". In work [182] Krull characterized a eudoxic semigroup as an archimedean semigroup, for each pair of elements a and b of which there exists a reflection P such that $Pa = b$. Here
by a reflection P of an archimedean semigroup we mean a mapping which
satisfies the following: 1) $Pa = n(P(na))$ for each natural number n; 2) $Pa <
Pb$ if $a > b$.

2. Lattice ordered semigroups. Birkhoff [86] stressed the applications of
such semigroups to algebraic number theory, algebraic geometry, the theory of
semisimple algebras, the algebra of binary relations, Brouwerian logic and
general topology.

Let the unit and zero of an l-semigroup L be denoted by e and 0. An
element $x \in L$ is called integral if $x \leq e$. The set C of all integral elements
of an l-semigroup L forms an l-semigroup. If all elements of L are integral,
then L is called an integral l-semigroup. Let $I(a)$ be the set of all $x \in L$
such that $axa \leq a$. If $I(a)$ is nonempty and has a greatest element, this

element is called the inverse to a and is denoted a^{-1}. $a \in L$ is called a quasi-divisor of $b \in L$ if $a^{-1} \le b^{-1}$. Quasi divisibility in integral l-semi-groups was studied by Kowalski [178]. He indicated applications to ideal theory; in particular, he established a connection with the relation of quasi divisibility introduced by Van der Waerden in ideal theory. He posed some problems.

P. D. Kruming [36] considered lattice ordered semigroups S such that 1) S is a complete lattice; 2) for each element a and subset A in S we have $\sup(aA) = a \sup A$, $\inf(aA) = a \inf A$, $\sup(Aa) = (\sup A)a$ and $\inf(Aa) = (\inf A)a$. His studies reflect mainly on the interval topology in such semi-groups.

A system $A = (A, +, \le, -)$ is called a dually lattice ordered semigroup with subtraction if: 1) $(A, +, \le)$ is a commutative lattice ordered semigroup with zero element 0; 2) for given $a, b \in A$ there exists a smallest element x such that $b + x \ge a$, and this element is denoted $a - b$ (it is uniquely de-fined); 3) $(a - b) \cup 0 + b \le a \cup b$; and 4) $(a - a) \ge 0$. Such systems were defined and their properties studied by Swamy [239]. A dually lattice ordered semigroup with subtraction is a general abstract notion including as special cases Brouwerian algebras, abelian l-groups and Boolean rings. It is used to partially solve problem 105 of Birkhoff [1].

3. **Partially ordered semigroups.** A subsemigroup A of a partially ordered semigroup P is called convex if $\alpha \ge \gamma \ge \beta$, $\alpha, \beta \in A$, $\gamma \in P$ always implies $\gamma \in A$. A semigroup P is called an l-semigroup if P is simultane-ously a lattice and the relations $(\alpha \cup \beta)\gamma = \alpha\gamma \cup \beta\gamma$ and $\gamma(\alpha \cup \beta) = \gamma\alpha \cup \gamma\beta$, and their duals, hold for any $\alpha, \beta, \gamma \in P$. A subsemigroup A of an l-group P is called an l-subsemigroup if A is a sublattice of P and convex in P. Ja. V. Hion [64] showed that any partially ordered semigroup (l-semigroup) with no proper convex subsemigroups (l-subsemigroups) consists only of one element. Every partially ordered semigroup in which no two distinct proper convex subsemigroups meet either has no more than two elements or is a trivially ordered cyclic group of prime order.

Consider a semigroup S and denote the set of all of its partial orders by $\Gamma = \Gamma(S)$. Γ is partially ordered by inclusion: $\sigma \le \rho$ if $(a, b) \in \sigma$ always implies $(a, b) \in \rho$. For any partial order $\tau \in \Gamma$ there exists a maximal ele-ment $\sigma \in \Gamma$ such that $\tau \le \sigma$. The paper [42] by E. S. Ljapin is devoted to characterizing the maximal elements of Γ. In [43] Ljapin studied partial orderings on semigroups of linear transformations.

Burgess and MacFadden [96] studied the question of embedding a partially ordered semigroup in an l-semigroup while preserving joins, meets, and left and right quotients in all cases when the partial operations are defined in the initial partially ordered semigroup.

We will have in mind a partially ordered semigroup S with unique left unit which is a maximal element, and possibly with a zero, 0, which is a minimal element. An element $p \in S$, $0 < p < e$, is called prime if $ab \le p$ implies $a \le p$ or $b \le p$. Fuchs and Steinfeld [136] showed that under certain conditions in S any element $a \in S$, $0 < a < e$, is uniquely expressable in the form $a = p_1^{k_1} \cdots p_r^{k_r}$, where the p_i are distinct primes.

A stable equivalence θ on a partially ordered semigroup S with left unit is called a Dubreil-Jacotin congruence if all its classes $\theta(a)$ are convex subsets in S and $\theta(a) \le \theta(e)$ implies $a \le e$ for any $a \in S$. Dubreil-Jacotin congruences on a semigroup S were considered by Fuchs [134]. In [135] Fuchs studies the problem of proving which are the equivalences on a partially ordered semigroup whose factor semigroups are ordered groups.

The report [67] of B. M. Šain is devoted to questions of representation of quasi-ordered semigroups in the form of subsemigroups of semigroups of transformations of a set.

P. G. Kontorovič and L. A. Prutkina [33] studied the ideal structure of a semigroup G^+ which is the semigroup of positive elements of a partially ordered torsion free group G.

4. **Ordered groupoids.** Blyth [89] considered partially ordered groupoids (groupoids with quotients) in which any two elements have a left and right quotient.

Lévy-Bruhl [187] extended the Jordan-Hölder Theorem to partially ordered groupoids as follows: if in a partially ordered groupoid G we have 1) $xy \le x$, 2) $xy = x$ if and only if $x \le y$, and 3) $x(c)y$ implies $xz(c)yz$, where $a(c)b$ denotes that $a = b$ or b covers a, then there exists a maximal chain of finite length between the elements a, $b \in G$, and all maximal chains between a and b have the same finite length. There is also given a correspondence between the elements of maximal chains between one and the same pair of elements.

Ordered semigroupoids were studied in the dissertation of Blyth [90]. In report [91] an l-groupoid is defined as a partially ordered groupoid which is a lattice. In the set $M(G)$ of n-square matrices over an l-groupoid G one defines in a special way a relation of partial order, and it is found when, in this case, $M(G)$ will be a divisible groupoid. There are solved certain other questions about the set $M(G)$ under various hypotheses put on the order of G.

5. **Betweenness semigroups.** A relation of "betweenness" on groups was considered by Shepperd, and on rings by A. I. Čeremisin. Gilder [138] introduced such a relation on semigroups. A semigroup S is called a betweenness semigroup if there is defined on S a ternary relation of "betweenness" (a, b, c) with the same axioms as in defining a betweenness group. If the "betweenness" relation is induced in the semigroup by a linear order, then the semigroup is called a linear betweenness semigroup, and if the "betweenness" relation is cyclic then the semigroup is called betweenness cyclic. All betweenness cyclic semigroups are found. He also studied linear betweenness semigroups.

BIBLIOGRAPHY

[1] G. Birkhoff, *Lattice theory*, Amer. Math. Soc. Colloq. Publ., vol. 25, rev. ed., Amer. Math. Soc., Providence, R. I., 1948; Russian transl., IL, Moscow, 1952.   MR 10, 673.

[2] B. M. Bredihin, *On the asymptotic distribution law of the generating elements in ordered semigroups*, Proc. Second Sci. Conf. Math. Dept. Ped. Inst. Volga Region, no. 1, Kuĭbyšev. Gos. Ped. Inst., Kuybyshev, 1962, pp. 3–6. (Russian)   MR **34** #1417.

[3] N. Bourbaki, *Algèbre*. Chap. VI: *Groupes et corps ordonnés*, Actualités Sci. Indust., no. 1179, Hermann, Paris, 1952; Russian transl., "Nauka", Moscow, 1965.   MR **14**, 237; **32** #5642.

[4] Wang Shih-Chiang, *Representation of ordered abelian groups and ordered rings of finite degree*, Acta Math. Sinica **5** (1955), 425–432. (Chinese) MR **17**, 710.

[5] A. A. Vinogradov, *Partially ordered locally nilpotent groups*, Ivanov. Gos. Ped. Inst. Učen. Zap. Fiz.-Mat. Nauki **4** (1953), 3–18. (Russian) MR **17**, 823.

[6] ———, *On the theory of partially ordered nilpotent groups*, Ivanov. Gos. Ped. Inst. Učen. Zap. Fiz.-Mat. Nauki **5** (1954), 61–64. (Russian)

[7] ———, *Remarks on the theory of partially ordered groups and semigroups*, Algebra i Logika Sem. **1** (1962), no. 2, 22–29. (Russian) MR **27** #2567.

[8] A. A. Vinogradov, *Metabelian partially ordered groups*, Ivanov. Gos. Ped. Inst. Učen. Zap. 34 (1963), 20–26. (Russian)   MR 33 #5757.

[9] E. Gabovič, *Partially ordered groups with no non-trivial convex sub-groups*, Tartu. Riikl. Ül. Toimetised No. 102 (1961), 289–293. (Russian)   MR 26 #1373.

[10] ———, *On Archimedean ordered Ω-groups*, Tartu Riikl. Ül. Toimetised No. 129 (1962), 19–22. (Russian)   MR 27 #2566.

[11] ———, *Endormorphisms of certain ordered semigroups*, Litovsk. Mat. Sb. 3 (1963), no. 2, 69–76. (Russian)   MR 30 #175.

[12] L. M. Gluskin, *Semigroups*, Itogi Nauki (Algebra. Topology. 1962) Akad. Nauk SSSR Inst. Naučn. Informacii, Moscow, 1963, pp. 33–58. (Russian)   MR 30 #176.

[13] Ju. Š. Gurevič, *Elementary equivalence of ordered abelian groups*, Uspehi Mat. Nauk 19 (1964), no. 2 (116), 222–223. (Russian)

[14] ———, *Elementary properties of ordered Abelian groups*, Algebra i Logika Sem. 3 (1964), no. 1, 5–39. (Russian)   MR 28 #5004.

[15] Ju. Š. Gurevič and A. I. Kokorin, *Universal equivalence of ordered Abelian groups*, Algebra i Logika Sem. 2 (1963), no. 1, 37–39. (Russian)   MR 27 #5687.

[16] M. I. Zaĭceva, *On the set of ordered Abelian groups*, Uspehi Mat. Nauk 8 (1953), no. 1 (53), 135–137. (Russian)   MR 14, 721.

[17] ———, *Right ordered groups*, Učen. Zap. Šuisk. Gos. Ped. Inst. 6 (1958), 205–226. (Russian)

[18] L. V. Kantorovič, *Lineare halbgeordnete Räume*, Mat. Sb. 2 (44) (1937), 121–168.

[19] L. V. Kantorovič, B. Z. Vulih and A. G. Pinsker, *Partially ordered groups and partially ordered linear spaces*, Uspehi Mat. Nauk 6 (1951), no. 3 (43), 31–98; English transl., Amer. Math. Soc. Transl. (2) 27 (1963), 51–124.   MR 13, 361; 27 #1517.

[20] M. I. Kargapolov, *Completely ordered groups*, Algebra i Logika Sem. 1 (1962), no. 2, 16–21. (Russian) MR 27 #2569.

[21] ———, *Classification of ordered Abelian groups by their elementary properties*, Algebra i Logika Sem. 2 (1963), no. 2, 31–46. (Russian) MR 27 #5823.

[22] ———, *Orderable groups. I*, Algebra i Logika Sem. 2 (1963), no. 6, 5–14. (Russian)   MR 30 #3156.

[23] A. I. Kokorin, *On uniquely orderable groups*, Abstracts of the 2nd Siberian Conference on Mathematics and Mechanics, 1962, Tomsk, Tomsk. Univ., 1962, pp. 87–88. (Russian)

[24] ———, *On a class of lattice-ordered groups*, Ural. Gos. Univ. Mat. Zap. **3** (1962), no. 3, 37–38. (Russian)   MR 29 #5936.

[25] ———, *Methods for the lattice-ordering of a free Abelian group with a finite number of generators*, Ural. Gos. Univ. Mat. Zap. **4** (1963), no. 1, 45–48. (Russian)   MR 29 #5937.

[26] ———, *On linearly orderable groups*, Dokl. Akad. Nauk SSSR **151** (1963), 31–33 = Soviet Math. Dokl. **4** (1963), 905–908.   MR 27 #1516.

[27] ———, *On the theory of orderable groups*, Algebra i Logika Sem. **2** (1963), no. 6, 15–20. (Russian)   MR **30** #3157.

[28] ———, *On the theory of completely ordered groups*, Ural. Gos. Univ. Mat. Zap. **4** (1963), no. 3, 25–29. (Russian)   MR 32 #1271.

[29] ———, *Ordering a direct product of ordered groups*, Ural. Gos. Univ. Mat. Zap. **4** (1963), no. 3, 95–96. (Russian)   MR 29 #5938.

[30] A. I. Kokorin and V. M. Kopytov, *Certain classes of ordered groups*, Algebra i Logika Sem. **1** (1962), no. 3, 21–23. (Russian)   MR 27 #5840.

[31] P. G. Kontorovič, *Questions on linear and lattice ordered groups*, Spisy Přírod. Fak. Univ. Brno **1964**, 472–473. (Russian)

[32] P. G. Kontorovič and A. I. Kokorin, *A type of partially ordered group*, Ural. Gos. Univ. Mat. Zap. **3** (1962), no. 3, 39–44. (Russian) MR **30** #3158.

[33] P. G. Kontorovič and L. A. Prutkina, *Strictly isolated ideals*, Abstracts of the 3rd Siberian Conference on Mathematics and Mechanics, 1964, Tomsk, Tomsk. Univ., 1964, pp. 227–228. (Russian)

[34] V. M. Kopytov, *On the completion of the center of an ordered group*, Ural. Gos. Univ. Mat. Zap. **4** (1963), no. 3, 20–24. (Russian) MR 29 #5934.

[35] ———, *The completion of completely ordered groups*, Ural. Gos. Univ. Mat. Zap. **4** (1963), no. 3, 76–78. (Russian) MR 32 #1272.

[36] P. D. Kruming, *Lattice-ordered semigroups*, Izv. Vysš. Učebn. Zaved. Matematika **1964**, no. 6 (43), 78–87. (Russian)   MR **30** #1198.

[37] K. M. Kutyev, *SL-isomorphisms of partially ordered locally nilpotent groups*, Uspehi Mat. Nauk **11** (1956), no. 2 (68), 193–198. (Russian) MR 17, 1184.

[38] K.M. Kutyev, SL-*isomorphism of ordered groups*, Izv. Akad. Nauk SSSR Ser. Mat. 24 (1960), 807–824. (Russian)   MR 23 #A231.

[39] ———, SL-*isomorphism of an ordered group*, Dokl. Akad. Nauk SSSR 135 (1960), 1326–1329 = Soviet Math. Dokl. 1 (1960), 1387–1390. MR 23 #A1736.

[40] ———, *On the SL-isomorphism of certain classes of R-groups*, Izv. Akad. Nauk SSSR Ser. Mat. 27 (1963), 701–722. (Russian) MR 27 #3707.

[41] Ja. B. Livčak, *On orderable groups*, Učen. Zap. Ural. Gos. Univ. 1959, no. 23, 11–12. (Russian)   MR 29 #5935.

[42] E. S. Ljapin, *On maximal two-sided stable orderings in semigroups*, Izv. Vysš. Učebn. Zaved. Matematika 1963, no. 3 (34), 88–94. MR 27 #5843.

[43] ———, *Orderings of linear transformations consistent with superposition*, Mat. Sb. 63 (105) (1964), 122–136. (Russian) MR 28 #2167.

[44] A. I. Mal'cev, *On ordered groups*, Izv. Akad. Nauk SSSR Ser. Mat. 13 (1949), 473–482. (Russian)   MR 11, 323.

[45] ———, *On the completion of group order*, Trudy Mat. Inst. Steklov. 38 (1951), 173–175. (Russian)   MR 14, 13.

[46] ———, *Remark on partially ordered groups*, Ivanov. Gos. Ped. Inst. Uč. Zap. Fiz.-Mat. Nauki 10 (1956), 3–5. (Russian)   MR 19, 530.

[47] ———, *On partially ordered nilpotent groups*, Algebra i Logika Sem. 1 (1962), no. 2, 5–9. (Russian)   MR 27 #5841.

[48] A. G. Pinsker, *Extensions of semiordered groups and spaces*, Leningrad. Gos. Ped. Inst. Uč. Zap. 86 (1949), 285–315. (Russian)

[49] ———, *Locally ordered groups*, Proc. 3rd All-Union Math. Conference, vol. 1, Izdat. Akad. Nauk SSSR, Moscow, 1956, pp. 32–33. (Russian)

[50] B. I. Plotkin, *On the theory of solvable groups without torsion*, Dokl. Akad. Nauk SSSR 84 (1952), 665–668. (Russian)   MR 14, 15.

[51] ———, *Generalized soluble and generalized nilpotent groups*, Uspehi Mat. Nauk 13 (1958), no. 4 (82), 89–172; English transl., Amer. Math. Soc. Transl. (2) 17 (1961), 29–115.   MR 21 #686; 23 #A1713.

[52] V. D. Podderjugin, *Conditions for orderability of a group*, Izv. Akad. Nauk SSSR Ser. Mat. 21 (1957), 199–208. (Russian)   MR 20 #911.

[53] K. S. Sibirskiĭ and A. M. Stahi, *On the question of partial orderability of groups*, Studies in Algebra and Math. Anal., Izdat. "Karta Moldovenjaske", Kishinev, 1965, pp. 73–78. (Russian) MR 34 #259.

[54] L. A. Skornjakov, *Rings*, (Algebra. Topology. 1962) Akad. Nauk SSSR Inst. Naučn. Informacii. Itogi Nauki, Moscow, 1963, pp. 59–79. (Russian) MR 30 #110.

[55] D. M. Smirnov, *Infrainvariant subgroups*, Ivanov. Gos. Ped. Inst. Uč. Zap. Fiz.-Mat. Nauki 4 (1953), 92–96. (Russian) MR 17, 823.

[56] ———, *On reduced free multi-operator groups*, Dokl. Akad. Nauk SSSR 150 (1963), 44–47 = Soviet Math. Dokl. 4 (1963), 600–603. MR 27 #5838.

[57] ———, *Groups of automorphisms of group rings of right-orderable groups*, Algebra i Logika Sem. 4 (1965), no. 1, 31–45. (Russian) MR 31 #5906.

[58] A. A. Terehov, *Completely orderable groups*, Dokl. Akad. Nauk SSSR 129 (1959), 34–36. (Russian) MR 22 #734.

[59] ———, *The structure of locally solvable, completely ordered groups*, Algebra i Logika Sem. 1 (1962), no. 2, 10–15. (Russian) MR 27 #2568.

[60] L. Fuchs, *Partially ordered algebraic systems*, Pergamon Press, New York; Addison-Wesley, Reading, Mass., 1963; Russian transl., "Mir", Moscow, 1965. MR 30 #2090.

[61] V. I. Frenkel', *Algorithm problems in partially ordered groups*, Uspehi Mat. Nauk 17 (1962), no. 4 (106), 173–179. (Russian) MR 31 #1194.

[62] ———, *Algorithmic problems in partially ordered groups*, Dokl. Akad. Nauk SSSR 152 (1963), 67–70 = Soviet Math. Dokl. 4 (1963), 1266–1269. MR 27 #3715.

[63] Ja. V. Hion, *Archimedean ordered rings*, Uspehi Mat. Nauk 9 (1954), no. 4 (62), 237–242. (Russian) MR 16, 442.

[64] ———, *Partially ordered semigroups in which the proper convex sub-semigroups do not intersect*, Izv. Akad. Nauk SSSR Ser. Mat. 27 (1963), 67–74. (Russian) MR 27 #4874.

[65] A. I. Čeremisin, *Betweenness rings*, Izv. Vysš. Učebn. Zaved. Matematika 1962, no. 3 (28), 158–163. (Russian) MR 25 #2100.

[66] ———, *Separation rings*, Ivanov. Gos. Ped. Inst. Učen. Zap. 34 (1963), 62–73. (Russian)

[67] B. M. Šain, *The representation of ordered semigroups*, Mat. Sb. 65 (107) (1964), 188–197. (Russian) MR 30 #2093.

[68] L. N. Ševrin, *The lattice-subsemigroup characteristic of orderable groups*, Uspehi Mat. Nauk 19 (1964), no. 5 (119), 157–161. (Russian) MR 30 #2094.

[69] E. P. Šimbireva, *On direct factorizations of partially ordered groups*, Moskov. Oblast. Ped. Inst. Učen. Zap. 110 (1962), 347–350. (Russian)

[70] A. L. Šmel'kin, *Free polynilpotent groups*, Dokl. Akad. Nauk SSSR 151 (1963), 73–75 = Soviet Math. Dokl. 4 (1963), 950–953. MR 27 #5820.

[71] ――――, *Free polynilpotent groups*, Izv. Akad. Nauk SSSR Ser. Mat. 28 (1964), 91–122; English transl., Amer. Math. Soc. Transl. (2) 55 (1966), 270–304. MR 29 #161.

[72] Ja. Jakubík, *Representation and extension of l-groups*, Czechoslovak Math. J. 13 (88) (1963), 267–283. (Russian) MR 30 #2091.

[73] M. Jakubíková, *On some subgroups of l-groups*, Mat.-Fyz. Časopis Sloven. Akad. Vied 12 (1962), 97–107. (Russian) MR 29 #173.

[74] N. L. Alling, *On ordered divisible groups*, Trans. Amer. Math. Soc. 94 (1960), 498–514. MR 25 #4013.

[75] ――――, *A characterization of abelian η_α-groups in terms of their natural valuation*, Proc. Nat. Acad. Sci. U.S.A. 47 (1961), 711–713. MR 31 #259.

[76] ――――, *On the existence of real-closed fields that are η_α-sets of power $\aleph_\alpha$*, Trans. Amer. Math. Soc. 103 (1962), 341–352. MR 26 #3615.

[77] F. W. Anderson, *On f-rings with the ascending chain condition*, Proc. Amer. Math. Soc. 13 (1962), 715–721. MR 26 #1336.

[78] J. F. Andrus and A. T. Butson, *Ordered groups*, Amer. Math. Monthly 70 (1963), 619–628. MR 27 #1518.

[79] K. E. Aubert, *Caractérisations des anneaux des valuations à l'aide de la théorie des r-idéaux*, Séminaire Dubreil et Pisot 1956/57, Secrétariat mathématique, Paris, 1958. MR 21 #7221.

[80] R. Mallol Balmaña, *Note*, Rev. Mat. Hisp.-Amer. (4) 12 (1952), 137. (Spanish) MR 14, 616.

[81] B. Banaschewski, *Über die Vervollständigung geordneter Gruppen*, Math. Nachr. 16 (1957), 51–71. MR 19, 388.

[82] ――――, *On lattice-ordered groups*, Fund. Math. 55 (1964), 113–122. MR 29 #5930.

[83] H. Bauer, *Geordnete Gruppen mit Zerlegungseigenschaft*, S.-B. Bayer. Akad. Wiss. Math.-Nat. Kl. 1958, 25–36. MR 21 #7255.

[84] F. A. Behrend, *A characterization of vector spaces over fields of real numbers*, Math. Z. 78 (1962), 298–304.   MR 25 #1162.

[85] G. Birkhoff, *Lattice-ordered groups*, Ann. of Math. (2) 43 (1942), 298–331.   MR 4, 3.

[86] ————, *Lattice-ordered demigroups*, Séminaire P. Dubreil, M.-L. Dubreil-Jacotin et Pisot, 1960/61, fasc. 2, exposé 19, Secrétariat mathématique, Paris, 1963.   MR 28 #3911.

[87] M. N. Bleicher and S. Bourne, *On the embeddability of partially ordered halfrings*, J. Math. Mech. 14 (1965), 109–116.   MR 30 #2047.

[88] M. N. Bleicher and H. Schneider, *The decomposition of cones in modules over ordered rings*, J. Algebra 1 (1964), 233–258.   MR 29 #5866.

[89] T. S. Blyth, *La forme générale des structures algébriques résiduées*, C. R. Acad. Sci. Paris 254 (1962), 2506–2508.   MR 24 #A3221.

[90] ————, *Contribution à la théorie de la résiduation dans les structures algébriques ordonnées*, Doctoral Thesis, Fac. Sci. Univ. Paris, Paris, 1963.   MR 28 #156.

[91] ————, *Matrices over ordered algebraic structures*, J. London Math. Soc. 39 (1964), 427–432.   MR 29 #4711.

[92] B. Brainerd, *On a class of Φ-algebras with zero dimensional structure spaces*, Arch. Math. 12 (1961), 290–297.   MR 26 #6095.

[93] ————, *On the normalizer of an f-ring*, Proc. Japan Acad. 38 (1962), 438–443.   MR 27 #3671.

[94] J. L. Britton and J. A. H. Shepperd, *Almost ordered groups*, Proc. London Math. Soc. (3) 1 (1951), 188–199.   MR 13, 320.

[95] D. C. J. Burgess, *Generalized intervals in partially ordered groups*, Proc. Cambridge Philos. Soc. 55 (1959), 165–171.   MR 21 #4191.

[96] D. C. J. Burgess and R. McFadden, *Systems of ideals in partially ordered semigroups*, Math. Z. 79 (1962), 439–450.   MR 25 #5117.

[97] F. Busulini, *Sui gruppi non regolarmente ordinati*, Rend. Sem. Mat. Univ. Padova 33 (1963), 285–296.   MR 28 #1518a.

[98] C. G. Chehata, *On a relation on ordered groups*, Proc. Math. Phys. Soc. U. A. R. (Egypt) No. 25 (1961), 79–82.   MR 29 #3557.

[99] T.-H. Choe, *Notes on the lattice-ordered groups*, Kyungpook Math. J. 1 (1958), 37–42.   MR 20 #5809.

[100] ————, *The interval topology of a lattice ordered group*, Kyungpook Math. J. 1 (1958), 69–74.   MR 21 #3492.

[101] T.-H. Choe, *On a quasi-ordered group*, Kyungpook Math. J. 2 (1959), 47–
52.   MR 22 #4782.

[102] ———, *Erratum: Notes on lattice-ordered groups*, Kyungpook Math.
J. 2 (1959), 73.   MR 22 #1625.

[103] S. Ciampa, *Osservazioni sull'ordinabilità dei gruppi*, Ann. Scuola
Norm. Sup. Pisa (3) 18 (1964), 111–136.   MR 29 #5931.

[104] A. H. Clifford, *A class of partially ordered abelian groups related to
Ky Fan's characterizing subgroups*, Amer. J. Math. 74 (1952), 347–
356.   MR 13, 912.

[105] L. W. Cohen and C. Goffman, *The topology of ordered Abelian groups*,
Trans. Amer. Math. Soc. 67 (1949), 310–319.   MR 11, 324.

[106] P. F. Conrad, *Embedding theorems for abelian groups with valuations*,
Amer. J. Math. 75 (1953), 1–29.   MR 14, 842.

[107] ———, *The group of order preserving automorphisms of an ordered
abelian group*, Proc. Amer. Math. Soc. 9 (1958), 382–389.
MR 21 #1340.

[108] ———, *A note on valued linear spaces*, Proc. Amer. Math. Soc. 9
(1958), 646–647.   MR 20 #5808.

[109] ———, *A correction and improvement of a theorem on ordered groups*,
Proc. Amer. Math. Soc. 10 (1959), 182–184.   MR 21 #3490.

[110] ———, *Non-abelian ordered groups*, Pacific J. Math. 9 (1959), 25–41.
MR 21 #3491.

[111] ———, *Right-ordered groups*, Michigan Math. J. 6 (1959), 267–275.
MR 21 #5684.

[112] ———, *The structure of a lattice-ordered group with a finite number of
disjoint elements*, Michigan Math. J. 7 (1960), 171–180. MR 22 #6854.

[113] ———, *Some structure theorems for lattice-ordered groups*, Trans.
Amer. Math. Soc. 99 (1961), 212–240.   MR 22 #12143.

[114] ———, *Regularly ordered groups*, Proc. Amer. Math. Soc. 13 (1962),
726–731.   MR 26 #3794.

[115] ———, *The relationship between the radical of a lattice-ordered group
and complete distributivity*, Pacific J. Math. 14 (1964), 493–499.
MR 29 #3556.

[116] ———, *The lattice of all convex l-subgroups of a lattice-ordered
group*, Czechoslovak Math. J. 15 (90) (1965), 101–123. MR 30 #3926.

[117] P. F. Conrad and A. H. Clifford, *Lattice-ordered groups having at most two disjoint elements*, Proc. Glasgow Math. Assoc. 4 (1960), 111–113. MR 22 #9532.

[118] P. F. Conrad, J. Harvey and C. Holland, *The Hahn embedding theorem for abelian lattice-ordered groups*, Trans. Amer. Math. Soc. 108 (1963), 143–169. MR 27 #1519.

[119] R. Cristescu, *La notion de composantes dans un groupe dirigé*, C. R. Acad. Sci. Paris 247 (1958), 1700–1702. MR 20 #6467.

[120] ———, *Sur les groupes dirigés*, Czechoslovak Math. J. 10 (85) (1960), 17–26. MR 22 #4783.

[121] A . Doneddu, *Mesure des grandeurs archimédiennes*, Bull. Assoc. Professeurs Math. Enseignment Publ. 43 (1964), 225–235.

[122] O. Endler, *Über multiplikative Strukturen und eudoxische Hüllenvon archimedischen totalgeordneten Gruppen*, Math. Z. 77 (1961), 339–358. MR 28 #4044.

[123] E. Linés Escardó and R. Mallol Balmaña, *On l-groups*, Rev. Mat. Hisp.-Amer. (4) 12 (1952), 129–136. MR 14, 616.

[124] C. J. Everett, *Sequence completion of lattice modules*, Duke Math. J. 11 (1944), 109–119. MR 5, 169.

[125] C. J. Everett and S. Ulam, *On ordered groups*, Trans. Amer. Math. Soc. 57 (1945), 208–216. MR 7, 4.

[126] Ky Fan, *Partially ordered additive groups of continuous functions*, Ann. of Math. (2) 51 (1950), 409–427. MR 11, 525.

[127] I. Fleischer, *Sur les espaces normés non-archimédiens*, Nederl. Akad. Wetensch. Proc. Ser. A 57 = Indag. Math. 16 (1954), 165–168. MR 15, 964.

[128] ———, *Functional representation of partially ordered groups*, Ann. of Math. (2) 64 (1956), 260–263. MR 18, 136.

[129] ———, *A characterization of lexicographically ordered η_α-sets*, Proc. Nat. Acad. Sci. U.S.A. 50 (1963), 1107–1108.

[130] A Fouques, *a-systèmes S-dedekindiens*, C. R. Acad. Sci. Paris 260 (1965), 1525–1527. MR 30 #3047b.

[131] L. Fuchs, *The extension of partially ordered groups*, Acta Math. Acad. Sci. Hungar. 1 (1950), 118–124. MR 13, 436.

[132] ———, *On the ordering of quotient rings and quotient semigroups*, Acta. Sci. Math. (Szeged) 22 (1961), 42–45. MR 23 #A2445.

[133] L. Fuchs, *Partially ordered algebraic systems*, Pergamon Press, New York; Addison-Wesley, Reading, Mass., 1963.   MR 30 #2090.

[134] ———, *On group homomorphic images of partially ordered semigroups*, Acta Sci. Math. (Szeged) 25 (1964), 139–142.   MR 29 #3554.

[135] ———, *Über homomorphe Gruppenbilder teilweise geordneter Halbgruppen*, Spisy Přírod.   Fak. Univ. Brno 1964, 462–463.

[136] L. Fuchs and O. Steinfeld, *Principal components and prime factorization in partially ordered semigroups*, Ann. Univ. Sci. Budapest. Eötvös Sect. Math. 6 (1963), 103–111.   MR 29 #5941.

[137] G. Gemignani, *Digressione sui campi ordinati*, Ann. Scuola Norm. Sup. Pisa (3) 16 (1962), 143–157.   MR 27 #141.

[138] J. Gilder, *Betweenness and order in semigroups*, Proc. Cambridge Philos. Soc. 61 (1965), 13–28.   MR 30 #1195.

[139] C. Goffman, *Remarks on lattice ordered groups and vector lattices. I. Carathéodory functions*, Trans. Amer. Math. Soc. 88 (1958), 107–120. MR 20 #3800.

[140] W. Greve, *Partial betweenness groups*, Math. Z. 78 (1962), 305–318. MR 25 #1163.

[141] H. Hahn, *Über die nichtarchimedischen Grössensysteme*, S.-B. Math.-Nat. Kl. Akad. Wiss. Wien Abt. IIa 1907, 601–655.

[142] J. Harvey, *Complete holomorphs*, Pacific J. Math. 11 (1961), 961–970. MR 24 #A3211.

[143] A. Hayes, *A characterization of f-rings without non-zero nilpotents*, J. London Math. Soc. 39 (1964), 706–707.   MR 29 #4774.

[144] ———, *A representation theory for a class of partially ordered rings*, Pacific J. Math. 14 (1964), 957–968.   MR 29 #5867.

[145] M. Henriksen and J. R. Isbell, *Lattice-ordered rings and function rings*, Pacific J. Math. 12 (1962), 533–565.   MR 27 #3670.

[146] M. Henriksen, J. R. Isbell and D. G. Johnson, *Residue class fields of lattice-ordered algebras*, Fund. Math. 50 (1961/62), 107–117. MR 24 #A3184.

[147] M. Henriksen and D. G. Johnson, *On the structure of a class of archimedean lattice-ordered algebras*, Fund. Math. 50 (1961/62), 73–94.   MR 24 #A3524.

[148] C. Holland, *Extensions of ordered groups and sequence completion*, Trans. Amer. Math. Soc. 107 (1963), 71–82.   MR 26 #3795.

[149] C. Holland, *The lattice-ordered group of automorphisms of an ordered set*, Michigan Math. J. 10 (1963), 399–408.   MR 28 #1237.

[150] ———, *Transitive lattice-ordered permutation groups*, Math. Z.87 (1965), 420–433.   MR 31 #2310.

[151] ———, *A class of simple lattice-ordered groups*, Proc. Amer. Math. 16 (1965), 326–329.   MR 30 #3927.

[152] ———, *The interval topology of a certain l-group*, Czechoslovak Math. J. 15(90) (1965), 311–314.   MR 31 #1308.

[153] I. D. Ion, *On valuation groups*, Acad. R. P. Romîne Stud. Cerc. Mat. 14 (1963), 689–696. (Romanian)   MR 31 #5869.

[154] P. Jaffard, *Groupes archimédiens et para-archimédiens*, C. R. Acad. Sci. Paris 231 (1950), 1278–1280.   MR 12, 480.

[155] ———, *Contribution à l'étude des groupes ordonnés*, J. Math. Pures Appl. (9) 32 (1953), 203–280.   MR 15, 284.

[156] ———, *Sur les groupes réticulés associés à un groupe ordonné*, Séminaire P. Dubreil de la Faculté des Sciences de Paris, 1954/55, exposé 25, Secrétariat mathématique, Paris, 1955.   MR 17, 451.

[157] ———, *Extensions des groupes ordonnés*, Séminaire A. Châtelet et P. Dubreil 1953/54, Secrétariat mathématique, Paris, 1956. MR 18, 464.

[158] ———, *Réalisation des groupes complètement réticulés*, Bull. Soc. Math. France 84 (1956), 295–305.   MR 18, 790.

[159] ———, *Sur certains groupes réticulés*, Proc. Internat. Congress Math., 1954, Amsterdam, vol. 2, Noordhoff, Groingen, 1954, pp. 28–29.

[160] ———, *Extension des groupes réticulés et applications*, Publ. Sci. Univ. Alger. Sér. A. 1 (1954), 197–222.   MR 17, 346.

[161] ———, *Sur les groupes réticulés associés à un groupe ordonné*, Publ. Sci. Univ. Alger. Sér. A. 2 (1955), 173–203.   MR 19, 13.

[162] ———, *Sur le spectre d'un groupe réticulé et l'unicité des réalisations irréductibles*, Ann. Univ. Lyon. Sect. A (3) 22 (1959), 43–47. MR 22 #6853.

[163] ———, *Sur la théorie algébrique de la croissance*, C. R. Acad. Sci. Paris 243 (1956), 1383–1385.   MR 18, 464.

[164] J. Jakubík, *On a class of l-groups*, Časopis Pešt. Mat. 84 (1959), 150–161. (Russian)   MR 21 #7254.

[165] J. Jakubík, *Principal ideals in l-groups*, Czechoslovak Math. J. 9 (84) (1959), 528–543. (Russian)   MR 22 #12144.

[166] ——, *Konvexe Ketten im halbgeordneten Gruppen*, Mat.-Fyz. Časopis. Solvensk. Akad. Vied 9 (1959), 236–242. (Slovak)   MR 23 #A2473.

[167] ——, *Über eine Klasse von l-Gruppen*, Acta Fac. Nat. Univ. Comenian. 6 (1961), 267–273.   MR 28 #2156.

[168] ——, *The interval topology of an l-group*, Mat.-Fyz. Časopis Sloven. Akad. Vied 12 (1962), 209–211.   MR 29 #174.

[169] ——, *Über Teilbünde der l-Gruppen*, Acta Sci. Math. (Szeged) 23 (1962), 249–254.   MR 26 #2375.

[170] ——, *Über ein Problem von Paul Jaffard*, Arch. Math. 14 (1963), 16–21.   MR 32 #4195.

[171] ——, *Interval topology of an l-group*, Colloq. Math. 11 (1963), 65–72.   MR 28 #4041.

[172] ——, *Über halbgeordnete Gruppen mit verallgemeinerter Jordanscher Zerlegung*, Rev. Roumaine Math. Pures Appl. 9 (1964), 187–190. MR 33 #4161.

[173] ——, *Über Verbandsgruppen mit zwei Erzeugenden*, Czechoslovak Math. J. 14 (89) (1964), 444–454.   MR 30 #1196.

[174] ——, *Verbandsgruppen mit zwei Erzeugenden*, Spisy Přírod. Fak. Univ. Brno 1964, 468.

[175] D. G. Johnson, *On a representation theory for a class of Archimedean lattice-ordered rings*, Proc. London Math. Soc. (3) 12 (1962), 207–225.   MR 25 #5082.

[176] Y. Katznelson, *Sur les algèbres dont les éléments non négatifs admettent des racines carrées*, Ann. Sci. École Norm. Sup. (3) 77 (1960), 167–174.   MR 22 #12403.

[177] J. Kist, *Representations of Archimedean function rings*, Illinois J. Math. 7 (1963), 269–278.   MR 26 #6083.

[178] O. Kowalski, *Zum Begriff der Quasiteilbarkeit in ganzen l.-Halbgruppen*, Časopis Pešt. Mat. 89 (1964), 53–77.   MR 33 #7434.

[179] D. H. Krantz, *Conjoint measurement the Luce-Tukey axiomatization and some extensions*, J. Math. Psychology 1 (1964), 248–277.

[180] W. Krull, *Über geordnete Gruppen von reellen Funktionen*, Math. Z. 64 (1955), 10–40.   MR 17, 582.

[181] W. Krull, *Über die Endomorphismen von total geordneter Archimedischen Abelschen Gruppen*, Math. Z. 74 (1960), 81–90.   MR 22 #12021.

[182] ———, *Automorphismen and Speigelungen eudoxischer Halbgruppen*, Math. Z. 79 (1962), 53–68.   MR 26 #1381.

[183] V. Kudláček, *On some types of l-rings*, Sb. Vysoké. Učení Tech. Brno 1962, no. 1/2, 179–181. (Czech)   MR 32 #2353.

[184] G. Lallement, *Sur les homomorphismes d'un demi-groupe sur un demi-groupe complètement-0-simple*, C. R. Acad. Sci. Paris 258 (1964), 3609–3612.   MR 28 #4046.

[185] A. Lavis, *Groupes topologiques ordonnés*, Bull. Soc. Roy. Sci. Liège 31 (1962), 497–503.   MR 25 #4022.

[186] ———, *Sur les quotients totalement ordonnés d'un groupe linéairement ordonné*, Bull. Soc. Roy. Sci. Liège 32 (1963), 204–208. MR 26 #5080.

[187] J. Lévy-Bruhl, *Le théorème de Jordan-Hölder dans certains groupoïdes ordonnés*, C. R. Acad. Sci. Paris 258 (1964), 1114–1116. MR 29 #5939.

[188] F. Loonstra, *L'extension du groupe ordonné des entiers rationnels par le même groupe*, Nederl. Akad. Wetensch. Proc. Ser. A 58 = Indag. Math. 17 (1955), 41–49.   MR 17, 12.

[189] K. Lorenz, *Über Strukturverbände von Verbandsgruppen*, Acta Math. Acad. Sci. Hungar. 13 (1962), 55–67.   MR 25 #2979.

[190] P. Lorenzen, *Über halbgeordnete Gruppen*, Math. Z. 52 (1949), 483–526.   MR 11, 497.

[191] ———, *Die Erweiterung halbgeordneter Gruppen zu Verbandsgruppen*, Math. Z. 58 (1953), 15–24.   MR 15, 7.

[192] H. Lugowski, *Über die Vervollständigung geordneter Halbringe*, Publ. Math. Debrecen 9 (1962), 213–222.   MR 27 #4840.

[193] ———, *Über gewisse geordnete Halbmoduln mit negativen Elementen*, Publ. Math. Debrecen 11 (1964), 23–31.   MR 31 #260.

[194] ———, *Über die Vervollständigung gewisser geordneter Halbmoduln mit negativen Elementen*, Publ. Math. Debrecen 11 (1964), 135–138. MR 31 #261.

[195] G. Maltese, *Convex ideals and positive multiplicative forms in partially ordered algebras*, Math. Scand. 9 (1961), 372–382.   MR 29 #5117.

[196] S. Matsushita, *Sur la puissance des ordres dans un groupe libre*, Nederl. Akad. Wetensch. Proc. Ser. A 56 = Indag. Math. 15 (1953), 15–16. MR 15, 284.

[197] T. Michiura, *On a definition of lattice ordered groups*, J. Osaka Inst. Sci. Tech. 1 (1949), part 1, 27. MR 11, 497.

[198] ———, *On a definition of lattice-ordered groups*. II, J. Osaka Inst. Sci. Tech. 1 (1949), part 1, 117–119. MR 12, 389.

[199] ———, *On simply ordered groups*, Portugal. Math. 10 (1951), 89–95. MR 13, 320.

[200] ———, *Sur les groupes ordonnés*. II, III, C. R. Acad. Sci. Paris 234 (1952), 1422–1423, 1521–1522. MR 14, 19.

[201] ———, *On partially ordered groups without proper convex subgroups*, Nederl. Akad. Wetensch. Proc. Ser. A 56 = Indag. Math. 15 (1953), 231–232. MR 15, 8.

[202] D. Müller, *Verbandsgruppen und Durchschitte allgemeiner Bewertungsringe*, Dissertation, Math.-Nat. Fak. Bonn, 1960.

[203] T. Nakano, *A theorem on lattice ordered groups and its applications to the valuation theory*, Math. Z. 83 (1964), 140–146. MR 28 #3106.

[204] B. H. Neumann, *On ordered groups*, Amer. J. Math. 71 (1949), 1–18. MR 10, 428.

[205] E. S. Northam, *The interval topology of a lattice*, Proc. Amer. Math. Soc. 4 (1953), 824–827. MR 15, 244.

[206] F. Papangelou, *Order convergence and topological completion of commutative lattice-groups*, Math. Ann. 155 (1964), 81–107. MR 30 #4699.

[207] G. Pickert, *Nachbarschaftsfilter und Verbandshalbgruppen*, Arch. Math. 13 (1962), 151–159. MR 29 #5214.

[208] J. Querré, *Equivalences de fermeture dans un demi-groupe résidutif*, Séminaire P. Dubreil, M.-L. Dubreil-Jacotin et C. Pisot 1961/62, fasc. 1, exposé 3, Secrétariat mathématique, Paris, 1963. MR 28 #3912.

[209] W. Rautenberg, *Beweis des Kommutativgesetzes in elementar-archimedisch geordneten Gruppen*, Z. Math. Logik Grundlagen Math. 11 (1965), 1–4. MR 30 #1937.

[210] R. Ree, *On ordered, finitely generated, solvable groups*, Trans. Roy. Soc. Canada. Sect. III (3) 48 (1954), 39–42. MR 16, 792.

[211] ———, *The existence of outer automorphisms of some groups*. II, Proc. Amer. Math. Soc. 9 (1958), 105–109. MR 20 #912.

[212] P. Ribenboim, *Un théorème de réalisation de groupes réticulés*, Pacific J. Math. 10 (1960), 305–308.   MR 22 #1624.

[213] ————, *Théorie des groupes ordonnés*, Monografías Mat., Universidad Nacional del Sur, Bahía Blanca, 1963.   MR 31 #5905.

[214] L. S. Rieger, *On ordered and cyclically ordered groups*, I–III, Vestnik Královské České Společnosti. Nauk Třída Matemat. Přírodověd. No. 6 (1946); No. 1 (1947); No. 1 (1948).   MR 9, 7; 10, 99.

[215] A. Robinson and E. Zakon, *Elementary properties of ordered abelian groups*, Trans. Amer. Math. Soc. 96 (1960), 222–236.   MR 22 #5673.

[216] T. Saitô, *Ordered idempotent semigroups*, J. Math. Soc. Japan 14 (1962), 150–169.   MR 26 #2533.

[217] ————, *Regular elements in an ordered semigroup*, Pacific J. Math. 13 (1963), 263–295.   MR 27 #2574.

[218] N. Sankaran and R. Venkataraman, *A generalization of the ordered group of integers*, Math. Z. 79 (1962), 21–31.   MR 25 #1224.

[219] P. Scherk, *On ordered geometries*, Canad. Math. Bull. 6 (1963), 27–36.   MR 26 #5454.

[220] J. A. H. Shepperd, *Betweenness groups*, J. London Math. Soc. 32 (1957), 277–285.   MR 19, 729.

[221] ————, *Separation groups*, Proc. London Math. Soc. (3) 7 (1957), 518–548.   MR 20 #910.

[222] F. Šik, *Zur Theorie der halbgeordneten Gruppen*, Czechoslovak Math. J. 6 (81) (1956), 1–25. (Russian)   MR 18, 465.

[223] ————, *Über Summen einfach geordneter Gruppen*, Czechoslovak Math. J. 8 (83) (1958), 22–53.   MR 20 #5810.

[224] ————, *Über subdirekte Summen geordneter Gruppen*, Czechoslovak Math. J. 10 (85) (1960), 400–424.   MR 23 #A951.

[225] ————, *Erweiterungen teilweise geordneter Gruppen*, Spisy Přírod. Fak. Univ. Brno 1960, 65–80.   MR 24 #A775.

[226] ————, *Zwei Konstruktionen quasilinearer Erweiterungen der Anordnung einer abelschen Gruppe mit Hilfe additiver und isotoner Funktionale*, Z. Math. Logik Grundlagen Math. 7 (1961), 39–45. MR 24 #A2620.

[227] ————, *Über die algebraische Charakterisierung der Gruppen von reellen Funktionen*, Ann. Mat. Pura Appl. (4) 54 (1961), 295–299. MR 24 #A1848.

[228] F. Šik, *Über die Beziehungen zwischen eigenen Spitzen und minimalen Komponenten einer l-Gruppe*, Acta Math. Acad. Sci. Hungar. 13 (1962), 171–178.   MR 26 #242.

[229] ——, *Über additive und isotone Funktionale auf geordneten Gruppen*, Czechoslovak Math. J. 12 (87) (1962), 611–621.   MR 29 #172.

[230] ——, *Kompakt erzeugte vollständige l-Gruppen*, Bull. Inst. Politehn. Iaşi 8 (12) (1962), fasc. 3/4, 5–8.   MR 32 #2496.

[231] ——, *Zum Disjunktivitätsproblem auf geordneten Gruppen*, Math. Nachr. 25 (1963), 83–93.   MR 27 #3716.

[232] ——, *Über direkte Zerlegungen gerichteter Gruppen*, Math. Nachr. 25 (1963), 95–110.   MR 27 #3717.

[233] ——, *Über Fortsetzung additiver und isotoner Funktionale auf geordneten Gruppen*, Czechoslovak Math. J. 13 (88) (1963), 24–36. MR 28 #146.

[234] ——, *Compactness of a class of spaces of ultra-antifilters*, Mem. Fac. Ci. Univ. Habana Ser. Mat. 1 (1963/64), no. 1, fasc. 1, 19–25. (Spanish)   MR 31 #3363.

[235] ——, *Structure and realizations of lattice-ordered groups*, Mem. Fac. Ci. Univ. Habana Ser. Mat. 1(1963/64), no. 3, fasc. 2/3, 1–29. (Spanish)   MR 32 #145.

[236] O. Steinfeld, *Über Hauptkomponenten und Primfaktorisation in halbgeordneten Halbgruppen*, Spisy Přírod. Fak. Univ. Brno 1964, 492.

[237] W. Strodt, *On the algebraic closure of certain partially ordered fields*, Trans. Amer. Math. Soc. 105 (1962), 229–250.   MR 25 #3934.

[238] K. L. Narasimha Swamy, *Autometrized lattice ordered groups*. I, Math. Ann. 154 (1964), 406–412.   MR 29 #171.

[239] ——, *Dually residuated lattice ordered semigroups*, Math. Ann. 159 (1965), 105–114.   MR 32 #1273.

[240] S. Swierczkowski, *On cyclically ordered groups*, Fund. Math. 47 (1959), 161–166.   MR 22 #1627.

[241] H. H. Teh, *A note on L-groups*, Proc. Edinburgh Math. Soc. (2) 13 (1962), 123–124.   MR 26 #56.

[242] J. R. Teller, *On the extensions of lattice-ordered groups*, Pacific J. Math. 14 (1964), 709–718.   MR 29 #1269.

[243] ——, *On ordered algebraic structures*, Doctoral Dissertation, Tulane University, New Orleans, La., 1964; Dissertation Abstracts 25 (1965), 4182–4183.

[244] G. Thierrin, *Sur les anneaux partiellement ordonnés*, Canad. Math. Bull. 5 (1962), 123–128.   MR 25 #3067.

[245] G. Trevisan, *Classificazione dei semplici ordinamenti di un gruppo libero commutativo con generatori*, Rend. Sem. Mat. Univ. Padova 22 (1953), 143–156.   MR 15, 8.

[246] E. J. Tully, Jr., *The existence of a total order on a group*, Proc. Amer. Math. Soc. 13 (1962), 217–219.   MR 24 #A3216.

[247] K. Urbanik, *Remarks on ordered absolute-valued algebras*, Colloq. Math. 11 (1963), 31–39.   MR 28 #5350.

[248] D. Vaida, *Sur les sous-groupes isolés d'un groupe réticulé non commutatif*, Com. Acad. R. P. Romîne 10 (1960), 935–939. (Romanian)   MR 23 #A1738.

[249] ———, *Un problème de G. Birkhoff*, C. R. Acad. Bulgare Sci. 15 (1962), 801–803.   MR 29 #3530.

[250] ———, *Groupes ordonnés dont les éléments admettent une décomposition jordanienne généralisée*, C. R. Acad. Sci. Paris 257 (1963), 2053–2055.   MR 27 #4875.

[251] ———, *Groupes ordonnés dont les éléments admettent une décomposition jordanienne généralisée*, C. R. Acad. Sci. Paris 257 (1963), 2222–2223.   MR 27 #5842.

[252] ———, *Groupes ordonnés dont les éléments admettent une décomposition jordanienne généralisée*, Rev. Roumaine Math. Pures Appl. 9 (1964), 929–948.   MR 31 #5907.

[253] ———, *Groupes ordonnés dont les éléments admettent une décomposition jordanienne généralisée*, Spisy Přírod. Fak. Univ. Brno 1964, 494.

[254] E. C. Weinberg, *Completely distributive lattice-ordered groups*, Pacific J. Math. 12 (1962), 1131–1137.   MR 26 #5054.

[255] ———, *Higher degrees of distributivity in lattices of continuous functions*, Trans. Amer. Math. Soc. 104 (1962), 334–346.   MR 25 #2013.

[256] ———, *Free lattice-ordered abelian groups*, Math. Ann. 151 (1963), 187–199.   MR 27 #3720.

[257] H. J. Weinert, *Über Halbringe und Halbkörper.* I, Acta Math. Acad. Sci. Hungar. 13 (1962), 365–378.   MR 26 #3634.

[258] ———, *Über Halbringe und Halbkörper.* II, Acta Math. Acad. Sci. Hungar. 14 (1963), 209–227.   MR 26 #6219.

[259] W. Strodt, *On the algebraic closure of certain partially ordered fields*, Trans. Amer. Math. Soc. 105 (1962), 229–250.   MR 25 #3934.

[260] E. S. Wolk, *On the interval topology of an l-group*, Proc. Amer. Math. Soc. 12 (1961), 304–307.   MR 23 #A232.

[261] I. Yakabe, *Equivalence of the Krull-Müller-Jaffard theorem and Ribenboim's approximation theorem*, Mem. Fac. Sci. Kyushu Univ. Ser. A 17 (1963), 145–152.   MR 28 #4042.

[262] E. Zakon, *Generalized archimedean groups*, Trans. Amer. Math. Soc. 99 (1961), 21–40.   MR 22 #11049.

[263] M. Zamansky, *Groupes de Riesz*, C. R. Acad. Sci. Paris 248 (1959), 2933–2934.   MR 21 #3489.

Translated by:
J. Ellis

Amer. Math. Soc. Transl.
(2) Vol. 96, 1970

ORDERED ALGEBRAIC SYSTEMS [*]

A. A. VINOGRADOV

A brief summary of works in the theory of ordered algebraic systems reviewed in the Referativnyĭ Žurnal Matematika up to 1966 was made in the report [4]. A summary of works reviewed in the same place in 1966 is done below.

§1. Linearly ordered groups

Several general questions on the theory of linearly ordered groups are listed in the book of Bourbaki [1].

1. o- and o^*-groups. Linearly ordered groups are called briefly o-groups, and groups every maximal order of which is linear are called o^*-groups.

Fuchs [47] pointed out a problem of B. Neumann: if the number of linear orders of an o-group is finite, then must it be a power of 2? M. I. Kargapolov, A. I. Kokorin and V. M. Kopytov [11] showed that there are groups the number of linear orders of which is finite but not a power of 2. They also showed that there exist o-groups which are not o^*-groups whose central factorgroups are o^*-groups. A free n-solvable group with k generates for $n \geq 3$ and $k \geq 2$ is not an o^*-group. In particular, a free group with more than one generator is not an o^*-group. Note that M. I. Kargapolov [10] and Fuchs and Sąsiada [50] established earlier that a free group with more than two generators is not an o^*-group. We call a subgroup of an o-group relatively convex if it is convex in some ordering of the group. In the three-author work mentioned above it is also shown that the free product of o-groups with amalgamated subgroup relatively convex in each of the factors may not be an o-group. The normalizer and centralizer of a relatively convex subgroup are not always relatively convex.

We call a subgroup H of a group G weakly pure relative to a system of subgroups G_α, $\alpha \in M$, of G if, for every $\alpha \in M$, from the solvability of the equation $x^n = h$, $h \in H$, in the subgroup G_α and in the subgroup G^α generated by all the G_β, $\beta \neq \alpha$, α, $\beta \in M$, there follows the solvability of this equation in H. V. M. Konytov [16] proved that in order that the generalized

---

[*] Translation of *Algebra. Topology. Geometry.* 1966, Akad. Nauk SSSR Inst. Naučn. Informacii, Moscow, 1968, pp. 91–108.

direct product of o-groups (o^*-groups) with amalgamated subgroup be an o-group (o^*-group) it is necessary and sufficient that the amalgamated subgroup be weakly pure relative to the system of generalized direct factors. He also investigated conditions for the orderability of the nilpotent product of o-groups.

A subgroup H of a group G is called Γ-orderable if under any maximal order of G the subgroup H is linearly ordered. A. I. Kokorin [12] gave a connection between Γ-orderable and relatively convex subgroups of orderable groups and pointed out the value of Γ-orderable subgroups in the theory of orderable groups as well as one new necessary and sufficient condition for the orderability of a group. The following open questions were raised. 1) Is a subgroup of an o^*-group also an o^*-group? 2) Is a torsion-free group whose central factorgroup is an o^*-group also an o^*-group?

The reports of Keimel [61] and Dubreil-Jacotin [42] were devoted to finding conditions for the orderability of groups.

The long report of D. M. Smirnov [18] is devoted to the investigation of criteria of orderability of a partial Ω-group and a distributive Ω-group.

2. **Hereditary discreteness.** A discrete linearly ordered group is called hereditarily discrete if each of its subgroups is discrete. Sankaran [71] established a connection between discreteness and hereditary discreteness for abelian groups.

3. **The interval topology.** It is known that a linearly ordered group considered as a topological space under the interval topology is zero dimensional or one dimensional. In [71] there is derived a characterization of the one-dimensional linearly ordered groups, and specially defined hereditary one-dimensional locally discrete and locally one-dimensional linearly ordered groups are considered.

4. **η_α-groups.** Let us recall the definition. Let T be a linearly ordered set and H and K subsets of T. We write $H < K$ if $h < k$ for any $h \in H$ and $k \in K$. Suppose that α is an ordinal. T is called an η_α-set if for any subsets H and K of T such that $|H| + |K| < \aleph_\alpha$ and $H < K$, there exists an element $t \in T$ such that $H < \{t\} < K$. A linearly ordered group G is called an η_α-group if when viewed as a linearly ordered set it is an η_α-set. Ribenboim [68] proved the theorem of Alling [31] on the existence of abelian η_α-groups of power $\aleph_\alpha$ if $\alpha > 0$, $\aleph_\alpha$ is regular and $\Sigma_{\delta < \alpha} 2^{\aleph_\delta} \leq \aleph_\alpha$.

5. **Special divisions in the theory of linearly ordered groups.** Ucsnay [82] considered well-ordered power series with exponents in a linearly ordered group and coefficients in a given field.

§2. Lattice ordered groups (l-groups)

Certain general questions in the theory of lattice ordered groups are listed in the book [1].

1. **Convergence.** Papangelou [67] extended his investigations [66] on convergence in abelian l-groups. He introduced two new notions of convergence of a directed net (x_i) in the abelian l-group: α-convergence and L^*-convergence. He established the equivalence of certain distinct forms of convergence and gave various characterizations of the types of convergence.

2. **Linear operators.** By a linear operator on an abelian l-group G we mean a mapping $E\colon G \to G$ such that $E(a + b) = Ea + Eb$. The report of Langford [64] is devoted to the clarification of certain properties of such operators.

3. **Archimedean complete abelian l-groups.** Recall that a maximal l-ideal of an l-group is an l-ideal which is maximal among all the l-ideals not containing a given element. If G is an l-subgroup of an l-group G' we say that G' is an archimedean extension of the group G if the mapping $L \to L \cap G$ is a one-to-one mapping of the set of maximal l-ideals of the l-group G' onto the set of maximal l-ideals of the l-group G. A group which possesses no proper archimedean extension is called archimedean complete. The study of archimedean complete extensions was taken up by Wolfenstein [89].

4. **Free lattice ordered abelian groups.** Let $[G, P]$ be a partially ordered abelian group with set of positive elements P. In [86] Weinberg called the abelian l-group $[G(P), G(P)^+]$ the free abelian l-group over $[G, P]$ if the following conditions are satisfied: 1) there exists an isomorphism ψ of the partially ordered group $[G, P]$ into $[G(P), G(P)^+]$; 2) $\psi([G, P])$ generates $[G(P), G(P)^+]$; 3) if $\phi\colon [G, P] \to [L, L^+]$ is an isomorphism of the partially ordered group into an l-group, then there exists an l-homomorphism $\phi'\colon [G(P), G(P)^+] \to [L, L^+]$ such that $\phi = \phi'\psi$. In [87] Weinberg extended the study of free lattice ordered abelian groups and defined the free abelian l-groups over a distributive lattice. In particular, it was proved by him that any free abelian l-group is a subdirect sum of copies of the group. Note that the last result was reproved by H. G. Hisamiev [24].

5. **Compactly generated l-groups.** We say that an l-group G satisfies condition $k(\alpha)$, where α is a cardinal number, if whenever $a = \bigvee a_j$ $(j \in J)$, where the power of J is $\leq \alpha$, there exists a finite subset $J_1 \subset J$ such that $a = \bigvee a_j$ $(j \in J_1)$. An l-group is called compactly generated if for any $\alpha \leq$ the power of G,

condition $k(\alpha)$ is satisfied. The structure of compactly generated l-groups was determined by Šik [72]. Jakubík [59] extended the study of compactly generated l-groups, and in particular proved that conditions $k(\alpha_1)$ and $k(\alpha_2)$ are not always equivalent for two infinite cardinals α_1 and α_2.

6. **Prime subgroups and prime ideals.** In the representation theory of l-groups, of basic importance are the notions of prime subgroups and l-ideals. A convex l-subgroup (respectively l-ideal) I of an l-group G is called a prime subgroup (respectively l-ideal) if $a \in I$ or $a' \subseteq I$ for every $a \in G$, where a' denotes the set of all elements of G which are disjoint from a. Such subgroups and l-ideals were studied by Šik [73]. Each prime subgroup contains a minimal element in the partially ordered set (with respect to set-theoretic inclusion) of prime subgroups of the group G. Each such minimal element is called a minimal prime subgroup. One defines similarly the notation of minimal prime l-ideal. In [73] it is noted that until now nobody has determined the minimal prime subgroups L such that $N(L)$ is a minimal prime ideal, where $N(L)$ denotes the normal subgroup of G generated by the set L.

7. **Representations of l-groups.** By a representation $G = (G_x : x \in M)$ of an l-group we mean a subdirect sum of linearly ordered groups G_x isomorphic to the given groups. A representation of G is called reduced if for any $x, y \in M$, $x \neq y$, there exists an $f \in G$ such that $f(x) > 0$ and $f(y) < 0$. A representation is called completely regular if for $f \in G$, $x \in M$ and $f(x) = 0$ there exists a $g \in G$ such that $|f| \wedge |\epsilon| = 0$ and $g(x) \neq 0$. A regulator kernel (representing kernel) of an l-group is a set of prime subgroups (prime l-ideals) of the given group with zero intersection. Every representing kernel permits one to construct, in a special way, a representation of the group, a so-called canonical representation. Reduced completely regular canonical representations were studied by Šik [74]. The reports of Bigard [36] and Lloyd [65] are devoted to the investigation of certain other types of representations of l-groups.

Fuchs [49] considered the problem of embedding one l-group into another using a notion of approximation, while Bernau [35] resolved an analogous problem by means of a notion of orthocompletion.

8. **Filets in l-groups.** On the positive cone P of an l-group define the following equivalence relation:

$$x \sim y \Leftrightarrow [x \wedge z = e \Leftrightarrow y \wedge z = e, \ \forall z \in P].$$

The classes of this relation are called filets of the l-group. Let $\bar{x}$ denote the class containing all elements equivalent to x. The set F of all filets of the l-group G can be partially ordered as follows: $\bar{x} \geq \bar{y}$ if and only if $x \wedge z = e \Rightarrow y \wedge z = e$, for all $z \in P$. Jaffard [54] proved that F forms a distributive lattice with the above order.

Recall that a lattice is called completely distributive if

$$\bigwedge_{i \in I} \bigvee_{j \in J} a_{ij} = \bigvee_{\sigma \in J^I} \bigwedge_{i \in I} a_{i\,\sigma(i)},$$

provided the indicated sup's and inf's all exist.

Bigard [36] gave necessary and sufficient conditions under which the set F is completely distributive. He noted that if F is completely distributive then so is G. In particular he gave conditions for the complete distributivity of F and G for a representable l-group G.

The atoms of the lattice F are called minimal filets of the l-group.

The structure of abelian l-groups for which F is finite were studied by Jaffard [54]. T. È. Kaminskiĭ [9] proved that if an l-group H has a finite number of filets and G is a lexicographic extension of H by a totally ordered group Q, then the lattices of filets of the l-groups H and G are isomorphic, while if G is a direct sum of the l-groups G_1 and G_2 having n_1 and n_2 filets respectively, then G has $n_1 n_2$ filets. The number of minimal filets of a direct sum of two l-groups G_1 and G_2 is equal to the sum of the number of minimal filets of the factors. If the l-group G has n filets among which k are minimal, then $n = 2^k$. In [9] a connection is established between the notions of filet and index of orderability. Under the mapping $g \to |g| = g \vee (-g)$ of the l-group G let the number of preimages of the element $x \in G^+$ be equal to n and let m of them be comparable to zero $(n, m \neq \infty)$. The quotient m/n is called the index of the element x, and the smallest of the indices of the elements $x \in G^+$, if it exists (under the hypothesis that the index of x is defined for all $x \in G^+$), is the index of the order of G. The above-mentioned connection expresses, in particular, the fact that the classes of l-groups having a finite number of filets, or containing only finite disjoint sets of elements or possessing an order index, all coincide.

A. I. Kokorin and N. G. Hisamiev [14] found a necessary and sufficient condition under which lattice ordered abelian groups with a finite number of filets are elementarily equivalent and also when they are universally equivalent. It follows that the universal theory of lattice ordered abelian groups with a finite number of filets is solvable.

By the orthogonal rank of a lattice ordered abelian group we will mean the maximal number α of pairwise orthogonal elements of the group. If α is infinite, we write $\alpha = \infty$. N. G. Hisamiev [24] proved that l-groups are universally equivalent if and only if they have equal orthogonal ranks, and that the universal theory of abelian l-groups is solvable.

9. l-groups with projections. The polar $M^\perp$ of a subset M of an l-group G is the set of all elements of the group which are disjoint from every element of M: $M^\perp = \{x \in G;\ x \perp m(m \in M)\}$. We say that for a given element x of G its projection into the polar $M^\perp$ exists if there are elements $x \in M^\pm$ and $x'' \perp M^\pm$ such that $x = x' + x''$. An l-group in which the projection of any element into every polar exists is called an l-group with projections. Jakubík [55] raised the following question: Let G be an l-group with projections. Is the factorgroup G/I modulo an l-ideal I always again an l-group with projections? A negative answer to this question was given by A. I. Veksler [2]. Let us just mention here that A. I. Veksler also considered abelian l-groups in [3].

10. The interval topology in l-groups. Jakubík [58] obtained certain results on lattice ordered groups which are Hausdorff spaces in their interval topology.

Fuchs [49] used the notion of interval topology to investigate embeddings of l-groups.

§3. Partially ordered groups

Certain general questions in the theory of partially ordered groups are listed in the book of Bourbaki [1] and in the report of Vrabec [84].

1. How to specify a partially ordered group. V. I. Frenkel' [22] showed how one can specify a partially ordered group by a system of generators and a particular system of inequalities. One composes words from the generators, and then performs elementary transformations on the words, by which means one defines a relation of inequality and equality of words. This allows the introduction of a notion of equivalence of words and equivalence classes. On the equivalence classes one performs an operation of multiplication, under which the set of classes becomes a group, while the transferal of the relation of inequality to the classes converts the group into a partially ordered group. It is proved that every partially ordered group is order isomorphic to some partially ordered group of classes; thus inequalities for classes determine inequalities for the elements of the partially ordered group. In [21] V. I. Frenkel' has given proofs of previously published results on algorithmic problems in partially ordered groups.

Fuchs [48] observed that the class of partially ordered groups is not a manifold. By introducing a new operation in a partially ordered group one gets a class of partially ordered groups, which is now a manifold. One considers, as a manifold, the class of semigroups which are the positive parts of partially ordered groups.

2. **Riesz groups.** A partially ordered group G is called a Riesz group if it is directed and possesses the property that for any $a_1, a_2, b_1, b_2 \in G$ such that $a_i \le b_j$ $(i = 1, 2; j = 1, 2)$ there exists an element $c \in G$ such that $a_i \le c \le b_j$ $(i = 1, 2; j = 1, 2)$. Such groups were thoroughly studied by Fuchs [45,49]. He gave examples of Riesz groups which are not l-groups and considered the Schreier extension problem for commutative Riesz groups. From the set of all Riesz groups he singled out a subset called the antilattices. A Riesz group H is called an antilattice if it has the following property: if $a \wedge b$ exists in H then either $a \wedge b = a$ or $a \wedge b = b$. The structure of commutative antilattices was clarified, and the possibility of representing commutative Riesz groups by means of antilattices was studied.

Teller [81] studied when a commutative extension of a commutative Riesz group by a commutative Riesz group is a Riesz group. He also introduced some nontrivial examples of Riesz groups.

3. **The interval topology.** Conrad [40], Jakubík [56,57], Wolk [90], and Holland [52] studied conditions under which an l-group in its interval topology is a topological group, or a topological space. In [58] Jakubík did the same for partially ordered groups.

Consider an arbitrary partially ordered group $(G, +, \le)$ and let $G^+ = \{x, x \in G, x > 0\}$. The open interval topology $\mathfrak{T}$ on G is defined by means of the sub-basis $\sigma = \{x \pm G^+ : x \in G\}$. Note the following five properties: 1) if $0 \in T \in \mathfrak{T}$ then there exist neighborhoods $U_1, U_2 \in \mathfrak{T}$ such that $0 \in U_1 \cap U_2$ and $U_1 - U_2 \subseteq T$; 2) $\mathfrak{T}$ is a T_0-space; 3) $\mathfrak{T}$ is not discrete; 4) if $x \in G$ and $G^+ + \nu x = G^+$ for each integer ν, then $x = 0$; 5) if $a, b \in G$ and $a < b$, then there exists an $x \in G$ such that $a < x < b$. Properties 1) and 2) together are equivalent to $(G, +, \mathfrak{T})$ being a topological group. Fuchs [44] indicated that 4) is a necessary and sufficient condition that $(G, +, \mathfrak{T})$ be a topological group and, in proving the theorem, noted that 3) implies 5). Ward [85] gave examples of a group with 2), 3) and 4) but without 1) and 5); of a group with 1)–4) but without 5); and of a group with 1), 2) and 4) but without 3) and 5).

4. **Representation of partially ordered groups.** Holland studied the

problem of representing l-groups by l-subgroups of groups of order automorphisms of totally ordered sets. Fried [43] studied a similar problem for partially ordered groups. He established that a partially ordered group can be represented by order automorphisms of an ordered set if and only if it is order isomorphic to a subgroup of an l-group.

Šmarda [75] solved two problems: 1) He found necessary and sufficient conditions for the existence of a nonzero additive isotone functional on partially ordered groups (i.e. a weakly monotone homomorphism of a partially ordered group into the totally ordered additive group of real numbers with the usual order). 2) Let H be a subgroup of a partially ordered group G and let f be a nonzero additive isotone functional on H. He found conditions for the extension of an additive isotone of f from H to G.

5. Generalized partially ordered groups. Suppose that G is at once a group and a partially ordered set. An element $a \in G$ is called left (right) conservative if $x \le y$ $(x, y \in G) \Rightarrow ax \le ay$ $[xa \le ya]$ and left (right) invertive if $x \le y$ $(x, y \in G) \Rightarrow ax \ge ay$ $[xa \ge ya]$. It is called conservative (invertive) if it is at once left and right conservative (invertive). If each element of G is conservative then G is a partially ordered group in the usual sense; in this case we call G a partially ordered group of the first kind. If each element in G is conservative or invertive and not every element in G is conservative then G is called by Clifford [39] a partially ordered group of the second kind. Such groups in work [15] of P. G. Kontorovič and A. I. Kokorin are called semihomogeneous partially ordered groups. Totally ordered semigroups of the second kind were considered in the commutative case by Clifford [38] and in the general case by Gilder [51]. A group G having a nontrivial partial order such that each element in G is either left conservative or left invertive, and also either right conservative or right invertive, and also such that G contains at least one element which is conservative on one side and invertive on the other, is called a group of the third kind. Clifford reduced the structure problem of partially ordered groups of the second and third kinds to the structure problem of partially ordered groups of the first kind.

§4. Ordered semigroups

1. Totally ordered semigroups. Suppose that a totally ordered semigroup S is mapped order homomorphically onto a totally ordered group G. Bigard [36] established the properties of the full preimage H of the negative cone G_- of G under such a homomorphism and showed that every complex $H \le S$ satisfying these properties determines an isotone homomorphism ϕ

of the semigroup S onto some ordered group G where $\phi^{-1}G_- = H$.

The work of Dubreil-Jacotin [41] is devoted to the study of homomorphic images of a totally ordered semigroup.

Kowalski [62] studied a totally ordered semigroup with the following properties: 1) S is positively ordered, i.e. for any a, $b \in S$, $ab \geq a$, and $ab \geq b$; 2) S is archimedean, i.e. $a^n < b$ for any natural number n implies that a is a unit in S.

Saito [69, 70] considered totally ordered semigroups. Let S be an inverse semigroup and let σ be the largest congruence on S such that S/σ is a group. S is called regular if the unit of the group S/σ contains only idempotents of S. In [70] totally ordered regular inverse semigroups are studied.

2. Lattice ordered semigroups. A lattice ordered semigroup T (l-semigroup) is called quasi-integral if $ab \leq b$ and $ab \leq a$ for any a, $b \in T$. Ion [52] built up the theory of the Baer-McCoy and Jacobson radicals for complete quasi-integral l-semigroups, unifying the part of the theory of radicals common to associative rings, ideals of a semigroup with zero and submodules of some module.

Several results were obtained by Steinfeld [77] on lattice ordered semigroups with division, containing an element 0 less than any other element and an element e which is a unit and greater than any other element.

A dually residuated lattice ordered semigroup is a system $A = (A, +, \leq, -)$ such that: 1) $(A, +, \leq)$ is a commutative l-semigroup with zero element 0; 2) for any given a, $b \in A$ there exists a smallest element x such that $b + x \geq a$, and it is denoted $a - b$ (for given a and b it is uniquely defined); 3) $(a - b) \cup a + b \leq a \cup b$; and 4) $(a - a) \geq 0$. Semigroups of this type were first studied by Swamy [78]. He continued the study of these semigroups in [79]. In particular, let 1 be an element of the semigroup such that $a + (1 - a) = 1 + 1$. If the subgroup has a 1, then such an element is unique; such a semigroup is an l-group if and only if $1 = 0$. If in the semigroup $(a + b) - (c + c) \geq (a - c) + (b + c)$ and 1 satisfies the equality $1 - (1 - a) = a$ for all a, then such a semigroup is called a Boolean l-group. The geometry of Boolean l-groups is studied where the distance between two points a and b is $a * b = (a - b) \cup (b - a)$.

3. Partially ordered semigroups. Steinfeld [77] brought out certain properties of a partially ordered semigroup S satisfying the following conditions: 1) $ab \leq a$ and $ab \leq b$ for any a, $b \in S$, and if $a < b$, then $a = bx = yb$ for some x, $y \in S$; 2) it contains an element 0 which is smaller than any other

element; 3) it contains an element s which is greater than any other element; 4) it contains an element z $(0 < z < s)$ which is not a zero divisor; 5) each $a \in S$ $(0 < a < s)$ can be factored uniquely up to the order of the factors into a product of prime elements $a = p_1 p_2 \cdots p_k$ (an element $p \in S$ $(0 \leq p < s)$ is called prime if $ab \leq p$ implies $a \leq p$ or $b \leq p$).

The class of semigroups which are the positive parts of partially ordered groups were considered as manifolds by Fuchs [48].

4. **Generalized ordered semigroups.** We call a semigroup a generalized ordered semigroup if it is a totally ordered set, where no connection is assumed between the operation of addition and the order relation. An element a of a generalized ordered semigroup is called positive if $x + a > x$ and $a + x > x$ for any element x of the semigroup. A system of generalized measure of a generalized ordered semigroup is a single-valued monotone additive mapping (i.e. homomorphic relative to the addition of the order $\leq$) of the semigroup into the set of real numbers.

N. E. Domošnickaja [5] found necessary and sufficient conditions for the existence of a nontrivial system of generalized measure on a generalized ordered semigroup with positive elements. In [6] there were previously found weakened conditions, and the independence of these weakened conditions was proved.

5. **Partially ordered groupoids.** Ján Jakubík [58] generalized certain results of A. I. Mal'cev [17] on lexicographic decompositions of totally ordered groups into (partially) ordered groupoids with particular special restrictions. An additive groupoid Y is called a u_1-groupoid if 1) Y is a partially ordered set, 2) $x < y$ implies $x + z < y + z$ and $z + x < z + y$, 3) $x \parallel y$ implies $x + z \parallel y + z$ and $z + x \parallel z + y$, and 4) there exists an element $0 \in Y$ such that $0 + x = x + 0 = x$ for any $x \in Y$. Suppose that I is a totally ordered set and that a u_1-groupoid Y_i is given for each $i \in I$. The set of all elements f of the full direct product of these u_1-groupoids for which the totality of indices at nonzero components is well ordered forms a u_1-groupoid, the so-called lexicographic direct product of the Y_i. Using this notion one defines a u-groupoid, and a result similar to that of Mal'cev is proved, namely, that any two lexicographic decompositions of a u-groupoid admit isomorphic extensions.

§5. Ordered fields, rings, skew-fields and algebras

1. **Ordered fields.** Some questions in the theory of ordered fields are listed in the book of Bourbaki [1].

Aguiló-Fuster [29] indicated how to construct a completion of a non-archimedean totally ordered field similar to the Dedekind construction of the reals from the rationals. He also [30] established the fact that among the archimedean subfields contained in a non-archimedean ordered field there exist maximal ones (in the sense of the partial order of inclusion).

2. **Ordered rings.** Yang Tsung-p'an and Tung K'o-ch'eng [28] showed that every complete totally ordered ring is either order isomorphic to the ring of integers Z, or contains Z as a subring, or is isomorphic to the ring of real numbers, or to the additive group of real numbers with the trivial multiplication.

Let A be a commutative ring with unit containing the field of rationals Q and such that $1 + x_1^2 + \cdots + x_n^2 \neq 0$ for any $x_1, \cdots, x_n \in A$. A subset Ω of A satisfying the conditions 1) $A^2 \subseteq \Omega$, 2) if $x, y \in \Omega$ then $x + y \in \Omega$ and $xy \in \Omega$, and 3) $-1 \notin \Omega$, is called a preorder. Rings with preorders were studied by Krivine [63].

Vaida [83] introduced the notion of isolated radical in a totally ordered ring and brought out its connection with the notions of isolated radicals introduced by other authors.

3. **Fully orderable rings.** A ring, any total order of any subring of which can be extended to a total order of the whole ring, is called a fully orderable ring (V-ring). V-rings were first studied by A. A. Terehov [19] who proved the local character of property V for rings. The study of the structural properties of V-rings was carried on by A. I. Čeremisin [25, 26]. There were also formulated the following open questions: 1) Is it possible to obtain a complete structural description of the V-subfields of the field of all real algebraic numbers? In particular, is it possible to find necessary and sufficient conditions that a simple real algebraic extension of the field of rationals of even degree be a V-field? (In [26] it is proved that a simple real algebraic extension of the field of rationals of odd degrees is a V-field and that this assertion is not true generally for simple real algebraic extensions of the field of rationals of even degree.) 2) Is it possible to obtain a complete structural description of the o^*-subfields of the field of real algebraic numbers?

4. **Totally ordered semirings and semi-sfields.** A semiring is a system which is a semigroup relative to an addition and a semigroup relative to a multiplication, the two being connected by the two distributive laws. A semi-sfield (semi-skewfield) is a semiring whose nonzero elements form a group under multiplication. A semiring R is called totally ordered if there is a total order $<$ defined on R satisfying the two requirements: 1) $a < b$

implies $a + c < b + c$ and $c + a < c + b$ for all $c \in R$; 2) $a < b$ and $c \neq 0$ implies $ac < bc$ and $ca < ab$ or $ac > bc$ and $ca > cb$. Totally ordered semirings and semi-sfields were studied by Weinert [88].

5. **Lattice ordered rings.** A lattice ordered ring (l-ring) is an l-group and a ring in which $a \wedge b \geq 0$ implies $ab \geq 0$. A lattice ordered ring is called an f-ring if $a \wedge b = 0$ and $c \geq 0$ implies that $ca \wedge b = 0 = ac \wedge b$. For a nonempty set M of an l-group we set $M^{\perp} = \{x: \; |x| \wedge |m| = 0 \; (m \in M)\}$. If $a \wedge b \geq 0$ implies $(ab)^{\perp\perp} \subset a^{\perp\perp} \cap b^{\perp\perp}$, then the l-ring is called seminormal. Seminormal l-rings were studied by Bernau [34,35]. f-rings were considered in [3] by A. I. Veksler. Anderson [32] studied rings of quotients of f-rings. Johnson [60] showed that every f-ring can be embedded in a specially defined complete f-ring.

Lattice ordered rings were considered also by Yang Tsung-p'an and Tung K'o-ch'eng [28], while Smith [76] studied the structure of lattice ordered semirings.

6. **Partially ordered rings.** The class of partially ordered rings and the class of positive parts of partially ordered rings were considered as manifolds by Fuchs in [48].

7. **Partially ordered algebras.** Swamy [80] studied a system which he called an autométric algebra. This is a system $A = \;<A, +, \leq, *>\;$ such that 1) $<A, +>$ is a commutative algebra with distinguished element 0, 2) $\leq$ is an antisymmetric reflexive order relation on A, and 3) the operation $*$ on A satisfies the metric axioms: $a * b \geq 0$, $a * b = 0 \Longleftrightarrow a = b$, $a * b = b * a$, and $a * c \leq a * b + b * c$ for any $a, b, c \in A$. The class of autometric algebras contains, in particular, the commutative l-groups.

A. I. Veksler [3] studied associative algebras over the field of reals which contain no nilpotent elements. He found conditions for when a commutative algebra can be made into an f-algebra.

The reports of Fuchs [46,47] are devoted to the study of the basic notions common to the theory of partially ordered algebras. Here by a partially ordered algebra is meant an algebra (A, F) with underlying partially ordered set A and with a family F of finitary operations f defined in A and connected with the order relation $\leq$ on A by monoticity laws.

BIBLIOGRAPHY

[1] N. Bourbaki, *Algèbre.* Chap. VI: *Groupes et corps ordonnés,* Actualités Sci. Indust., no. 1179, Hermann, Paris, 1952; Russian transl., "Nauka", Moscow, 1965. MR 14, 237; 32 #5642.

[2] A. I. Veksler, *Two problems in the theory of semi-ordered spaces,* Sibirsk. Mat. Ž. 5 (1964), 952–954. (Russian) MR 29 #5096.

[3] ———, *Lattice orderability of algebras and rings,* Dokl. Akad. Nauk SSSR 164 (1965), 259–262 = Soviet Math. Dokl. 6 (1965), 1201–1204. MR 32 #7609.

[4] A. A. Vinogradov, *Ordered algebraic systems,* Algebra. Topology. Geometry. 1965, Akad. Nauk SSSR Inst. Naučn. Tehn. Informacii, Moscow, 1967, pp. 83–131. (Russian) MR 35 #6596.

[5] N. E. Domošnickaja, *Generalized measure of elements of an almost monotonically ordered semigroup,* Perm. Politehn. Inst. Naučn. Trud. 7 (1960), no. 1, 41–56. (Russian)

[6] ———, *On the axiomatics of almost monotone ordered semigroups,* Sibirsk. Mat. Ž. 5 (1964), 804–814. (Russian) MR 29 #3555.

[7] L. D. Zybina, *On relations determining the order of certain endomorphisms of an ordered set,* Leningrad. Gos. Ped. Inst. Učen. Zap. 274 (1965), 106–121. (Russian) MR 33 #5526.

[8] ———, *Aggregate of relations determining the ordering of partial endomorphisms,* Leningrad. Gos. Ped. Inst. Učen. Zap. 274 (1965), 122–142. (Russian) MR 34 #110.

[9] T. È. Kaminskiĭ, *Direct sums and lexicographical extensions of lattice ordered groups,* Moskov. Oblast. Ped. Inst. Učen. Zap. 150 (1964), 189–200. (Russian) MR 34 #258.

[10] M. I. Kargapolov, *Orderable groups.* I, Algebra i Logika Sem. 2 (1963), no. 6, 5–14. (Russian) MR 30 #3156.

[11] M. I. Kargapolov, A. I. Kokorin and V. M. Kopytov, *On the theory of orderable groups,* Algebra i Logika Sem. 4 (1965), no. 6, 21–27. (Russian) MR 23 #4162.

[12] A. I. Kokorin, *G-fully orderable and relatively convex subgroups of orderable groups,* Sibirsk. Mat. Ž. 7 (1966), 713–717. (Russian) MR 36 #89.

[13] A. I. Kokorin and V. M. Kopytov, *Relative convexity of the generalized centers of orderable groups,* Ural. Gos. Univ. Mat. Zap. 5 (1965), no. 1 49–53. (Russian) MR 35 #1525.

[14] A. I. Kokorin and N. G. Hisamiev, *An elementary classification of lattice-ordered abelian groups with a finite number of fibers*, Algebra i Logika Sem. 5 (1966), no. 1, 41–50. (Russian) MR 33 #7433.

[15] P. G. Kontorovic and A. I. Kokorin, *A type of partially ordered groups*, Ural. Gos. Univ. Mat. Zap. 3 (1962), no. 3, 39–44. (Russian) MR 30 #3158.

[16] V. M. Kopytov, *Certain forms of products of orderable groups*, Ural. Gos. Univ. Mat. Zap. 5 (1966), no. 3, 75–81. (Russian) MR 34 #4389.

[17] A. I. Mal'cev, *On ordered groups*, Izv. Akad. Nauk SSSR Ser. Mat. 13 (1949), 473–482. (Russian) MR 11, 323.

[18] D. M. Smirnov, *Ordered multi-operator groups*, Sibirsk. Mat. Ž. 6 (1965), 433–458. (Russian) MR 31 #2337.

[19] A. A. Terehov, *Completely orderable groups*, Dokl. Akad. Nauk SSSR 129 (1959), 34–36. (Russian) MR 22 #734.

[20] V. I. Frenkel', *Algorithmic problems in partially ordered groups*, Dokl. Akad. Nauk SSSR 152 (1963), 67–70 = Soviet Math. Dokl. 4 (1963), 1266–1269. MR 27 #3715.

[21] ——, *On the effective partial ordering of finitely presented groups*, Sibirsk. Mat. Ž. 5 (1964), 651–670. (Russian) MR 29 #2290.

[22] ——, *On the representation of partially ordered groups by defining inequalities*, Uspehi Mat. Nauk 20 (1965), no. 6 (126), 164–168. (Russian) MR 32 #7653.

[23] L. Fuchs, *Partially ordered algebraic systems*, Pergamon Press, New York; Addison-Wesley, Reading, Mass., 1963; Russian transl., "Mir", Moscow, 1965. MR 30 #2090.

[24] N. G. Hisamiev, *Universal theory of lattice-ordered Abelian groups*, Algebra i Logika Sem. 5 (1966), no. 3, 71–76. (Russian) MR 34 #2727.

[25] A. I. Čeremisin, *On completely orderable rings*, Algebra i Logika Sem. 4 (1965), no. 2, 67–85. (Russian) MR 33 #7380.

[26] ——, *Completely orderable rings*, Algebra i Logika Sem. 4 (1965), no. 6, 29–46. (Russian) MR 33 #7381.

[27] J. Jakubík, *Lexicographic products of partially ordered groupoids*, Czechoslovak Math. J. 14 (89) (1964), 281–305. (Russian) MR 29 #4830.

[28] Yang Tsung-p'an and Tung K'o-ch'eng, *Two theorems on lattice-ordered rings*, Acta Math. Sinica 15 (1965), 574–581 = Chinese Math.– Acta 7 (1965), 296–303. MR 32 #7610.

[29] R. Aguiló-Fuster, *Application of Dedekind's method to a non-archimedean ordered field,* Collect. Math. 15 (1963), 77–90. (Spanish) MR 29 #4758.

[30] ———, *Maximal archimedean fields contained in a non-archimedean ordered field,* Fourth Annual Meeting Spanish Mathematicians, Univ. Salamanca, Salamanca, 1965, pp. 51–60. (Spanish) MR 36 #2603.

[31] N. L. Alling, *On the existence of real-closed fields that are η_α-sets of power $\aleph_\alpha$,* Trans. Amer. Math. Soc. 103 (1962), 341–352. MR 26 #3615.

[32] F. W. Anderson, *Lattice-ordered rings of quotients,* Canad. J. Math. 17 (1965), 434–448.   MR 30 #4801.

[33] S. J. Bernau, *On semi-normal lattice rings,* Proc. Cambridge Philos. Soc. 61 (1965), 613–616.   MR 32 #1136.

[34] ———, *Unique representation of Archimedean lattice groups and normal Archimedean lattice rings,* Proc. London Math. Soc. (3) 15 (1965), 599–631.   MR 32 #144.

[35] ———, *Orthocompletion of lattice groups,* Proc. London Math. Soc. (3) 16 (1966), 107–130.   MR 32 #5554.

[36] A. Bigard, *Sur les images homomorphes d'un demi-groupe ordonné,* C. R. Acad. Sci. Paris 260 (1965), 5987–5988.   MR 31 #2338.

[37] ———, *Étude de certaines réalisations des groupes réticulés,* C. R. Acad. Sci. Paris Sér. A–B 262 (1966), A853–A855. MR 33 #2736.

[38] A. H. Clifford, *Ordered commutative semigroups of the second kind,* Proc. Amer. Math. Soc. 9 (1958), 682–687.   MR 20 #3223.

[39] ———, *Partially ordered groups of the second and third kinds,* Proc. Amer. Math. Soc. 17 (1966), 219–225.   MR 33 #7431.

[40] P. F. Conrad, *Some structure theorems for lattice-ordered groups,* Trans. Amer. Math. Soc. 99 (1961), 212–240.   MR 22 #12143.

[41] M.-L. Dubreil-Jacotin, *Sur les images homomorphes d'un demi-groupe ordonné,* Bull. Soc. Math. France 92 (1964), 101–115.   MR 29 #5933.

[42] ———, *Quelques propriétés des groupes ordonnés liées à la considération de fuseaux,* C. R. Acad. Sci. Paris Sér. A–B 262 (1966), A989–A992.   MR 37 #119.

[43] E. Fried, *Representation of partially ordered groups,* Acta Sci. Math. (Szeged) 26 (1965), 15–18.   MR 31 #1307.

[44] L. Fuchs, *Partially ordered algebraic systems,* Pergamon Press, New York; Addison-Wesley, Reading, Mass., 1963.   MR 30 #2090.

[45] L. Fuchs, *Riesz groups*, Ann. Scuola Norm. Sup. Pisa (3) 19 (1965), 1–34,   MR 31 #4843.

[46] ——, *On partially ordered algebras*. I, Colloq. Math. 14 (1966), 115–130.   MR 32 #2359.

[47] ——, *On partially ordered algebras*. II, Acta Sci. Math. (Szeged) 26 (1965), 35–41.   MR 31 #4749.

[48] ——, *Note on the class of partially ordered groups*, Bull. Acad. Polon. Sci. Sér. Sci. Math. Astronom. Phys. 13 (1965), 757–759. MR 33 #203.

[49] ——, *Approximation of lattice-ordered groups*, Ann. Univ. Sci. Budapest. Eötvös Sect. Math. 8 (1965), 187–203.

[50] L. Fuchs and E. Sąsiada, *Note on orderable groups*, Ann. Univ. Sci. Budapest. Eötvös Sect. Math. 7 (1964), 13–17.   MR 30 #3925.

[51] J. Gilder, *Betweenness and order in semigroups*, Proc. Cambridge Philos. Soc. 61 (1965), 13–28.   MR 30 #1195.

[52] C. Holland, *The interval topology of a certain l-group*, Czecholovak Math. J. 15 (90) (1965), 311–314.   MR 31 #1308.

[53] D. Ion Ion, , *Baer-McCoy radical and Jacobson radical for semigroups*, Stud. Cerc. Mat. 17 (1965), 1321–1331. (Romanian)   MR 34 #7684.

[54] P. Jaffard, *Contribution à l'étude des groupes ordonnés*, J. Math. Pures Appl. (9) 32 (1953), 203–280.   MR 15, 284.

[55] J. Jakubík, *On a class of l-groups*, Časopis Pešt. Mat. 84 (1959), 150–161. (Russian)   MR 21 #7254.

[56] ——, *The interval topology of a certain l-group*, Mat.-Fyz. Časopis Sloven. Akad. Vied 12 (1962), 209–211.   MR 29 #174.

[57] ——, *Interval topology of an l-group*, Colloq. Math. 11 (1963), 65–72.   MR 28 #4041.

[58] ——— *Über die Intervalltopologie auf einer halbgeordneten Gruppe*, Mat.-Fyz. Časopis Sloven. Akad. Vied 15 (1965), 257–272. MR 34 #4387.

[59] ——, *Kompakt erzeugte Verbandsgruppen*, Math. Nachr. 30 (1965), 193–201.   MR 33 #7432.

[60] D. C. Johnson, *The completion of an archimedean f-ring*, J. London Math. Soc. 40 (1965), 493–496.   MR 31 #1279.

[61] K. Keimel, *Fuseaux généralisés dans la théorie des groupes ordonnés*, C. R. Acad. Sci. Paris Sér. A–B 262 (1966), A993–A995. MR 36 #1372.

[62] O. Kowalski, *On archimedean positively fully ordered semigroups*, Ann. Univ. Sci. Budapest. Eötvös. Sect. Math. 8 (1965), 97–99.

[63] J. L. Krivine, *Anneaux préordonnés*, J. Analyse Math. 12 (1964), 307–326.   MR 31 #214.

[64] E. S. Langford, *Some results on linear operators on lattice groups*, Amer. Math. Monthly 72 (1965), 841–846.   MR 33 #4163.

[65] J. T. Lloyd, *Representations of lattice-ordered groups having a basis*, Pacific J. Math. 15 (1965), 1313–1317.   MR 32 #2489.

[66] F. Papangelou, *Order convergence and topological completion of commutative lattice-groups*, Math. Ann. 155 (1964), 81–107. MR 30 #4699.

[67] ———, *Some considerations on convergence in abelian lattice-groups*, Pacific J. Math. 15 (1965), 1347–1364.   MR 32 #7655.

[68] P. Ribenboim, *On the existence of totally ordered Abelian groups which are η_α-sets*, Bull. Acad. Polon. Sci. Sér. Sci. Math. Astronom. Phys. 13 (1965), 545–548.   MR 33 #5756.

[69] T. Saitô, *Ordered completely regular semigroups*, Pacific J. Math. 14 (1964), 295–308.   MR 28 #4047.

[70] ———, *Proper ordered inverse semigroups*, Pacific J. Math. 15 (1965), 649–666.   MR 33 #204.

[71] N. Sankaran, *Classification of totally ordered abelian groups*, J. Indian Math. Soc. 29 (1965), 9–29.   MR 32 #5748.

[72] F. Šik, *Kompakt erzeugte vollständige l-Gruppen*, Bul. Inst. Politehn. Iaşi 8 (12) (1962), no. 3/4, 5–8.   MR 32 #2496.

[73] ———, *Sous-groupes simples et idéaux simples des groupes réticulés*, C. R. Acad. Sci. Paris 261 (1965), 2791–2793.   MR 33 #2738.

[74] ———, *Types spéciaux de réalisations des groupes réticulés*, C. R. Acad. Sci. Paris 261 (1965), 4948–4949.   MR 32 #7656.

[75] B. Šmarda, *Additive and isotone functionals on partially ordered groups*, Mem. fasc. Ci. Univ. Habana Ser. Mat. 1 (1963/64), no. 3, facs. 2/3, 51–62.   MR 32 #146.

[76] F. E. A. Smith, *A structure theory for a class of lattice-ordered semirings*, Doctoral Dissertation, Purdue University, Lafayette, Ind., 1965; Dissertation Abstracts 26 (1965), 1675.

[77] O. Steinfeld, *On residuals in partially ordered semigroups*, Publ. Math. Debrecen 12 (1965), 107–116.   MR 32 #4201.

[78] K. L. N. Swamy, *Dually residuated lattice ordered semigroups*, Math. Ann. 159 (1965), 105–114.   MR 32 #1273.

[79] K. L. N. Swamy, *Dually residuated lattice ordered semigroups*. II, Math. Ann. 160 (1965), 64–71.   MR 33 #78.

[80] ——— , *A general theory of autometrized algebras*, Math. Ann. 157 (1964), 65–74.   MR 30 #1077.

[81] J. R. Teller, *On partially ordered groups satisfying the Riesz interpolation property*, Proc. Amer. Math. Soc. 16 (1965), 1392–1400. MR 32 #5749.

[82] P. Ucsnay, *Wohlgeordnete Untermengen in totalgeordneten Gruppen. Mit einer Anwendung auf Potenzreihenkörper*, Math. Ann. 160 (1965), 161–170.   MR 32 #147.

[83] D. Vaida, *Idéaux isolés gauches nilpotents et radical isolé*, Bull. Math. Soc. Sci. Math. Phys. R. P. Roumaine 6 (54) (1962), 257–260. MR 33 #75.

[84] J. Vrabec, *Partially ordered algebraic structures*, Obzornik Mat. Fiz. 12 (1965), 15–31. (Slovenian)   MR 34 #4171.

[85] M. A. Ward, *Remarks on a theorem of Fuchs*, Acta Math. Acad. Sci. Hungar. 17 (1966), 5–8.   MR 32 #7657.

[86] E. C. Weinberg, *Free lattice-ordered abelian groups*. I, Math. Ann. 151 (1963), 137–199.   MR 27 #3720.

[87] ——— , *Free lattice-ordered abelian groups*. II, Math. Ann. 159 (1965), 217–222.   MR 31 #5895.

[88] H. J. Weinert, *Über Halbringe and Halbkörper*. III, Acta Math. Acad. Sci. Hungar. 15 (1964), 177–194.   MR 28 #4012.

[89] S. Wolfenstein, *Sur les groupes réticulés archimédiennement complets*, C. R. Acad. Sci. Paris Sér. A–B 262 (1966), A813–A816. MR 33 #2739.

[90] E. S. Wolk, *On the interval topology of an l-group*, Proc. Amer. Math. Soc. 12 (1961), 304–307.   MR 23 #A232.

Translated by:
J. Ellis

Amer. Math. Soc. Transl.
(2) Vol. 96, 1970

o-RINGS AND LA-RINGS[1]

UDC 519.48

B. M. ŠAĬN

In the present article we shall use the notation and definitions of our previous article [1]. A ring whose multiplicative semigroup is regular is called a regular ring [2]. In §1 of [1] we considered a special class of regular semigroups, the generalized groups. There arises in a natural way the question of whether a [given] generalized group is the multiplicative semigroup of some ring and what properties this sort of ring possesses. Part of the present work is devoted to resolving this question. A series of results, then in the process of being published, were communicated by the author in a report to the Fourth All-Union Mathematical Congress on July 11, 1961.

In the future, a ring whose multiplicative semigroup is a generalized group will be called an o-ring. The center of a ring A, defined in the usual manner, will be denoted by $Z(A)$. Elements in the center are said to be central. An element a of A is said to be nilpotent if there exists a natural number n such that $a^n = 0$. A nilpotent element different from zero is said to be a nontrivial nilpotent element. An element a is said to be annihilating if $\{a\}A = A\{a\} = 0.$[2] The set of all annihilating elements of A is denoted by $N(A)$ and is called the annihilator of A. Clearly $N(A)$ is an ideal of A, $Z(A)$ is a subring of A, and, moreover, $N(A) \subset Z(A)$.

The principal right (left) ideal generated by an element a of A will be denoted by $(a)_r [(a)_l]$. The set of all principal right [left] ideals of a ring A will be denoted by $\Re_A [\mathfrak{L}_A]$. The principal (two-sided) ideal generated by an element a of A will be denoted by $(a)_f$. The set of all principal ideals of A is denoted by $\mathfrak{F}_A$. The sets $\Re_A$, $\mathfrak{L}_A$ and $\mathfrak{F}_A$ are ordered by set-theoretic inclusion.

We shall need the following proposition in the future.

Proposition 1. *If a regular ring contains a unique idempotent, then this ring consists of exactly one element.*

Proof. Since every ring contains an idempotent, namely the zero element,

---

1) Translation of Izv. Vysš. Učebn. Zaved. Matematika 1966, no. 2 (51), 111-122.

2) As usual, $\{a\}$ denotes the set of consisting of just the element a.

a regular ring satisfying the conditions of the proposition has the zero element as its unique idempotent. Now let a be any element of this ring and
let $\bar{a}$ be one of its generalized inverses. Then $\overline{a}a$ is idempotent and consequently $\overline{a}a = 0$. Then $a = a\overline{a}a = 0$ as desired.

Proposition 2. *For a ring to be a division ring, it is necessary and sufficient that this ring be regular and contain exactly two idempotents.*

Proof. The necessity of the conditions is obvious.

Let A be a regular ring containing exactly two idempotents: 0 and e.
Let a be any element of A other than 0 and let $\bar{a}$ be one of its generalized
inverses. Then $a\bar{a} = \overline{a}a = e$, from which it is clear that A is a division ring
since e will be the identity of A (actually $ea = a\overline{a}a = a$ for $a \neq 0$ and
$e0 = 0$, i.e. e is a left identity; one shows analogously that e is a right
identity).

This proposition is a slight generalization of an assertion from $[3]$
(Russian page 172).

Proposition 3 $[4]$. *The following three conditions are equivalent:*

a) *A is regular;*

b) *$\mathfrak{R}_A$ is a Dedekind lattice with respect to complementation;*

b*) *$\mathfrak{L}_A$ is a Dedekind lattice with respect to complementation.* [1]

Proposition 4 ($[5]$, **page 75**).[2] *Let i and i_1 be idempotents of a regular
ring A. Then $(i)_r = (i_1)_r$ $[(i)_l = (i_1)_l]$ holds if and only if $i_1 = i + ia - iai$
$[i_1 = i + ai - iai]$, where a is in A.*

Theorem 1. *In a regular ring A the following conditions are equivalent
and are characteristic of o-rings:*

a) *any two idempotents commute under multiplication;*

b) *for any element a and any of its generalized inverses $\bar{a}$ the elements
$a\overline{a}$ and $\overline{a}a$ commute under multiplication;*

c) *there exists no more than one generalized inverse for each element
of the ring A;*

d) *there exists no more than one generalized inverse for each idempotent
of the ring A;*

---

    1) The asterisk indicates that condition b*) is the dual of condition b), i.e.
that b*) turns into b) if in the ring A a new multiplication $\times$ is introduced by setting $a_1 \times a_2 = a_2 a_1$.

    2) Regular rings with identity were considered in $[5]$. The reference to $[5]$ is
to indicate that the arguments introduced there carry over to arbitrary regular rings
without difficulty.

f) *any idempotent commutes with any of its generalized inverses under multiplication;*

g) *for any idempotent i and any of its generalized inverses $\bar{i}$ the elements $\bar{i}i$ and $i\bar{i}$ commute under multiplication;*

h) *every idempotent of A is central;*

i) *for any element a of the ring A one can find a generalized inverse $\bar{a}$ such that $a\bar{a} = \bar{a}a$;*

j) *the ring A contains no nontrivial nilpotent elements;*

k)[1] *in the ring A, $a_1a_2 = 0$ follows from $a_2a_1 = 0$;*

l) *if i is an idempotent, then, for every a in A, $ia = iai$ or, for every a in A, $ai = iai$;*

m) *every right ideal is two-sided;*

m*) *every left ideal is two-sided;*

n) *for any elements a_1 and a_2 it follows from $(a_1)_r \subset (a_2)_r$ that $(a_1)_l \subset (a_2)_l$;*

n*) *for any elements a_1 and a_2 it follows from $(a_1)_l \subset (a_2)_l$ that $(a_1)_r \subset (a_2)_r$;*

o) *for any elements a_1 and a_2 it follows from $(a_1)_r = (a_2)_r$ that $(a_1)_l = (a_2)_l$.*

o*) *for any elements a_1 and a_2 it follows from $(a_1)_l = (a_2)_l$ that $(a_1)_r = (a_2)_r$;*

p) *for any two idempotents i_1 and i_2, $i_1i_2 = i_2$ implies that $i_2i_1 = i_2$;*

p*) *for any two idempotents i_1 and i_2, $i_1i_2 = i_1$ implies that $i_2i_1 = i_1$;*

q) *for any two idempotents i_1 and i_2, $i_1i_2 = i_2$ and $i_2i_1 = i_1$ together imply that $i_1 = i_2$;*

q*) *for any two idempotents i_1 and i_2, $i_1i_2 = i_1$ and $i_2i_1 = i_1$ together imply that $i_1 = i_2$;*

r) *if for two idempotents i_1 and i_2 we have $(i_1)_r = (i_2)_r$ or $(i_1)_l = (i_2)_l$, then $i_1 = i_2$;*

s) *the lattice $\Re_A$ is distributive;*

s*) *the lattice $\mathfrak{L}_A$ is distributive;*

t) *the lattice $\Re_A$ is a lattice with unique relative complements;*

t*) *the lattice $\mathfrak{L}_A$ is a lattice with unique relative complements.*

---

[1] If the ring A contains an identity or if it is biregular (the definition of biregularity is introduced later), then k) can be replaced by a weaker condition, namely: when a_1 and a_2 are idempotents then $a_2a_1 = 0$ follows from $a_1a_2 = 0$.

Proof. The equivalence of conditions a)-g) was proved for regular semigroups in [1] Corollary 1.3, since conditions a)-g) coincide with conditions A1–A4, A6 and A7 of [1]. The implications a) $\rightarrow$ h) and h) $\rightarrow$ i) were proved by A. Clifford [6] and A. E. Liber [7] for regular semigroups. Suppose condition i) holds and $a^n = 0$. Then $a\bar{a} = \bar{a}a$, where a is some generalized inverse for a. Hence, $0 = a^n\bar{a}^n a = a^{n-1}a\bar{a}\bar{a}^{n-1}a = a^{n-1}\bar{a}\bar{a}a a^{n-1}a = a^{n-1}\bar{a}^{n-1}a = \cdots = a\bar{a}a = a$. This means j) holds. Now suppose j) holds and $a_1 a_2 = 0$. Then $(a_2 a_1)^2 = a_2 a_1 a_2 a_1 = 0$, whence $a_2 a_1 = 0$, and thus k) holds. Since $(ai - a)\, i = 0$ for any a in A, we obtain $i(ai - a) = 0$ or $ia = iai$. Analogously, $ai = iai$, so l) holds. Suppose l) holds and $iai = ia$ for all a in A. Then if a_1 is a generalized inverse for the element $(ia - a)i$, we have $(ia - a)\,i = (ia - a)\,ia_1(ia - a)\,i = (ia - a)\,ia_1 i(ia - a)\,i = 0$ since $ia_1 i = ia_1$. Therefore $iai = ai$. Analogously one can show that if $iai = ai$ for all a in A, the $iai = ia$. Consequently, $ai = ia$ and condition h) is fulfilled. Condition a) clearly follows from h). We have shown that conditions a)-l) are equivalent.

Suppose condition h) holds and that i is an idempotent. Then $(i)_r = \{i\}A = A\{i\} = (i)_l$, i.e. every principal right ideal is a left ideal and so is two-sided. Noting that every right ideal of a ring is generated by principal right ideals, we see that condition m) holds. m*) follows from m) in a similar fashion. h) follows from m) ([5], page 74) and analogously from m*). Thus conditions a)–m*) are equivalent. Now suppose condition i) holds and that $(a_1)_r \subset (a_2)_r$. It is not difficult to see that $(a_1)_r = (a_1\bar{a}_1)_r$ and $(a_2)_r = (a_2\bar{a}_2)_r$, where $\bar{a}_1$ and $\bar{a}_2$ are generalized inverses for the elements a_1 and a_2. If these generalized inverses are so chosen that $a_i\bar{a}_i = \bar{a}_i a_i$ for $i = 1, 2$, then $(\bar{a}_2 a_2)_r = (\bar{a}_2 a_2)_l$ and $(\bar{a}_1 a_1)_l = (\bar{a}_1 a_1)_r$, whence it follows that $(a_1)_r = (a_1)_l$ and $(a_2)_r = (a_2)_l$. Therefore $(a_1)_l \subset (a_2)_l$, and condition n) holds. In a similar fashion n*) follows from i). Condition o) follows from n) in an obvious fashion, and o*) follows from n*) analogously.

For the future we shall need

Lemma 1. *If i_1 and i_2 are idempotents of a regular ring A, then $(i_1)_r \subset (i_2)_r$ if and only if $i_2 i_1 = i_1$, and $(i_1)_l \subset (i_2)_l$ if and only if $i_1 i_2 = i_1$.*

Proof. Let $(i_1)_r \subset (i_2)_r$. Then $i_1 \in (i_2)_r$. Consequently $i_1 = i_2 a$ for some a in A. Multiplying this equation on the left by i_2 we obtain $i_2 i_1 = i_2 a$, whence it follows that $i_2 i_1 = i_1$. Conversely, if $i_2 i_1 = i_1$, then clearly $i_1 \in (i_2)_r$ and $(i_1)_r \subset (i_2)_r$. The second part of the lemma is proved analogously.

We return to the proof of our theorem. Suppose condition n) holds. Then, according to Lemma 1, $(i_1)_r \subset (i_2)_r$ means that $i_2 i_1 = i_1$; hence we get $(i_1)_l \subset (i_2)_l$ or $i_1 i_2 = i_1$, so that p) holds. Condition p*) follows from n*) analogously. Condition q) follows from p) in an obvious manner, as does q*) from p*). We note that condition q) is equivalent to condition n) and condition q*) is equivalent to n*) in view of Lemma 1. We also note that, by virtue of this same lemma, r) follows from q) [from q*)] and q) and q*) from r). By virtue of Proposition 4 1) follows from r). Thus conditions a)—r) are equivalent. From h) it follows that all elements of the lattice $\mathcal{L}_A$ are central (compare with proposition 98-a of [3]). Consequently the lattice $\mathcal{L}_A$ is distributive, so condition s*) is satisfied. Similarly s) follows from h) ([8], Russian page 53).

Conditions s) and t) are equivalent ([8], Russian page 192), just as are conditions s*) and t*). Finally, h) follows from t) [from t*)] (compare with [5], page 76). Thus conditions a)—t*) are equivalent. It remains to note that conditions a)—g) are characteristic of generalized groups [1], and thus of o-rings. Theorem 1 is proved.

We note that regular rings satisfying condition j) were considered in [9]. Regular rings satisying condition k) were considered by Thierrin [10] (Thierren called condition k) reflectivity). It is clear that a commutative regular ring is an o-ring, since such rings satisfy condition a).

Corollary 1. *In order that a ring A be an o-ring, it is necessary and sufficient that A satisfy either of the conditions t) or t*) of Theorem 1.*

Proof. The necessity follows from Theorem 1. To show the sufficiency it is only necessary to note that it follows from t) that the lattice $\mathfrak{R}_A$ will be distributive ([8], Russian page 192) and consequently Dedekind ([8], Russian page 192), and to apply Proposition 3 and Theorem 1.

A ring A is said to be strongly regular if for every a in A there exists an $\bar{a}$ such that $a^2 \bar{a} = a$ [11].

Corollary 2. *The class of strongly regular rings coincides with the class of o-regular rings.*

Proof. It is known that every strongly regular ring is a regular ring without nontrivial nilpotent elements (see, for example, [3], Russian page 177), and so is an o-ring by condition j) of Theorem 1. On the other hand, in an o-ring A, for every element a we have $a^2 \bar{a} = a$, where $\bar{a}$ is the generalized inverse for a (this follows from conditions c) and i) of Theorem 1). Therefore every o-ring is strongly regular.

It follows from Corollary 2 that a ring which is anti-isomorphic to a strongly regular ring (i.e. a ring in which for every element a there exists an element $\bar{a}$ with the property that $\bar{a}a^2 = a$) is strongly regular, since a ring which is anti-isomorphic to an o-ring is an o-ring. Thus out of Corollary 2 comes the following extension of Theorem 1.

Corollary 3. *The following properties are equivalent for regular rings and are characteristic of o-rings:*

u) *for any element a of A one can find a generalized inverse $\bar{a}$ such that $a\bar{a}$ is a central element;*

u*) *for any element a of A one can find a generalized inverse $\bar{a}$ such that $\bar{a}a$ is a central element.*

Proof. Condition u) follows from condition h). Conversely, a ring satisfying u) is strongly regular, and consequently an o-ring by Corollary 2.

We note that we have strengthened a result of Arens and Kaplansky ([11], Theorems 3.3 and 3.4).

Corollary 4. *A regular ring is an o-ring if and only if it is isomorphic to a dense subdirect sum ([12], Russian page 31) of division rings.*

Proof. Clearly any regular ring which is a subdirect sum of some family of division rings is an o-ring, since it satisfies condition j) of Theorem 1. A homomorphic image of an o-ring is evidently an o-ring (since a homomorphic image of a generalized group is a generalized group [13]). Therefore every o-ring can be decomposed into a subdirect product of subdirectly indecomposable o-rings. A subdirectly indecomposable o-ring does not contain idempotents other than 0 and the identity (for otherwise the ring admits a nontrivial Pierce decomposition). From Propositions 1 and 2 it follows that a subdirectly indecomposable o-ring is a division ring. Thus every o-ring A is isomorphic to a subdirect sum of some family $\{A_i\}_{i \in I}$ of division rings. Let $\{B_i\}_{i \in I}$ be a family of ideals of the ring A which are the kernels of the homomorphisms of A onto the division rings of the family $\{A_i\}_{i \in I}$. We can assume that all the ideals of $\{B_i\}_{i \in I}$ are different. Since all the ideals $\{B_i\}_{i \in I}$ are maximal, the ring A will be isomorphic to the dense subdirect sum of the division rings $\{A_i\}_{i \in I}$ ([12], Russian page 32).

Corollary 5. *Every o-ring containing a finite number of idempotents is isomorphic to a direct sum of a finite number of division rings. Moreover, if k is the number of idempotents, then $k = 2^n$, where n is the number of direct summands. Conversely, a direct sum of a finite family of division rings is an o-ring and contains a finite number of idempotents.*

Proof. It is clear that a ring which is the direct sum of n division rings contains 2^n idempotent elements. If a ring is isomorphic to a dense sub-direct sum of an infinite family of division rings, then the number of idempotents of such a ring is greater than 2^n for all n, as easily follows from the definition of dense subdirect sum ([12], Russian page 31). Thus such a ring contains an infinite number of idempotents. Consequently an o-ring with a finite number of idempotents is, by Corollary 4, a dense subdirect sum of a finite number of division rings, and this means it is a direct sum of a finite number of division rings, since the notions of dense subdirect sum and direct sum coincide for a finite family of rings. The second part of our assertion also follows from Corollary 4.

Corollary 6. *Every finite o-ring is isomorphic to a direct sum of fields, and consequently is commutative.*

The proof follows from Corollary 5 and the fact that every finite division ring is a field.

We note that certain of the connections existing between the conditions a)–t*) of Theorem 1 were considered earlier. For example A. I. Gerčikov [14] showed that m), m*) and h) follow from j); certain other implications were proved in books by John von Neumann [5] and L. A. Skornjakov [3].

For the remainder of this article we turn our attention to the structure of the centers of full matrix rings. Let $(a_{ij}) \in Z(A_n)$ and $(b_{ij}) \in A_n$, where $b_{ij} \neq 0$ only for $i = i_0$, $j = j_0$. From the equality $(a_{ij})(b_{ij}) = (b_{ij})(a_{ij})$ we obtain $a_{i_0 i_0} b_{i_0 j_0} = b_{i_0 j_0} a_{j_0 j_0}$ and $a_{i_0 i_0} b_{i_0 j_0} = b_{i_0 j_0} a_{j_0 j_0} = 0$ for $i \neq i_0$, $j \neq j_0$. In particular, for $i_0 \neq j_0$ we have $a_{j_0 i_0} b_{i_0 j_0} = b_{i_0 j_0} a_{j_0 i_0} = 0$, whence it follows that $a_{j_0 i_0} \in N(A)$. Setting $i_0 = j_0$, we obtain $a_{i_0 i_0} b_{i_0 i_0} = b_{i_0 i_0} a_{i_0 i_0}$; that is, $a_{i_0 i_0} \in Z(A)$. Therefore $b_{i_0 j_0} a_{i_0 i_0} = b_{i_0 j_0} a_{j_0 j_0}$ or $b_{i_0 j_0}(a_{i_0 i_0} - a_{j_0 j_0}) = 0$. In exactly the same way $(a_{i_0 i_0} - a_{j_0 j_0}) b_{i_0 j_0} = 0$. Consequently $a_{i_0 i_0} - a_{j_0 j_0} \in N(A)$.

A matrix $(a_{ij}) \in A_n$ is called a scalar matrix if all its main diagonal elements are equal and all the remaining elements are zero. The set of all scalar matrices of order n will be denoted by $S(A_n)$. It is clear that $S(A_n)$ constitutes a subring of A_n which is isomorphic to the ring A. The center of this ring, $ZS(A_n)$, is the set of all those scalar matrices whose main diagonal elements are elements from $Z(A)$.

We have shown that all matrices $(a_{ij}) \in Z(A_n)$ can be represented in the form $(a_{ij}) = (b_{ij}) + (c_{ij})$, where $(b_{ij}) \in ZS(A_n)$ and $(c_{ij}) \in N(A)_n$ (for (b_{ij}) one may take the scalar matrix all of whose main diagonal elements are equal to a_{11}, while for (c_{ij}) one may choose that matrix such that $c_{ij} = a_{ij}$

for $i \neq j$ and $c_{ii} = a_{ii} - a_{11}$). On the other hand, it is clear that $ZS(A_n) \subset Z(A_n)$ and $N(A)_n \subset Z(A_n)$, whence $ZS(A_n) + N(A)_n \subset Z(A_n)$. Therefore we have proved

Theorem 2. $Z(A_n) = ZS(A_n) + N(A)_n$.

We shall say that A is a ring with trivial annihilator if $N(A) = \{0\}$.

Corollary 1. *If A is a ring with trivial annihilator, then $Z(A_n) = ZS(A_n)$.*

Since, as we have already noted, the rings $ZS(A_n)$ and $Z(A)$ are isomorphic, we have

Corollary 2. *If A is a ring with trivial annihilator, then for any n the centers of the rings A and A_n are isomorphic.*

As is not difficult to see, every ring which is semisimple in the sense of Jacobson is a ring with trivial annihilator. Therefore we have

Corollary 3. *If A is semisimple in the sense of Jacobson, then the centers of the rings A and A_n are isomorphic.*

A ring is said to be biregular if every principal ideal of the ring is generated by some central idempotent. From conditions h) and m) of Theorem 1 and from the regularity of o-rings it follows that all o-rings are biregular. On the other hand, as we shall see later, not every regular ring, nor even every biregular ring, is an o-ring. As is well known, regular and biregular rings are semisimple in the sense of Jacobson. Therefore from Corollary 3 follows

Corollary 4. *If a ring A is regular or biregular, then the centers of A and A_n are isomorphic.*

The set I of central idempotents of a biregular ring A can be ordered by setting $i_1 \leqslant i_2$ if and only if $i_1 i_2 = i_1$. As is not difficult to verify, it follows from $(i_1)_f = (i_2)_f$ for i_1, $i_2 \in I$ that $i_1 i_2 = i_1$ and $i_2 i_1 = i_2$ and $i_1 = i_2$. Therefore the correspondence between an idempotent i and the principal ideal $(i)_f$ gives a one-to-one correspondence between the sets I and $\mathfrak{F}_A$. It is easy to see that this correspondence is isotonic, and therefore the ordered sets $\mathfrak{F}_A$ (of principal ideals of a biregular ring A) and I (the central idempotents of this ring) are isomorphic.

Let A be a biregular ring, $a \in Z(A)$, $i \in I$ and $(a)_f = (i)_f$. Then it is not difficult to see that $ia = a$ and $aa' = i$ for some $a' \in A$. If $na = i$, then $nai = i$, $ani = i$ and $a' = ni$. We verify without difficulty that $\bar{a} = a'aa'$ is a generalized inverse for a. Let $a_1 \in A$. Then $\bar{a}a_1 = \bar{a}a\bar{a}a_1 = \bar{a}aa'aa'a_1 = \bar{a}ia_1 = \bar{a}a_1 i = \bar{a}a_1 aa' = \bar{a}aa_1 a' = a'aa'aa_1 a' = ia_1 a' = a_1 ia' = a_1 a'aa' = a_1 \bar{a}$, i.e. $\bar{a} \in Z(A)$. Therefore $Z(A)$ is a commutative regular subring of A. By condition h) of Theorem 1 $Z(A)$ is an o-ring, and consequently is a biregular ring. Thus the center of a biregular ring is a biregular ring [15].

It is clear that $I \subset Z(A)$. Thus, as was shown earlier, the ordered set $\mathfrak{F}_A$ is isomorphic to the ordered set I and the latter is isomorphic to the ordered set $\mathfrak{F}_{Z(A)}$. Therefore $\mathfrak{F}_A$ and $\mathfrak{F}_{Z(A)}$ are isomorphic. But since $Z(A)$ is an o-ring, $\mathfrak{F}_{Z(A)}$ is isomorphic to $\mathfrak{R}_{Z(A)}$ by property m) of Theorem 1, while $\mathfrak{R}_{Z(A)}$ is a distributive lattice with relative complements, by properties s) and t) of Theorem 1. Finally we obtain the result that the lattice of principal ideals of a biregular ring is a distributive lattice with relative complements [16] and that the lattices of principal ideals of the biregular rings A and $Z(A)$ are isomorphic [15].

Theorem 3. *If A is a biregular ring, then for any n the lattices $\mathfrak{F}_A$ and $\mathfrak{F}_{A_n}$ are isomorphic.*

Proof. $\mathfrak{F}_A$ is isomorphic to $\mathfrak{F}_{Z(A)}$, and $Z(A)$ is isomorphic to $Z(A_n)$ by Corollary 4 of Theorem 2, so $\mathfrak{F}_A$ is isomorphic to $\mathfrak{F}_{A_n}$, since $\mathfrak{F}_{Z(A_n)}$ is isomorphic to $\mathfrak{F}_{A_n}$ and A_n is biregular [16].

Corollary 1. *If A is an o-ring, then the lattices $\mathfrak{F}_A$ and $\mathfrak{F}_{A_n}$ are isomorphic.*

We note that if A is an o-ring and $n > 1$, then A_n will not be an o-ring, although this ring will be regular and biregular. Indeed, consider a matrix $(a_{ij}) \in A_n$ with all elements except a_{11} equal to zero and a_{11} a nonzero idempotent of the ring A. Clearly (a_{ij}) is an idempotent, but $(a_{ij}) \notin Z(A_n)$, as follows from Theorem 2. Therefore A_n is not an o-ring, since condition h) of Theorem 1 is violated.

It is clear that the assertion of Theorem 3 does not hold for all rings. We next consider rings for which this assertion is valid. As a preliminary we turn our attention to a certain wider class of rings.

An element a of a ring A is called a lateral annihilating element if $A\{a\}A = 0$. The set of all lateral annihilating elements of a ring A, which we denote by $M(A)$, is clearly an ideal of A, and, moreover, it is nilpotent since $M(A)^3 = \{0\}$. We shall say that A is a ring with trivial lateral annihilator if $M(A) = \{0\}$. It is clear that if A is a ring with trivial lateral annihilator, then A is a ring with trivial annihilator, for, as is easy to see, $N(A) \subset M(A)$. It is also evident that every ring which is semisimple in the sense of Jacobson is a ring with trivial lateral annihilator, since such a ring does not contain a nontrivial nilpotent ideal. We also note that the lateral annihilator and the annihilator of a ring do not, generally speaking, coincide. For example, if A

is a ring such that $A^3 = \{0\}$ but $A^2 \neq \{0\}$, then it is not difficult to see that $M(A) = A$ but $N(A) \neq A$.

We will say that A is an LA-ring if every homomorphic image of A is a ring with trivial lateral annihilator. Clearly, if A is an LA-ring, then A is a ring with trivial lateral annihilator, because every ring is a homomorphic image of itself. Let A be some ring and B an ideal of A. Then B_n, as is easy to see, will be an ideal of the ring A_n. The converse is not in general true: one can find an ideal $\overline{B}$ of A_n for which there does not exist an ideal B of A satisfying the condition $B_n = \overline{B}$. On the other hand,

Theorem 4. *In order that every ideal of the ring A_n ($n > 1$) be of the form B_n, where B is some ideal of A, it is necessary and sufficient that A be an LA-ring.*

Before proving this theorem, we will prove a series of other propositions. Let B be some subset of the ring A. By B^+ we will denote the additive closure of this subset; that is, the set of all elements of the form $b_1 + b_2 + \cdots + b_n$, $b_i \in B$, $n \geq 1$.

Lemma 2. *If B is an ideal of A, then $(ABA)^+$ is also an ideal of A.*

Proof. It is clear that $-(ABA)^+ = (A(-B)A)^+$, so that $(ABA)^+$ is a sub-group of the additive group of the ring. It is no less clear that $(ABA)^+A \subset (ABA)^+$ and $A(ABA)^+ \subset (ABA)^+$, for this is a consequence of the distributivity of multiplication. Therefore $(ABA)^+$ is an ideal. We will denote the set $(ABA)^+$ by B^0. If B is an ideal, then clearly $B^0 \subset B$. However, in general $B^0 \neq B$. An ideal B of a ring A is said to be proper if $B^0 = B$.

Proposition 5. *In order that every ideal of a ring be proper[1] it is necessary and sufficient that it be an LA-ring.*

Necessity. Suppose every ideal of A is proper but A is not an LA-ring. Then one can find an ideal B of the ring A such that the factor ring A/B has a nontrivial lateral annihilator $M(A/B)$. The preimage of $M(A/B)$, under the canonical homomorphism from A onto A/B, is an ideal C such that $ACA \subset B$, as follows from the definition of lateral annihilator. Hence one gets $C^0 \subset B$. Since C is a proper ideal, $C \subset B$, i.e. the lateral annihilator $M(A/B)$ is trivial. The contradiction thus obtained completes the proof of necessity.

Sufficiency. Let A be an LA-ring and B some ideal of this ring. We

---

1) *Translator's note.* The reader should note the nonstandard use of the term "proper ideal". The author's definition of this term is to be found at the end of the proof of Lemma 2.

consider the factor ring A/B^0 and its lateral annihilator $M(A/B^0)$. It is clear that $B/B^0 \subset M(A/B^0)$. From the triviality of $M(A/B^0)$ it follows that $B = B^0$, i.e. the ideal B is proper. Proposition 5 is proved.

Proposition 6. *Let A be an LA-ring which is an ideal of a ring A'. Then every ideal of A is an ideal of A'.*

Sufficiency. Let B be an ideal of the ring A. Then $(ABA)^+ A' \subset (ABA)^+$, for A is an ideal of A'. Consequently B^0 is a right ideal of A'. One shows analogously that B^0 is a left ideal of A'. It remains to note that $B^0 = B$.

Let A be a ring having the following property: if A is an ideal of some ring A', then every ideal of A is an ideal of A'. Is A an LA-ring? In other words, is the converse of Proposition 6 valid? Apparently not, although the author has been unable to show this.

We introduce the following notation for the future: a matrix having the element a in the i, j position and zeroes everywhere else will be denoted by $(a)^{ij}$. Let A be a ring having an improper ideal B. Then there exists an element $a \in B \backslash B^0$. It is not difficult to see that for this element $(a)_f^0 \subset B^0$ and therefore $(a)_f \neq (a)_f^0$. Let n be some fixed natural number greater than 1. We consider the principal ideal of A_n generated by the matrix $(a)^{11}$. All elements of the ideal $((a)^{11})_f$ have the form

$$k\,(a)^{11} + \sum_l (a)^{11}(a_{ij}^l) + \sum_m (b_{ij}^m)\,(a)^{11} + \sum_{p,\,q} (c_{ij}^p)\,(a)^{11}(d_{ij}^q).$$

Only elements of the form

$$k \cdot 0 + \sum_{l,\,i} 0 \cdot a_{i\,2}^l + \sum_{m,\,j} b_{2j}^m \cdot 0 + \sum_{p,\,q} c_{21}^p a d_{12}^q = \sum_{p,\,q} c_{21}^p a d_{12}^q,$$

can occur in the i, j position of a matrix from $((a)^{11})_f$. Such an element is an element of $(a)_f^0$. Since $a \notin (a)_f^0$, the element a does not occur in the 2, 2 position of any matrix from $((a)^{11})_f$, but it does stand in the 1, 1 position of $(a)^{11}$ which belongs to $((a)^{11})_f$. Hence it follows that $((a)^{11})_f$ is not of the form B_n where B is some ideal of A. Actually, if $((a)^{11})_f = B_n$, then $(a)^{11} \in B_n$, whence it follows that $a \in B$. Therefore $(a)^{22} \in B_n = ((a)^{11})_f$, which we saw to be untrue. Thus we have proved the necessity of the conditions in Theorem 4. Actually, if every ideal of the ring A_n is of the form B_n where B is an ideal of A, then A, as we saw, cannot contain any improper ideals. By Proposition 5, A is an LA-ring.

We will show the sufficiency of the conditions of Theorem 4. Let A be an LA-ring and B' some ideal of the ring A_n. Finally, let B be the set of all those elements of A which occur in any of the matrices from B'. We will show that $B' = B_n$. First we note the obvious fact that $B' \subset B_n$. Second, we note that every matrix of B_n decomposes in an obvious manner into a sum of matrices of the form $(b)^{ij}$, where $b \in B$. Therefore we need only show that $(a)^{ij} \in B'$ for any $a \in B$ and any i, j such that $1 \leq i, j \leq n$. By hypothesis there exists a matrix $(a_{ij}) \in B'$ such that $a = a_{i_0 j_0}$. Since A is an LA-ring, the ideal $(a)_f$ is a proper one by Proposition 5, and consequently $(a)_f = (a)_f^0$ and $a \in (a)_f^0$; thus one can find elements $x_1, x_2, \cdots, x_m$ and $y_1, y_2, \cdots, y_m$ in A for which $a = x_1 a y_1 + \cdots + x_m a y_m$. Clearly, $\sum_{k=1}^m (x_k)^{ij0}(a_{ij})(y_k)^{i0j} \in B'$. Finally, direct computation assures us that $(a)^{ij} = \sum_{k=1}^m (x_k)^{ii0}(a_{ij})(y_k)^{j0j}$. Therefore $(a)^{ij} \in B'$, as required. In the future we shall need

Lemma 3. *If B_n is an ideal of the ring A_n, then B is an ideal of A.*

Proof. Let B_n be an ideal of A_n. Let $b_1, b_2 \in B$. Then $(b_1)^{11}$ and $(b_2)^{11}$ are elements of B_n, whence it follows that $(b_1 - b_2)^{11} = (b_1)^{11} - (b_2)^{11} \in B_n$. Therefore $b_1 - b_2 \in B$. If $b \in B$ and $a \in A$, then $(ab)^{11} = (a)^{11}(b)^{11} \in B_n$, so that $ab \in B$. Analogously $ba \in B$. Therefore B is an ideal of the ring A.

We have shown that if A is an LA-ring and B' is an ideal of A_n, then $B' = B_n$. From Lemma 3 it follows that B is an ideal of A. Consequently the conditions of Theorem 4 are sufficient. Theorem 4 is proved.

Corollary 1. *If for some $m > 1$ every ideal of the ring A_m has the form B_m, then for any n any ideal of A_n is of the form B_n, where B is a suitable ideal of the ring A.*

Proof. If every ideal of A_m has the form B_m, then by Lemma 3 and Theorem 4 A is an LA-ring. Corollary 1 now follows from Theorem 4 in an obvious manner.

Corollary 2. *A ring A is an LA-ring if and only if A_m is an LA-ring for some m. If A is an LA-ring, then for all m A_m will be an LA-ring.*

Proof. Let A be an LA-ring. The rings A_4 and $(A_2)_2$ are clearly isomorphic. Under this isomorphism the matrix

$$\begin{pmatrix} a_{11} & a_{12} & a_{13} & a_{14} \\ a_{21} & a_{22} & a_{23} & a_{24} \\ a_{31} & a_{32} & a_{33} & a_{34} \\ a_{41} & a_{42} & a_{43} & a_{44} \end{pmatrix}$$

corresponds to the matrix

$$\left(\begin{array}{cc} \begin{pmatrix} a_{11} & a_{12} \\ a_{21} & a_{22} \end{pmatrix} & \begin{pmatrix} a_{13} & a_{14} \\ a_{23} & a_{24} \end{pmatrix} \\ \begin{pmatrix} a_{31} & a_{32} \\ a_{41} & a_{42} \end{pmatrix} & \begin{pmatrix} a_{33} & a_{34} \\ a_{43} & a_{44} \end{pmatrix} \end{array}\right)$$

Every ideal of the ring A_4 has the form B_4, where B is an ideal of A (Theorem 4). Under the isomorphism between A_4 and $(A_2)_2$ the ideal B_4 corresponds to the ideal $(B_2)_2$. Therefore every ideal of $(A_2)_2$ has the form B_2', where B' is some ideal of A_2. By Theorem 4, A_2 is an LA-ring. In an analogous manner one shows that A_m is an LA-ring for any m.

Conversely, let A_m be an LA-ring. By Proposition 5 all ideals of A_m are proper. Let B be an ideal of A. Then B_m is an ideal of A_m and therefore is proper, i.e. $(A_m B_m A_m)^+ = B_m$. Let $a \in B$. Then $(a)^{11} \in B_m$, whence $(a)^{11} \in B_m^0$. Thus it follows that $a = x_1 a_1 y_1 + \cdots + x_n a_n y_n$, where $a_i \in B$. Therefore $a \in (ABA)^+ = B^0$, i.e. $B \subset B^0$. This means that the ideal B is proper. By Proposition 5 the ring A is an LA-ring. Corollary 2 is proved.

Proposition 7. *A homomorphic image of an LA-ring is an LA-ring.*

The proof follows from the definition of an LA-ring and from the fact that a homomorphic image of a homomorphic image of some ring is a homomorphic image of that ring.

It is not difficult to prove the discrete direct sum of a family of LA-rings is an LA-ring. On the other hand, a dense direct sum of a family of LA-rings need not be an LA-ring. From Theorem 4 it follows that the lattices of ideals of the rings A and A_n are isomorphic in the case that A is an LA-ring. Since from $M(A) = \{0\}$ it follows that $N(A) = \{0\}$, as was mentioned earlier, by Corollary 2 of Theorem 2 the centers of the rings A and A_n are isomorphic if A is an LA-ring.

Proposition 8. *The following rings are LA-rings:*

1) *rings in which for every element there is a left and a right identity;*

2) *rings with identity;*

3) *regular rings;*

4) *biregular rings;*

5) *o-rings;*

6) *f-regular rings of Blair* [17] (*i.e. the strongly idempotent rings of V. A. Andrunakievič* [18]).

Proof. If A is a regular ring, then $\overline{a}a$ is a right and $a\overline{a}$ is a left identity for the element a if $\overline{a}$ is any generalized inverse for a. If A is a biregular ring, then e will be a two-sided identity for a if e is a central idempotent such that $(a)_f = (e)_f$. Hence it is evident that the classes of rings 2)–5) are included in class 1). Let B be an ideal of the ring A, let $b \in B$, and let i_1 and i_2 be respectively right and left identities for the element b. Then $b = i_1 b i_2 \in B^0$. It follows that $B \subset B^0$ in a ring satisfying condition 1), and by Proposition 5 such a ring is an LA-ring. f-regular rings are characterized by the fact that if B is an ideal of such a ring then $B^{2+} = B$. Hence it is evident that $B^{3+} = B$, and since $B^{3+} \subset B^0$, we get that $B^0 = B$. By Proposition 5, f-regular rings are LA-rings.

Corollary 1. *Every ring can be embedded as an ideal in an LA-ring.*

The proof follows from case 2) of Proposition 8 and from the well-known fact that every ring can be imbedded as an ideal in a ring with identity.

Let A be an LA-ring. Then A will be a proper ideal of this ring, i.e. $A^{3+} = A$. From this, as is easy to see, it follows that if A is an LA-ring, then $A^{2+} = A$.

We also note that a homomorphic image of a full matrix LA-ring will be a full matrix LA-ring. In fact, let A_n be an LA-ring, B' an ideal of it. Then by Theorem 4 $B' = B_n$, where B is an ideal of A and the factor ring A_n/B' is clearly isomorphic to the ring $(A/B)_n$. By Proposition 7 and Corollary 2 of Theorem 4, $(A/B)_n$ is an LA-ring.

It is clear that if B is an ideal of A, then $B_n \cap S(A_n) = S(B_n)$. If the ring $S(A_n)$ of scalar matrices isomorphic to the ring A is identified with A, then we get $B_n \cap A = B$; that is, the ideal B_n of A_n induces the ideal B in A.

Proposition 9. *If an ideal B of an LA-ring A is generated by a subset C of A, then the ideal B_n of A_n is generated by the set $S(C_n)$ of all scalar matrices whose main diagonals consist of elements from C.*

Proof. Let D' be the ideal generated by $S(C_n)$. Then $D' = D_n$ by Theorem 4. Since $S(C_n) \subset D_n$, $C \subset D$. Since D is an ideal (Lemma 3), $B \subset D$. Since $C \subset B$, $S(C_n) \subset B_n$. Since B_n is an ideal, $D_n \subset B_n$. Consequently $D' = B_n$.

Corollary 1. *Every ideal of an LA-ring A_n is generated by some set of scalar matrices of this ring.*

Proposition 10. *If an ideal B' of an LA-ring A_n is generated by a family of matrices $\{(a_{ij}^k)\}_{k \in K}$, then $B' = B_n$, where B is the ideal of the ring A generated by the family of elements $\{a_{ij}^k\}$, $k \in K$, $1 \leq i, j \leq n$.*

We omit the virtually obvious proof of this proposition.

Professor L. A. Skornjakov read over the first part of the manuscript of this article and gave some valuable advice for which the author expresses his deep gratitude.

BIBLIOGRAPHY

[1] B. M. Šaĭn, *On the theory of generalized groups and generalized heaps*, Theory of Semigroups Appl. I, Izdat. Saratov. Univ., Saratov, 1965, pp. 286–324. (Russian)   MR 35 #283.

[2] J. von Neumann, *On regular rings*, Proc. Nat. Acad. Sci. U. S. A. 22 (1936), 707–713.

[3] L. A. Skornjakov, *Complemented modular lattices and regular rings*, Fizmatgiz, Moscow, 1961; English transl., Oliver and Boyd, London, 1964.   MR 29 #3404; 30 #42.

[4] K. K. Fryer and I. Halperin, *The von Neumann coordinatization theorem for complemented modular lattices*, Acta Sci. Math. 17 (1956), 203–249. MR 20 #12.

[5] J. von Neumann, *Continuous geometry*, Princeton Math. Series, no. 25, Princeton Univ. Press, Princeton, N. J., 1960.   MR 22 #10931.

[6] A. H. Clifford, *Semigroups admitting relative inverses*, Ann. of Math. (2) 42 (1941), 1037–1049.   MR 3, 199.

[7] A. E. Liber, *On the theory of generalized groups*, Dokl. Akad. Nauk SSSR 97 (1954), 25–28. (Russian)   MR 16, 9.

[8] G. Birkhoff, *Lattice theory*, Amer. Math. Soc. Colloq. Publ., vol 25, rev. ed., Amer. Math. Soc., Providence, R. I., 1948; Russian transl., IL, Moscow, 1952.   MR 10, 673.

[9] A. Forsythe and N. H. McCoy, *On the commutativity of certain rings*, Bull. Amer. Math. Soc. 52 (1946), 523–526.   MR 7, 509.

[10] G. Thierrin, *Contribution à la théorie des anneaux et des demi-groupes*, Comment. Math. Helv. 32 (1957), 93–112.   MR 19, 1158.

[11] R. F. Arens and I. Kaplansky, *Topological representation of algebras*, Trans. Amer. Math. Soc. 63 (1948), 457–481.   MR 10, 7.

[12] N. Jacobson, *Structure of rings*, Amer. Math. Soc. Colloq. Publ., vol. 37, Amer. Math. Soc., Providence, R. I., 1956; Russian transl., IL, Moscow, 1961.   MR 18, 373; 22 #11005.

[13] V. V. Vagner, *Generalized groups*, Dokl. Akad. Nauk SSSR 84 (1952), 1119–1122.   MR 14, 12.

[14] A. I. Gerčikov, *Rings decomposing into a direct sum of fields*, Mat. Sb. 7 (49) (1940), 591–597. (Russian)   MR 2, 121.

[15] D. R. Morrison, *Bi-regular rings and the ideal lattice isomorphisms*, Proc. Amer. Math. Soc. 6 (1955), 46–49.   MR 16, 669.

[16] V. A. Andrunakievič, *Biregular rings*, Mat. Sb. 39 (81) (1956), 447–464. (Russian)   MR 19, 244.

[17] R. L. Blair, *Ideal lattices and the structure of rings*, Trans. Amer. Math. Soc. 75 (1953), 136–153.   MR 15, 4.

[18] V. A. Andrunakievič, *Antisimple and strongly idempotent rings*, Izv. Akad. Nauk SSSR Ser. Mat. 21 (1957), 125–144. (Russian)   MR 19, 244.

Translated by:
G. Wagner

Amer. Math. Soc. Transl.
(2) Vol. 96, 1970

ON ČEBYŠEV AND ALMOST ČEBYŠEV SUBSPACES[*]

A. L. GARKAVI

Abstract. The subspaces of a Banach space having a unique element of best approximation for all of the elements of the space, as well as the subspaces for which the uniqueness of an element of best approximation is violated only for elements constituting a set of the first category, are studied. Questions concerning the existence of such subspaces are considered and some general propositions on isomorphic subspaces are established.

§1. Introduction

Let l be a subspace of a Banach space B. An element $y^* \in l$ is called an element of best approximation of $x \in B$ (or a closest element to x) if

$$\| x - y^* \| = E(l, x) = \inf_{y \in l} \| x - y \| .$$

The quantity $E(l, x)$ is called the best approximation of x by l.

The paper is devoted to questions connected with the problem of the uniqueness of an element of best approximation.

We will say that l is a *uniqueness subspace* for x (or has the property Ux) if it contains a unique element of best approximation of x. A subspace l is called a *Čebyšev subspace* if it has the property Ux for every $x \in B$. If B is strictly convex (i.e. strictly normed) then any finite-dimensional subspace of it is a Čebyšev subspace. But if B is not strictly convex then even if it is separable it cannot have a Čebyšev subspace of finite dimension. There also exist separable spaces which do not have a Čebyšev subspace of infinite dimension. The question concerning the existence of Čebyšev subspaces is taken up by us in more detail in §2. In particular, we give an example there of a (nonseparable) Banach space which does not have a Čebyšev subspace of either finite or infinite dimension.

In connection with this there arises the question of whether an arbitrary Banach space contains subspaces that are close to Čebyšev subspaces, i.e. subspaces for which a violation of the property Ux is in some definite

---

[*]Translation of Izv. Akad. Nauk SSSR Ser. Mat. 28 (1964), 799–818.

sense an exceptional phenomenon which occurs, for example, only for elements constituting a set of the first category in B.

A subspace l will be called an *almost Čebyšev subspace* if it has the
property Ux for all $x \in B$ except for elements constituting a set of the
first category in B.

It is clear that an investigation of questions concerning the existence
and uniqueness of elements of best approximation becomes essentially an
investigation of the geometric properties of the unit sphere of a Banach
space. Every notion of this circle of questions can be formulated in terms of
the geometry of the unit sphere. In particular, the property of a subspace l
being almost Čebyšev means that if the unit sphere is circumscribed by a
"cylinder" whose axis is the subspace l, then "almost all" of the "elements"
of the cylinder will have a unique point in common with the sphere.[*]

It is proved in the paper that for any reflexive subspace l_0 of a separable space B there exists an almost Čebyšev subspace that is isomorphic
to l_0.[**] It is proved that the conditions of reflexivity of l_0 and separability
of B cannot be dropped. It turns out in this connection that, in a certain
sense, "almost all" of the reflexive subspaces are almost Čebyšev.

In the case of a separable conjugate space B^* the condition of reflexivity of l_0 can be replaced by the requirement that it be regularly closed.

In a recent paper [14] S. B. Stečkin investigated the approximating properties of sets of very general form. He showed in particular that every closed
set in a uniformly convex space is almost Čebyšev. The question of the
existence of almost Čebyšev subspaces does not arise for such spaces since
every subspace in a uniformly convex space is a Čebyšev subspace.

In §2 we deal with questions on the existence of Čebyšev subspaces.

In §3 we establish a number of general propositions concerning isomorphic subspaces of a Banach space. The results of the section are used in
the sequel. Some of them may be of independent interest.

§4 contains theorems on the existence of almost Čebyšev subspaces.

The author intends to consider in another paper [15] finite-dimensional
almost Čebyšev subspaces in the space $C(Q)$ of continuous functions on a

---

[*] "Almost all" means with the exception of a set of the first category in the
space of cylinder elements metrized by means of the distance between them. The
meaning of the other words in quotes is obvious.

[**] A somewhat stronger result is actually obtained.

sequentially compact space Q. The main results obtained in this direction were formulated by the author in [3]. Here we note only that almost Čebyšev systems of functions (i.e. bases of almost Čebyšev subspaces in $C(Q)$) have a number of properties that are similar to properties of Čebyšev systems. In certain problems, for example, in interpolation problems on sequentially compact spaces, almost Čebyšev systems are to a certain extent capable of compensating for the absence of the Čebyšev systems that are normally used for interpolation on an interval.

The author wishes to express his gratitude to S. B. Stečkin for discussion on the proper formulation of the problems and on the results of the work in a seminar conducted by him on the theory of approximations.

§2. On the existence of Čebyšev subspaces

If a space B is strictly convex then any finite-dimensional subspace of it is a Čebyšev subspace. If B is not only strictly convex but also reflexive then any subspace of it is a Čebyšev subspace. In particular, every subspace of finite (deficiency) index in such a space is a Čebyšev subspace. The situation is different for spaces that are not strictly convex. In the accompanying table we give data on the existence of Čebyšev subspaces of finite dimension or of finite index in the classical Banach spaces that are not strictly convex. In the table we adopt the following notation, which is borrowed from [6].

$C(S)$ is the space of continuous functions on a compact space S.

$L_1(T, \Sigma, \mu)$ is the space of summable functions on a measure space (T, Σ, μ).

$L_\infty(T, \Sigma, \mu)$ is the space of essentially bounded functions on a measure space (T, Σ, μ).

The spaces l_1, c, c_0 and m are defined in the same way as by Banach [1].

N, S and NS respectively stand for necessary, sufficient and necessary and sufficient conditions for the existence of Čebyšev subspaces in a space of the given type. In this connection all of the sufficient conditions are essental, i.e. cannot be dropped. The sources are indicated in brackets.

We also note that if B^* is the conjugate of a separable space B then it always has Čebyšev subspaces of unit index (see [10, 4]). But it is not known whether such a space contains Čebyšev subspaces of index $n > 1$ (if B is not separable then B^* cannot have a Čebyšev subspace of finite index; see [4]). A little is known concerning the existence of Čebyšev subspaces of infinite dimension and of infinite index in spaces that are not strictly convex. We note in particular that a Čebyšev subspace $l \subset B$ with a reflexive factor space B/l can exist only if the dimension of the set of extremal points of a sphere in B is not less than the codimension of l (see [4]).

Space	Čebyšev subspaces of dimension n	Čebyšev subspaces of index n
1) $C(S)$	a) $n = 1$. Always exist. b) $n > 1$. NS: the space S is homeomorphic to a circular arc (see [8,12]).	a) $n = 1$. S: the space S is separable (see [4]). b) $n > 1$. S: the space S is countable (see [14]). N: S does not contain an infinite connected open set (see [4]).
in particular, c	exist	exist (see [14])
2) $L_1(T, \Sigma, \mu)$	NS: (T, Σ, μ) has $m \geq n$ atoms (see [7,11] for $m = 0$ and [14] for $m \geq 0$).	NS: (T, Σ, μ) has $m \geq n$ atoms (see [11] for $m = 0$ and [14] for $m \geq 0$).
in particular, l_1	exist (see [14])	exist (see [14])
3) c_0	exist (see [14])	do not exist (see [11])
4) $L_\infty(T, \Sigma, \mu)$	a) $n = 1$. Exist. b) $n > 1$. Do not exist (see [11]).	a) $n = 1$. S: μ is σ-finite (see [14]). b) $n > 1$. S: there exist n functions from $L_1(T, \Sigma, \mu)$ that are linearly independent to any set of measure (see [14]).
in particlar, m	a) $n = 1$. Exist. b) $n > 1$. Do not exist (see [11]).	exist (see [14])

To the above facts we add the following proposition.

Theorem I. *There exists a (nonseparable) Banach space B_0 which* *does not have a Čebyšev subspace.*[*]

Let T be a set of cardinality greater than the continuum (i.e. card $T > c$). We consider the space B_0 of bounded real functions $f(t)$ on T that are different from zero at not more than a countable number of points $t \in T$ and with norm

$$\| f \| = \sup_{t \in T} | f(t) |.$$

Let l be an arbitrary subspace of B_0. Two cases are possible: 1) card $l \leq c$ or 2) card $l > c$.

Since a continuum of countable sets has cardinality c, in the first case there exists a point $t_0 \in T$ at which all of the functions from l vanish. But then, clearly, an element of best approximation in l for the function

$$f_0(t) = \begin{cases} 1, & \text{if} \quad t = t_0, \\ 0, & \text{if} \quad t \neq t_0, \end{cases}$$

will not be unique.

Suppose now card $l > c$, $f \notin l$ and $p \in l$ is an element of best approximation of f. The difference $f(t) - p(t)$ is different from zero at not more than a countable number of points $A = \{t_1, t_2, \cdots\}$. But the cardinality of all bounded real functions defined on A is equal to c. Since card $l > c$ there exist two nonidentical functions p_1 and p_2 from l which coincide on A. But then the element $p(t) - \epsilon[p_1(t) - p_2(t)]$ will also be an element of best approximation of f for sufficiently small ϵ. Thus B_0 does not have a Čebyšev subspace.

As to separable Banach spaces, it can be seen from the table that even such classical spaces as $L[0, 1]$ and $C(P)$ (where P is a square) do not have Čebyšev subspaces of finite dimension (the latter with the exception of one-dimensional subspaces). To these well-known examples we can add the following proposition.

Theorem II. *There exists a separable Banach space which does not have a Čebyšev subspace of infinite dimension.*

An example of such a space is the space c_0 of sequences converging to zero.

For suppose l is an infinite subspace of c_0, $a = \{a_1, a_2, \cdots\} \notin l$ and

---

[*]Excluding, of course, the trivial subspaces: B_0 itself and the zero element.

$\beta^0 = \{\beta_1^0, \beta_2^0, \cdots\} \in l$ is an element of best approximation of α. It is clear that an N and a $\delta > 0$ exist such that

$$|\alpha_n - \beta_n^0| < \|\alpha - \beta^0\| - \delta$$

for $n > N$. Inasmuch as the set of all sequences $\{\beta_1, \beta_2, \cdots, \beta_N\}$ has dimension N while the subspace l is infinite dimensional, there exists a nonzero element $\beta' = \{\beta_1', \beta_2', \cdots\} \in l$ for which $\beta_n' = 0$ $(n = 1, 2, \cdots, N)$. But then the element $\beta^0 + \epsilon\beta'$ will also be an element of best approximation of α for sufficiently small ϵ.

§3. On isomorphic subspaces

As is well known, a natural generalization of the notion of dimension of finite-dimensional spaces is the notion of isomorphic spaces (in the sense of Banach [1]). But a direct application of this notion to subspaces of one and the same space leads to a number of difficulties. For example, two subspaces of different indices can turn out to be isomorphic. On the other hand, the annihilators[*] of isomorphic subspaces need not be isomorphic. In connection with this we introduce the following definition.

Two subspaces l_1 and l_2 of a Banach space B are said to be *B-isomorphic* if there exists an isomorphism T of B onto itself which maps l_1 onto l_2.

B-isomorphic subspaces are obviously also isomorphic in the usual sense. We also recall the following notions.

A subspace l of the conjugate space B^* is said to be regularly closed if it is closed in the topology $\sigma(B^*, B)$. An operator $\widetilde{T}$ which maps B^* into itself is said to be regularly continuous if it is continuous in the topology $\sigma(B^*, B)$. An isomorphism $\widetilde{T}$ of B^* onto itself is regularly continuous if and only if it is the adjoint of an isomorphism T of B onto itself, i.e. $\widetilde{T} = T^*$ (see [6]). It follows that a regularly continuous isomorphism is regularly bicontinuous.

Two subspaces l_1 and l_2 of the conjugate space B^* will be said to be *regularly B^*-isomorphic* if there exists a regularly continuous isomorphism T^* of B^* onto itself which maps l_1 onto l_2.

Lemma I. *Two subspaces l_1 and l_2 of B are B-isomorphic if and only if their annihilators $l_1^\perp$ and $l_2^\perp$ are regularly B^*-isomorphic.*

---

[*]The annihilator $l^\perp$ of a subspace $l \subset B$ is the set of all functionals $f \in B^*$ such that $f(x) = 0$ if $x \in l$.

For suppose an isomorphism T of B onto itself maps l_1 onto l_2. Then its adjoint T^* is an isomorphism of B^* onto itself. In this connection, if $f \in l_2^{\perp}$ then

$$T^* f(x) = f(Tx) = 0$$

for each $x \in l_1$. Consequently, $T^* f \in l_1^{\perp}$. And if $\phi \in l_1^{\perp}$ then

$$T^{*-1} \varphi(x) = \varphi(T^{-1}x) = 0$$

for each $x \in l_2$. Hence $T^{*-1} \phi \in l_2^{\perp}$ and $l_1^{\perp}$ and $l_2^{\perp}$ are regularly B^*-isomorphic. The converse assertion is proved analogously.

Lemma II. *Suppose l_1 and l_2 are subspaces of B and suppose the index (dimension) of one of them is finite. Then l_1 and l_2 are B-isomorphic if and only if their indices (dimensions) are the same.*

Suppose l_1 and l_2 are subspaces of index $n < \infty$ which are respectively given by the systems of equations

$$f_i(x) = 0 \qquad (i = 1, \ldots, n)$$

and

$$\varphi_i(x) = 0 \qquad (i = 1, \ldots, n),$$

where $x \in B$, $\phi_i \in B^*$ and $\|f_i\| = \|\phi_i\| = 1$. We assume that exactly $n - m$ $(1 \le m \le n)$ of the functionals f_i linearly depend on the functionals ϕ_i. It can be assumed without loss of generality that the functionals $f_1, \cdots, f_m$, $\phi_1, \cdots \phi_m$ are linearly independent. But it is then possible to choose m points $z_1, \cdots, z_m$ in B which satisfy the conditions

$$f_i(z_k) - \varphi_i(z_k) = 0, \\ \varphi_i(z_k) = \begin{cases} 0 & \text{for } i \neq k, \\ 1 & \text{for } i = k \end{cases} \quad (i, k = 1, 2, \ldots, m) \tag{1}$$

We define an operator

$$y = Tx \equiv x + [f_1(x) - \varphi_1(x)] z_1 + \ldots + [f_m(x) - \varphi_m(x)] z_m. \tag{2}$$

Using the conditions (1), it is not difficult to verify that the operator T is one-to-one and that $f_i(x) = 0$ $(i = 1, \cdots, n)$ if and only if $\phi_i(y) = 0$ $(i = 1, \cdots, n)$. Thus T is actually an isomorphism of B onto itself which takes l_1 into l_2.

Suppose now l_1 and l_2 are B-isomorphic subspaces and the index n of the first of them is finite (i.e. def $l_1 = n < \infty$). By Lemma I their annihilators $l_1^{\perp}$ and $l_2^{\perp}$ are isomorphic and

$$\dim l_1^{\perp} = \operatorname{def} l_1 = n.$$

But then

$$\dim l_2^{\perp} = \dim l_1^{\perp} = n$$

and hence

$$\operatorname{def} l_2 = \dim l_2^{\perp} = n.$$

The case of finite-dimensional subspaces l_1 and l_2 can be reduced to the case of subspaces of finite indices by considering their annihilators $l_1^{\perp}$ and $l_2^{\perp}$ and constructing a regularly continuous isomorphism in B^* that is analogous to (2).

Lemma III. *Suppose l_1 and l_2 are two hypersubspaces* of B given by the equations $f(x) = 0$ and $\phi(x) = 0$ ($\|f\| = \|\phi\| = 1$). There exists an isomorphism T of B onto itself which takes l_1 into l_2 and is such that*

$$\max \{ \|T\|,\ \|T^{-1}\| \} \leqslant 1 + 4 \|f - \varphi\|. \tag{3}$$

It is seen from the proof of Lemma II that the isomorphism

$$y = Tx \equiv x + [f(x) - \varphi(x)] z \tag{4}$$

maps l_1 onto l_2 if

$$\left. \begin{array}{r} f(z) - \varphi(z) = 0, \\ \varphi(z) = 1. \end{array} \right\} \tag{5}$$

Here $\|T\| \leq 1 + \|z\| \, \|f - \phi\|$. Let us show that, for any f and ϕ ($\|f\| = \|\phi\| = 1$), z can be chosen so that $\|z\| \leq 3 + \epsilon$ (where $\epsilon > 0$ is arbitrarily small). Since the functionals f and ϕ are defined except for sign and at least one of the following inequalities is always satisfied:

$$\|f + \lambda\varphi\| \geqslant \|f\| = \|\varphi\|, \quad \|f - \lambda\varphi\| \geqslant \|f\| = \|\varphi\| \quad (\lambda > 0),$$

we will assume that

---

*I.e. subspaces of unit index.

$$\| f + \lambda \varphi \| \geqslant \| f \| = \| \varphi \| \quad \text{for } \lambda > 0. \tag{6}$$

Using Helly's Theorem (see [6], p. 86 (Russian p. 100)), we conclude that the system (5) has a solution z whose norm satisfies the condition

$$\frac{1}{\| z \| - \varepsilon} = \min_{\alpha} \| \alpha f + (1 - \alpha) \varphi \|.$$

Since $\| f \| = \| \phi \|$ it is clear that this minimum is achieved for $0 \leq \alpha \leq 1$. By taking (6) into account it is easily verified that

$$\| \alpha f + (1 - \alpha) \varphi \| \geqslant \min \{ \alpha, | 1 - 2\alpha | \}$$

for $0 \leq \alpha \leq 1$. Hence

$$\frac{1}{\| z \| - \varepsilon} \geqslant \min_{0 \leqslant \alpha \leqslant 1} \min \{ \alpha, | 1 - 2\alpha | \} = \frac{1}{3},$$

so that $\| z \| \leq 3 + \epsilon \leq 4$. Under such a choice of z we have

$$\| T \| \leqslant 1 + 4 \| f - \varphi \|.$$

We go over to the inverse operator T^{-1}. Clearly,

$$x = T^{-1} y \equiv y - [f(x) - \varphi(x)] z.$$

But it is easily verified that by virtue of (5)

$$f(x) - \varphi(x) = f(y) - \varphi(y).$$

Thus

$$x = T^{-1} y \equiv y - [f(y) - \varphi(y)] z,$$

and we again obtain

$$\| T^{-1} \| \leqslant 1 + 4 \| f - \varphi \|,$$

which proves estimate (3).

Lemma III a. *Suppose under the conditions of Lemma* III *that* B *is a subspace of a Banach space* B_1 . *Then the subspaces* l_1 *and* l_2 *are* B_1*-isomorphic and estimate* (3), *in which* T *is an isomorphism of* B_1 *onto itself, remains valid.*

For the proof it suffices to extend the functional $f - \phi$ onto all of B_1

without increasing its norm.

Lemma III[*]. *Suppose* l_1 *and* l_2 *are regularly closed hyperspaces of* B^* *given by the equations* $x(f) = 0$ *and* $y(f) = 0$ $(x, y \in B, \|x\| = \|y\| = 1, f \in B^*)$. *There exists a regularly continuous isomorphism* T^* *of* B^* *onto itself which takes* l_1 *into* l_2 *and is such that*

$$\max\{\|T^*\|, \|T^{*-1}\|\} \leqslant 1 + 3\|x - y\|.$$

The proof is essentially analogous to the proof of Lemma III. The desired isomorphism has the form

$$\varphi = T^*f = f + [x(f) - y(f)]\,\psi, \tag{7}$$

where

$$\varphi, \psi \in B^*, \quad \psi(x - y) = 0,$$
$$\psi(y) = 1, \quad \|\psi\| \leqslant 3 \tag{8}$$

(the possibility of the latter inequality is proved by applying, in place of Helly's theorem, Hahn's theorem; see [6], p. 86 (Russian p. 100)).

Lemma III[*] **a.** *Suppose under the conditions of Lemma* **III**[*] *that* B^* *is a subspace of* B_1^* *and* $B = B_1/^\perp B^*$, *where* $^\perp B^*$ *is the annihilator of* B^* *in* B_1. *Then* l_1 *and* l_2 *are regularly* B_1^*-*isomorphic and there exists a regularly continuous isomorphism* T_1^* *of* B_1^* *onto itself which maps* l_1 *onto* l_2 *and is such that*

$$\max\{\|T_1^*\|, \|T_1^{*-1}\|\} \leqslant 1 + 4\|X - Y\|,$$

where $X, Y \in B = B_1/^\perp B^*$.

If $B = B_1/^\perp B^*$ then the isomorphism (7) takes the form

$$T^*f = f + [X(f) - Y(f)]\,\psi,$$

where $X, Y \in B_1/^\perp B^*$. The definition of a factor space and of the norm of its elements implies the existence of a $z \in B_1$ such that $X(f) - Y(f) = z(f)$ for all $f \in B^*$ and such that $\|z\| \leq \|X - Y\| + \epsilon$. Putting

$$T_1^*f = f + z(f)\,\psi \quad (f \in B_1^*) \tag{9}$$

we get

$$\|T_1^*\| \leqslant 1 + 3[\|X - Y\| + \varepsilon] \leqslant 1 + 4\|X - Y\|;$$

the same estimate is also valid for the inverse operator.

One can easily verify that T_1^* is one-to-one. In addition, it is regularly continuous since it is the adjoint of the operator

$$Tu = u + z \psi (u) \qquad (u \in B_1).$$

And, since T_1^* coincides with T^* on B^*, it maps l_1 onto l_2.

Let l_0 be a fixed subspace of a Banach space B. We consider the set $\{l\}$ of all of the subspaces of B that are B-isomorphic to l_0. We introduce in this set a metric ρ defined by the equality

$$\rho (l_1, l_2) = \inf_{T_{1,2}} \left\{ \sup_{x \neq \theta} \left\| \frac{x}{\|x\|} - \frac{T_{1,2}x}{\|T_{1,2}x\|} \right\| + \ln \max (\|T_{1,2}\|, \|T_{1,2}^{-1}\|) \right\}, \quad (10)$$

where the infimum is taken over all of the isomorphisms $T_{1,2}$ of B onto itself that map l_1 onto l_2.

The metric axioms are easily verified. The symmetry axiom follows from the fact that the set $\{T_{2,1}\}$ of isomorphisms taking l_2 into l_1 coincides with the set $\{T_{1,2}^{-1}\}$. The triangle axiom also follows from elementary considerations. One need only make use of the fact that if $T_{1,2}$ takes l_1 into l_2 and $T_{2,3}$ takes l_2 into l_3, then the operator $T_{1,3} = T_{2,3} T_{1,2}$ takes l_1 into l_3 and

$$\|T_{1,3}\| \leqslant \|T_{2,3}\| \|T_{1,2}\|.$$

The set $\{l\}$ of subspaces that are B-isomorphic to l_0, when equipped with the metric ρ, forms a metric space which we will denote by $L(B, l_0)$. A set of B-isomorphic subspaces can also be metrized with the use of the following metric:

$$\widetilde{\rho} (l_1, l_2) = \max \left\{ \sup_{x \in l_1 - \theta} \inf_{y \in l_2 - \theta} \left\| \frac{x}{\|x\|} - \frac{y}{\|y\|} \right\|, \quad \sup_{y \in l_2 - \theta} \inf_{x \in l_1 - \theta} \left\| \frac{y}{\|y\|} - \frac{x}{\|x\|} \right\| \right\}.$$

The resultant metric space will be denoted by $\widetilde{L}(B, l_0)$. The metric $\widetilde{\rho}$ was considered by I. C. Gohberg and A. S. Markus in [5], where it was introduced in the set of all subspaces of a space B. The resultant metric space $\widetilde{L}(B)$ turned out to be complete.

It is easily seen that

$$\widetilde{\rho} (l_1, l_2) \leqslant \rho (l_1, l_2). \qquad (11)$$

It can also be shown that ρ and $\tilde{\rho}$ are topologically equivalent when either the dimension or the index of l_0 is finite.

Theorem III. *The metric space $L(B, l_0)$ is complete.*

Let $l_1, l_2, \cdots$ be a fundamental sequence of subspaces. By virtue of (11) and the completeness of $\tilde{L}(B)$ it converges in the metric $\tilde{\rho}$ to a subspace l. We will show that l is B-isomorphic to l_1 (and hence to l_0). The fundamentalness of the sequence $\{l_n\}$ and the definition of the metric ρ imply the existence of isomorphisms $T_{n,n+1}$ of B onto itself which take l_n into l_{n+1} and are such that

$$\left.\begin{aligned} \|T_{n,n+1}\| &\leqslant 1 + \alpha_n, \\ \|T_{n,n+1}^{-1}\| &\leqslant 1 + \alpha_n, \end{aligned}\right\} \tag{12}$$

$$\sup_{x \neq \theta} \left\| \frac{x}{\|x\|} - \frac{T_{n,n+1}x}{\|T_{n,n+1}x\|} \right\| \leqslant \alpha_n, \tag{13}$$

where $\alpha_n \to 0$. It can be assumed without loss of generality that the α_n tend to 0 so fast that the following conditions are satisfied:

$$\sum_{k=n}^{\infty} \alpha_n = \beta_n \to 0 \quad \text{for} \quad n \to \infty, \tag{14}$$

$$\prod_{k=n}^{\infty} (1 + \alpha_n) = 1 + \gamma_n \to 1 \quad \text{for} \quad n \to \infty. \tag{15}$$

We consider the operator $U_n = \Pi_{k=n-1}^{1} T_{k,k+1}$ and show that there exists a constant C such that for all $x \in B$ and $n = 1, 2, \cdots$

$$\frac{1}{C} \|x\| \leqslant \|U_n x\| \leqslant C \|x\|. \tag{16}$$

In fact, by virtue of (12) and (15)

$$\|U_n\| \leqslant \prod_{k=1}^{n-1} \|T_{k,k+1}\| \leqslant \prod_{k=1}^{n-1} (1 + \alpha_k) \leqslant \prod_{k=1}^{\infty} (1 + \alpha_k) \leqslant C.$$

Analogously,

$$\|U_n^{-1}\| \leqslant C,$$

which implies (16).

Let us show that the limit

$$y = Ux = \lim U_n x$$

exists for each $x \in B$ and that $y \in l$ if $x \in l_1$.

Excluding the trivial case $x = \theta$ and putting $y_k = U_k x$, we get

$$\left\| \frac{y_n}{\|y_n\|} - \frac{y_m}{\|y_m\|} \right\| \leqslant \sum_{k=n}^{m-1} \left\| \frac{y_k}{\|y_k\|} - \frac{y_{k+1}}{\|y_{k+1}\|} \right\| \leqslant \sum_{k=n}^{m-1} \alpha_k \leqslant \beta_n \quad (m > n). \quad (17)$$

On the other hand,

$$y_m = \prod_{k=m-1}^{n} T_{k,k+1}\, y_n \quad (m > n),$$

and hence by virtue of (12) and (15)

$$\|y_m\| \leqslant \prod_{k=m-1}^{n} (1 + \alpha_n)\, \|y_n\| \leqslant (1 + \gamma_n)\, \|y_n\|.$$

Analogously,

$$\|y_n\| \leqslant \prod_{k=n}^{m-1} \|T_{k,k+1}^{-1}\|\, \|y_m\| \leqslant (1 + \gamma_n)\, \|y_m\|.$$

Therefore

$$\left\| \frac{y_n}{\|y_n\|} - \frac{y_m}{\|y_m\|} \right\| = \left\| \frac{y_n}{\|y_n\|} - \frac{y_m}{\|y_n\|(1 + \bar{\gamma}_n)} \right\|,$$

where $|\bar{\gamma}_n| \leq \gamma_n$. Hence, taking (15)–(17) into account,

$$\|y_n - y_m\| \leqslant C\,(\beta_n + \beta_n \gamma_n + \gamma_n)\, \|x\| \to 0 \quad (m > n,\ n \to \infty). \quad (18)$$

Thus the sequence $y_n = U_n x$ has a limit. In this connection, if $x \in l_1$ then $y_n \in l_n$. But l is the totality of the limits of all convergent sequences $\{z_n\}$, where $z_n \in l_n$ (see [5]). Hence $\lim y_n = y \in l$ if $x \in l_1$. Consequently the operator $Ux = \lim U_n x$ acting in B maps l_1 into l. Passing to the limit, we get

$$\frac{1}{C}\,\|x\| \leqslant \|Ux\| \leqslant C\,\|x\|,$$

which means that U is a one-to-one linear operator. Hence it takes l_1 into a closed subspace $\tilde{l} \subset l$. We show that $\tilde{l}$ coincides with l. Suppose $y_0 \in l$ but $y_0 \notin \tilde{l}$. Let r be the distance of the point $y_0/\|y_0\|$ from the subspace $l\,(r > 0)$. For sufficiently large n there exists by virtue of the definition of l a point $\bar{y}_n \in l_n$ such that

$$\left\| \frac{y_0}{\|y_0\|} - \frac{\bar{y}_n}{\|\bar{y}_n\|} \right\| < \frac{r}{2}.$$

It can be assumed in this connection that $\beta_n < r/2$. Let

$$\bar{x} = U_n^{-1}(\bar{y}_n) \quad (\bar{x} \in l_1),$$
$$\bar{y}_k = U_k \bar{x}, \quad \bar{y} = U\bar{x} \in \tilde{l},$$

Then

$$\left\| \frac{y_0}{\|y_0\|} - \frac{\bar{y}}{\|\bar{y}\|} \right\| \leq \left\| \frac{y_0}{\|y_0\|} - \frac{\bar{y}_n}{\|\bar{y}_n\|} \right\| + \left\| \frac{\bar{y}_n}{\|\bar{y}_n\|} - \frac{\bar{y}}{\|\bar{y}\|} \right\|$$

$$\leq \frac{r}{2} + \sum_{k=n}^{\infty} \left\| \frac{y_k}{\|y_k\|} - \frac{y_{k+1}}{\|y_{k+1}\|} \right\| \leq \frac{r}{2} + \beta_n < r,$$

which contradicts the definition of r, inasmuch as $\bar{y}/\|\bar{y}\| \in \tilde{l}$. Thus the iso-
morphism U of B onto itself maps l_1 onto l, and hence l_1 and l are B-
isomorphic. Finally, one can easily verify that the sequence $\{l_n\}$ also converges to l in the metric ρ. In order to see that $\rho(l, l_n) \to 0$ it suffices to
consider the operators $V_n = \lim_{m \to \infty} V_{n,m}$, where

$$V_{n,m} = \prod_{k=m}^{n} T_{k,k+1} \quad (m > n),$$

and again make use of estimates (12) and (13).

Lemma IV. *Suppose the conditions of Lemma* II *are satisfied. Then*

$$\rho(l_1, l_2) \leq K \|J - \varphi\|, \tag{19}$$

where K *is an absolute constant.*

Inequality (19) is proved by directly calculating the expression in braces
in the right side of (10) after substituting in it the isomorphism (4) extended
onto all of the space B_1 in the manner indicated in Lemma III a. In this connection one should make use of estimate (3).

Suppose now l_0 is a regularly closed subspace of B^*. We consider the
set $L^*(B^*, l_0)$ consisting of all of the subspaces of B^* that are regularly
B^*-isomorphic to l_0 (and hence regularly closed). We introduce in this set
a metric ρ^* defined by a relation analogous to (10), with the only difference

being that the infimum in (10) is now taken only over regularly continuous isomorphisms $T^*_{1,2}$. Clearly, the following inequality, which is analogous to (11), holds for the metrics ρ^* and $\tilde{\rho}$:

$$\tilde{\rho}\,(l_1,\, l_2) \leqslant \rho^*\,(l_1,\, l_2). \tag{20}$$

Theorem III*. *The metric space* $L^*(B^*,\, l_0)$ *is complete.*

The proof is carried out in the same way as for Theorem II. One should only note that the operator

$$U = \lim\, U_n = \lim\, \prod_{k=n}^{1} T_{k,k+1}$$

considered in this theorem, which maps l_1 onto the limit subspace l, will in this case be regularly continuous (and hence l will be regularly closed). The regular continuity of the isomorphism U follows from the fact that by virtue of (18) U is the limit in the uniform operator topology of a sequence of regularly continuous operators $\{U_n\}$.

Lemma IV*. *Suppose the conditions of Lemma* III* a *are satisfied. Then*

$$\rho^*\,(l_1,\, l_2) \leqslant K\|X - Y\| \quad (X,\, Y \in B == B_1/^{\perp}B^*), \tag{21}$$

where K *is an absolute constant.*

As in Lemma IV, the proof is carried out by calculating the expression in braces in the right side of (10) after substituting the operator (9) in it.

§4. On the existence of almost Čebyšev subspaces

Let $S_B(r)$ denote the sphere $\|x\| = r$ of a Banach space B. To each hyperplane H of B supporting $S_B(r)$ there corresponds the equation

$$f\,(x) = r, \quad \|f\| = 1 \quad (x \in B,\, f \in B^*).$$

The set of all hyperplanes supporting $S_B(r)$ can be metrized by putting

$$\rho_H\,(H_1,\, H_2) = \|\,f_1 - f_2\,\|\,. \tag{22}$$

Lemma V. *If* B *is separable and reflexive then the set of hyperplanes tangent to* $S_B(r)$ *contains an everywhere dense (in the sense of the metric* ρ_H*) subset of hyperplanes each of which is tangent to* $S_B(r)$ *at a unique point.*

The lemma is a consequence of the following well-known theorem of Mazur [9].

If a Banach space E is separable, then there exists on $S_E(r)$ an everywhere dense (in $S_E(r)$) set of points, for each of which there exists a unique support hyperplane that passes through this point.[*]

In order to obtain Lemma V from this theorem it suffices to put $B^* = E$ and consider the space $E^* = B^{**} = B$, using the correspondence existing between the points of $S_{B^*}(t)$ and the hyperplanes tangent to $S_{B^{**}}(1)$.

Theorem IV. *For any reflexive subspace l_0 of a separable Banach space B the totality of almost Čebyšev subspaces that are B-isomorphic to l_0 forms an everywhere dense set of second category in the metric space $L(B, l_0)$.*

For convenience we divide the proof into several parts. In each part we prove the assertion stated at the beginning of the part.

a) *For any element $x_0 \in B$ the totality of subspaces $\{l\}$ from $L(B, l_0)$ having the property U_{x_0} forms an everywhere dense set in $L(B, l_0)$.*

Suppose l is an arbitrary subspace of B that is B-isomorphic to l_0, and $x_0 \in B$. We will show that for any $\epsilon > 0$ there exists a subspace $\tilde{l} \in L(B, l_0)$ having the property U_{x_0} and such that $\rho(l, \tilde{l}) < \epsilon$. Assuming $x_0 \notin l$, we consider the space M spanned by l and the element x_0. Since l is reflexive, the space M is also reflexive.

The subspace l is obviously a hypersubspace of M. Suppose it is given by the equation

$$f(x) = 0 \quad (f \in M^*, \quad \|f\| = 1).$$

Putting $\bar{x} = x - x_0 \, (x \in M)$, we get that $f(\bar{x}) = -f(x_0)$ is a tangent hyperplane to the sphere $\|\bar{x}\| = |f(x_0)|$, and by Lemma V there exists for any $\epsilon' > 0$ a hyperplane

$$\tilde{f}(\bar{x}) = -f(x_0) \quad (\|\tilde{f}\| = 1, \quad \tilde{f} \in M^*),$$

which is tangent to this sphere at a unique point and such that $\|f - \tilde{f}\| < \epsilon'$. Going over again to $x = \bar{x} + x_0$, we get that the subspace $\tilde{f}(x) = 0$ is tangent to the sphere $\|x - x_0\| = |f(x_0)| \, (x \in M)$ at a unique point with $\|f - \tilde{f}\| < \epsilon'$.

Thus the subspace $\tilde{l}$ described by the equation $\tilde{f}(x) = 0 \, (x \in M)$ has the property U_{x_0}, and it is clear that it retains this property independently

[*]Mazur showed that this set is actually of type G_δ, but this fact is not required by us.

of whether it is considered as a subspace of M or as a subspace of B. Also, since l and $\tilde{l}$ are hypersubspaces of $M \subset B$, they are B-isomorphic according to Lemma III a, and by virtue of Lemma IV

$$\rho(l, \tilde{l}) \leqslant K \| f - \tilde{f} \| \leqslant K\varepsilon' < \varepsilon.$$

Part a) is proved.

b) *For any element* $x_0 \in B$ *the totality of subspaces* $\{l\}$ *from* $L(B, l_0)$ *having the property* $U(x_0)$ *forms an everywhere dense set* $N(x_0)$ *of type* G_δ *(and hence of the second category) in the metric space* $L(B, l_0)$.

Let $\Delta(l) = \Delta(l, x_0)$ denote the set of elements in a subspace l that are closest to x_0. Further, let $\{f_s\}$ $(s = 1, 2, \cdots)$ denote a total set of functionals from B^*, which exists by virtue of the separability of B. Let $G_{s,m}$ denote the set of those subspaces $\{l\}$ from $L(B, l_0)$ for which

$$\sup_{x \in \Delta(l)} f_s(x) - \inf_{x \in \Delta(l)} f_s(x) \geqslant \frac{1}{m}.$$

The set

$$M(x_0) = \bigcup_{s,m=1}^{\infty} G_{s,m}$$

coincides by virtue of the totalness of the sequence $\{f_s\}$ with the complement of $N(x_0)$ in $L(B, l_0)$. We will show that the sets $G_{s,m}$ are closed. Suppose a sequence of subspaces $\{l_n\}$ belonging to $G_{s,m}$ converges in the metric ρ to a subspace l. We will show that $l \in G_{s,m}$. In each l_n, from the elements that are closest to x_0 we choose two elements x_n and $\bar{x}_n$ such that

$$f_s(x_n) - f_s(\bar{x}_n) \geqslant \frac{1}{m} - \frac{1}{n}.$$

Let y_n and $\bar{y}_n$ be elements of l that are closest to x_n and $\bar{x}_n$ respectively. Clearly,

$$\left.\begin{array}{l} \| x_n - y_n \| \leqslant \| x_n \| \rho(l_n, l) \leqslant 2 \| x_0 \| \rho(l_n, l), \\ \| \bar{x}_n - \bar{y}_n \| \leqslant \| \bar{x}_n \| \rho(l_n, l) \leqslant 2 \| x_0 \| \rho(l_n, l). \end{array}\right\} \tag{23}$$

The sequences $\{y_n\}$ and $\{\bar{y}_n\}$ are weakly sequentially compact since they are bounded and belong to a reflexive space. We will assume without loss of generality that they converge weakly to elements y and $\bar{y}$ respectively $(y, \bar{y} \in l)$. But then, inasmuch as $\rho(l_n, l) \to 0$ we get from (23) that the sequences $\{x_n\}$ and $\{\overline{x}_n\}$ also converge weakly to the elements y and $\overline{y}$ respectively. Therefore

$$f_s(y) - f_s(\overline{y}) = \lim_{n\to\infty} f_s(x_n) - \lim_{n\to\infty} f_s(\overline{x}_n) \geqslant \frac{1}{m_l}. \tag{24}$$

We will prove that y is an element in l that is closest to x_0. Let y^* be an element in l that is closest to x_0 and let x_n^* be an element in l_n that is closest to y^*. We then have

$$E(l, x_0) = \|x_0 - y^*\| \leqslant \|x_0 - y_n\| \leqslant \|x_0 - x_n\| + \|x_n - y_n\|$$
$$\leqslant E(l_n, x_0) + 2\|x_0\| \, p(l_n, l). \tag{25}$$

On the other hand,

$$E(l_n, x_0) = \|x_0 - x_n\| \leqslant \|x_0 - x_n^*\| \leqslant \|x_0 - y^*\| + \|y^* - x_n^*\|$$
$$\leqslant E(l, x_0) + \|y^*\| \, p(l_n, l) \leqslant E(l, x_0) + 2\|x_0\| \, p(l_n, l). \tag{26}$$

It follows from (25) and (26) that

$$\lim_{n\to\infty} E(l_n, x_0) = E(l, x_0),$$

i.e.

$$\lim_{n\to\infty} \|x_0 - x_n\| = E(l, x_0).$$

But, since $\{x_n\}$ converges weakly to y,

$$\|x_0 - y\| \leqslant \lim_{n\to\infty} \|x_0 - x_n\| = E(l, x_0),$$

and hence y really is an element in l that is closest to x_0. An analogous assertion holds in regard to the element $\overline{y}$. But then relation (24) shows that $l \in G_{s,m}$. Consequently the sets $G_{s,m}$ are closed, and their complements $CG_{s,m}$ are open. And, since the set $N(x_0)$ is the complement of $M(x_0)$,

$$N(x_0) = CM(x_0) = \bigcap_{s,\,m=1}^{\infty} CG_{s,m}.$$

Thus $N(x_0)$ is a set of type G_δ. Taking into account that the space $L(B, l_0)$ is complete (Theorem III) while the set $N(x_0)$ is everywhere dense in it (part a)), we conclude that $N(x_0)$ is a set of the second category while its complement $M(x_0)$ is a set of the first category.

c) *For any countable everywhere dense set C in B the totality of sub-spaces from $L(B, l_0)$ having the property Ux for each $x \in C$ forms an every-where dense set of the second category in the space $L(B, l_0)$.*

For if C consists of elements $x_1, x_2, \cdots$ then the indicated totality of subspaces is the intersection of the sets $N(x_1), N(x_2), \cdots$. And since the complement of each $N(x_k)$ is a set of the first category, $\bigcap_{k=1}^{\infty} N(x_k)$ is an everywhere dense set of the second category.

d) *Every subspace $l \subset B$ having the property Ux for each x of an every-where dense set in B is almost Čebyšev.*

Let $N(l)$ be the set of those elements $x \in B$ with respect to which a subspace l has the property Ux, and let $M(l)$ be the complement of $N(l)$ in B. Let $F_{s,m}$ denote the set of elements $x \in B$ for which

$$\sup_{y \in \Delta(l,x)} f_s(y) - \inf_{y \in \Delta(l,x)} f_s(y) \geq \frac{1}{m},$$

where f_s and $\Delta(l, x)$ have the same values as in part b). Clearly

$$M(l) = \bigcup_{s,\, m=1}^{\infty} F_{s,\,m}.$$

We will show that the sets $F_{s,m}$ are closed. Suppose a sequence $\{x_n\}$ $(x_n \in F_{s,m})$ converges to x_0. For each x_n, from the elements in l that are closest to it we choose two elements y_n and $\bar{y}_n$ such that

$$f_s(y_n) - f_s(\bar{y}_n) \geq \frac{1}{m} - \frac{1}{n}. \tag{27}$$

Inasmuch as l is reflexive it can be assumed without loss of generality that $\{y_n\}$ and $\{\bar{y}_n\}$ converge weakly to points y and $\bar{y}$ respectively. We show that y is an element in l that is closest to x_0. In fact,

$$E(l, x_n) \leq \|x_0 - y_n\| \leq \|x_0 - x_n\| + \|x_n - y_n\| = E(l, x_0) + \|x_0 - x_n\|.$$

On the other hand,

$$E(l_n, x_n) = \|x_n - y_n\| \leq \|x_n - x_0\| + \|x_0 - y_n\| \leq E(l, x_0) + \|x_0 - x_n\|.$$

These relations imply

$$\lim_{n \to \infty} \|x_n - y_n\| = \lim_{n \to \infty} E(l, x_n) = E(l, x_0).$$

But since the sequence $\{x_n\}$ converges weakly to x_0 while the sequence $\{y_n\}$ converges weakly to y, the sequence $\{x_n - y_n\}$ converges weakly to $x - y$ and therefore

$$\|x - y\| \leqslant \lim_{n \to \infty} \|x_n - y_n\| = E\,(l,\,x_0).$$

Consequently, y really is an element in l that is closest to x_0. An analogous conclusion can be made concerning the element $\bar{y}$. Passing to the limit in (27), we get

$$f_s\,(y) - f_s\,(\bar{y}) \geqslant \frac{1}{m}$$

and hence $x_0 \in F_{s,m}$. Arguing further in the same way as in part b), we see that $N\,(l)$ is an everywhere dense set in B of the second category while its complement $M\,(l)$ is a set of the first category. But this means that the subspace l is almost Čebyšev.

Propositions c) and d) obviously constitute the assertion of the theorem.

Corollary I. *In every separable Banach space there exist almost Čebyšev subspaces of any finite dimension.*

Corollary II. *For any subspace l_0 of a reflexive space B there exist almost Čebyšev subspaces that are B-isomorphic to it.*

As the following theorem shows, neither the condition of separability of B nor the condition of reflexivity of l_0 can be dropped in Theorem IV.

Theorem V. a) *There exists a nonseparable Banach space which does not have an almost Čebyšev subspace.*

b) *There exists a (nonreflexive) separable Banach space which does not have a nonreflexive almost Čebyšev subspace.*

An example of a space described by assertion a) is provided by the space B_0 considered in Theorem I. This is easily discerned from the proof of Theorem I. A corresponding example for assertion b) is provided by the space c_0. In fact, it is seen from the proof of Theorem II that this space does not have an infinite-dimensional almost Čebyšev subspace.

The condition of reflexivity of the subspace l_0 can be weakened in the case of the conjugate space B^*.

Theorem IV*. *For any regularly closed subspace l_0 of a separable conjugate space B^* the totality of almost Čebyšev subspaces that are regularly B^*-isomorphic to l_0 forms an everywhere dense set of the second category* *in the metric space* $L^*(B^*, l_0)$.

The proof follows the plan of the proof of Theorem IV. One need only make use of the following additional remarks.

In part a):

1) If the subspace l is regularly closed and $f_0 \notin l$ ($f_0 \in B^*$), then the subspace M spanned by l and f_0 is also regularly closed (see [2]).

2) The subspace M, being regularly closed, is the conjugate of the factor space $B/^{\perp}M$.

3) Lemma V remains valid if the space B is assumed to be a conjugate space and the set of all hyperplanes tangent to $S_B(r)$ is replaced by the set of regularly closed tangent hyperplanes.

4) Instead of Lemmas III a and IV one should make use of Lemmas III*a and IV*, and instead of inequality (11), inequality (20).

In parts b) and d):

5) As the countable total set of functionals one should take a countable everywhere dense set of elements of B.

6) Inasmuch as l is a regularly closed subspace of a separable conjugate space, the sequences of its elements corresponding to the sequences $\{y_n\}$ and $\{\bar{y}_n\}$ of Theorem IV are weakly sequentially compact and their limit points belong to l.

7) The completeness of the space $L^*(B^*, l_0)$ is established in Theorem III*.

Taking Lemmas I and II into consideration, we obtain from Theorem IV* the following

Corollary. *In every separable conjugate space there exist almost Čebyšev subspaces of any finite index.*

We note that neither the condition of separability nor the condition of "conjugateness" in this corollary can be dropped. For example, the space D of bounded real functions on $[0, 1]$ is a conjugate space but is not separable. This space does not have an almost Čebyšev subspace of finite index (cf. Theorem 9 in [4]). The space c_0 is separable but is not a conjugate space. It also does not have almost Čebyšev subspaces of finite index.*

We consider, finally, a space B^* that is the conjugate of a separable space B. A subspace $l \subset B^*$ will be called *weakly almost Čebyšev* if it has

---

*The following assertion is valid: if the unit sphere of a space B does not have extremal points, then this space does not contain a subspace l of finite index which has the property Ux for even one $x \in B$ ($x \notin l$).

the property Ux for all $x \in B$ except for elements constituting a set of the first category in the sense of the weak topology $\sigma(B^*, B)$. In this connection it is necessary to bear in mind that if B is separable then the space B^* equipped with the topology $\sigma(B^*, B)$ is metrizable and all of the notions of metric space theory can be applied to it (see [6]).

Theorem IV**. *For any regularly closed subspace* l_0 *of a space* B^* *that is the conjugate of a separable space* B *the totality of weakly almost Čebyšev subspaces that are regularly* B^*-*isomorphic to* l_0 *forms an everywhere dense set of the second category in the metric space* $L^*(B^*, l_0)$.

The proof follows the plan of Theorem IV with due regard for the remarks made in Theorem IV*. Additional changes must be made only in part c). Namely, as the set C one must take a sequence $f_1, f_2, \cdots (f_n \in B^*)$ that is everywhere dense in the topology $\sigma(B^*, B)$ and which exists by virtue of the weak separability of B^* (see [2]).

BIBLIOGRAPHY

[1] S. Banach, *Théorie des opérations linéaires*, Monografie Mat., PWN, Warsaw, 1932; reprint, Chelsea, New York, 1955; Ukrainian transl., "Radjans 'ka Skola", Kiev 1948. MR 17, 175.

[2] N. Bourbaki, *Espaces vectoriels topologiques*, Actualités Sci. Indust., nos. 1189, 1229, Hermann, Paris, 1953–1955; Russian transl., IL, Moscow, 1959. MR 14, 880; 17, 1109.

[3] A. L. Garkavi, *On Čebyšev and almost Čebyšev subspaces*, Dokl. Akad. Nauk SSSR 149 (1963), 1250–1252 = Soviet Math. Dokl. 4 (1963), 532–534. MR 26 #6737.

[4] ———, *On best approximation by the elements of infinite-dimensional subspaces of a certain class*, Mat. Sb. 62 (104) (1963), 104–120. (Russian) MR 27 #6076.

[5] I. C. Gohberg and A. S. Markus, *Two theorems on the gap between subspaces of a Banach space*, Usephi Mat. Nauk 14 (1959), no. 5 (89), 135–140. (Russian) MR 22 #5880.

[6] N. Dunford and J. T. Schwartz, *Linear operators. I: General theory*, Pure and Appl. Math., vol. 7, Interscience, New York, 1958; Russian transl., IL, Moscow, 1962. MR 22 #8302.

[7] M. G. Kreĭn, *The L-problem in an abstract normed linear space*, in N. I. Ahiezer and M. G. Kreĭn, *Some questions in the theory of moments*,

Naučn. Tehn. Izdat. Ukrain., Kharkov, 1938; pp. 171–199; English transl., Transl. Math. Monographs, vol. 2, Amer. Math. Soc., Providence, R. I., 1962, pp. 175–204. MR 29 #5073.

[8] J. C. Mairhuber, *On Haar's theorem concerning Chebychev approximation problems having unique solutions*, Proc. Amer. Math. Soc. 7 (1956), 609–615. MR **18**, 125.

[9] S. Mazur, *Über konvexe Mengen in linearen normierten Räumen*, Studia Math. 4 (1933), 70–84.

[10] D. Mil'man, *Accessible points of a functional compact set*, Dokl. Akad. Nauk SSSR 59 (1948), 1045–1048. (Russian) MR **9**, 449.

[11] R. R. Phelps, *Uniqueness of Hahn-Banach extensions and unique best approximation*, Trans. Amer. Math. Soc. 95 (1960), 238–255. MR **22** #3964.

[12] K. Sieklucki, *Topological properties of sets admitting the Tschebycheff systems*, Bull. Acad. Polon. Sci. Sér. Sci. Math. Astr. Phys. 6 (1958), 603–606. MR 20 #6625.

[13] S. B. Stečkin, *Approximation properties of sets in normed linear spaces*, Rev. Math. Pures Appl. 8 (1963), 5–18. (Russian) MR 27 #5108.

[14] A. L. Garkavi, *On the uniqueness of the L-problem of moments*, Izv. Akad. Nauk SSSR Ser. Mat. 28 (1964), 553–570. (Russian) MR **29** #1525.

[15] ———, *Almost Čebyšev systems of continuous functions*, Izv. Vysš. Učebn. Zaved. Matematika 1965, no. 2 (45), 36–44; English transl., Amer. Math. Soc. Transl. (2) 96 (1970), 177–187. MR 32 #1550.

Translated by:
S. Smith

Amer. Math. Soc. Transl.
(2) Vol. 96, 1970

ALMOST ČEBYŠEV SYSTEMS OF CONTINUOUS FUNCTIONS[*]

A. L. GARKAVI

The present article is a continuation of our paper [1], in which we investigated almost Čebyšev subspaces of arbitrary Banach spaces. Let us recall some definitions and results from that article. Let L be a subspace of a Banach space B. We shall say that this subspace has property Tf, if for the element f in B there exists a unique element in L of best approximation, that is, an element p^* in L such that $E(L, f) = \inf_{p \in L} \|f - p\| = \|f - p^*\|$. If the subspace L has property Tf for every f in B, then it is said to be *Čebyšev*. Čebyšev subspaces of a given dimension (in particular, finite dimensional) do not exist in every Banach space. Subspaces L having property Tf for all f in B, except possibly for a set of the first category, we shall call *almost Čebyšev*. If L has property Tf for the elements of a dense subset of a separable space B, then it is almost Čebyšev [1].

In [1] it was shown that if L_0 is any reflexive subspace of a separable space B then there exists an almost Čebyšev subspace $L \subset B$ that is isomorphic to L_0. Thus in each separable space there exist almost Čebyšev subspaces of arbitrary finite dimension. In the present article we study finite-dimensional almost Čebyšev subspaces of the space $C(Q)$ of all continuous functions $f(q)$ on a compact metric space Q, with $\|f\| = \max |f(q)|$.

Let $S_n = \{\phi_1(q), \cdots, \phi_n(q)\}$ be a basis for the subspace $L \subset C(Q)$. As usual, an element p in L will be called a polynomial in the system S_n. A basis of a Čebyšev (almost Čebyšev) subspace is called a Čebyšev (almost Čebyšev) system of functions. An infinite system $\{\phi_1, \phi_2, \cdots\}$ is called a Markov (almost Markov) system if for each $n < \infty$ the system $\{\phi_1, \cdots, \phi_n\}$ is Čebyšev (almost Čebyšev). By a theorem of Haar, the system S_n is Čebyšev if and only if each nontrivial polynomial in this system vanishes at at most $n - 1$ points of Q. As Mairhuber [3] showed, this condition can only be fulfilled when Q is homeomorphic to some closed subset of a circle. In this connection the study of almost Čebyšev systems in the space $C(Q)$ is very relevant, since such systems exist on every compact space.

In the present paper we establish the characteristic properties of almost Čebyšev systems and give a constructive proof of the existence of almost

---

[*] Translation of Izv. Vysš. Učebn. Zaved. Matematika 1965, no. 2 (45), 36–44.

Čebyšev and almost Markov systems on arbitrary compact metric spaces. We also establish certain properties of almost Čebyšev systems that are close to the analogous properties of Čebyšev systems. We indicate, in particular, the possibility of using almost Čebyšev systems in interpolation problems on compact spaces. The basic results of the article were stated without proof in the note [2].

Let us introduce some notation and definitions. 1. $L_n = L(Q, S_n)$ denotes the subspace of all polynomials in the system S_n, defined on the compact metric space Q. 2. $E = E(Q, S_n, f)$ denotes the best approximation to the function f by means of polynomials in the system S_n on Q [Translator's note: that is, E denotes the distance from f to L_n]. 3. $\Delta(Q, S_n, f)$ denotes the set of polynomials $t(q)$ in $L(Q, S_n)$ that furnish a best approximation to $f(q)$ on Q. 4. $T(Q, S_n)$ denotes the class of functions f in $C(Q)$ with respect to which the subspace $L(Q, S_n)$ has the property Tf; we shall call it the uniqueness class. 5. $S_r(q_0)$ denotes the open ball in the metric space Q with center at the point q_0 and radius r; that is, the set $\{q: \rho(q_0, q) < r\}$. 6. $\mathfrak{M}(Q, S_n, f, t)$ denotes the set of points of maximal deviation of the polynomial t from the function f; that is, the set of those points q for which $|f(q) - t(q)| = \|f - t\|$. 7. $\mathfrak{M}(Q, S_n, f)$ denotes the minimal subset of points of maximal deviation of the function f, defined by $\mathfrak{M}(Q, S_n, f) = \bigcap_{t \in \Delta(Q, S_n, f)} \mathfrak{M}(Q, S_n, f, t)$, where the intersection is taken over all polynomials of best approximation. 8. To each subset $G \subset Q$ we associate the number $N_n(G)$ equal to the number of points of G, if this number does not exceed n, and equal to n otherwise. In what follows we shall always assume that the compact space Q has at least n points.

We present some simple lemmas.

Lemma I. *The minimal subset* $\mathfrak{M} = \mathfrak{M}(Q, S_n, f)$ *has the following properties:*

a) *All polynomials in* $\Delta(Q, S_n, f)$ *agree on* $\mathfrak{M}$.

b) *There exists a polynomial* $t^*(q)$ *in* $\Delta(Q, S_n, f)$ *for which the set* $\mathfrak{M}(Q, S_n, f, t^*)$ *coincides with the set* $\mathfrak{M}(Q, S_n, f)$.

c) *If* $f(q)$ *is not in the uniqueness class* $T(Q, S_n)$, *then at least* $n - N_n(\mathfrak{M}) + 1$ *linearly independent polynomials of* $L(Q, S_n)$ *vanish identically on the set* $\mathfrak{M}$.

Lemma II. *Let* $f \in C(Q)$. *To each* $\epsilon > 0$ *there corresponds some* $\delta > 0$ *such that for any function* $\tilde{f}$ *in* $C(Q)$, *satisfying the condition* $\|f - \tilde{f}\| < \delta$, *and for any polynomial* $\tilde{t}(q)$ *in* $\Delta(Q, S_n, \tilde{f})$, *there exists a polynomial* $t(q)$ *in* $\Delta(Q, S_n, f)$, *such that* $\|t - \tilde{t}\| < \epsilon$.

The proofs of these lemmas use the usual methods of the theory of uniform approximation, and we shall omit them. We only note that in the proof of c) of Lemma I one must use a) and b).

Lemma III. *If for the points* $q_1, \cdots, q_n$ *of* Q *the determinant*

$$\left| \begin{array}{c} \varphi_1, \cdots, \varphi_n \\ \hline q_1, \cdots, q_n \end{array} \right| = \left| \begin{array}{ccc} \varphi_1(q_1) & \cdots & \varphi_n(q_1) \\ \cdot \cdot \cdot & \cdot & \cdot \cdot \cdot \\ \varphi_1(q_n) & \cdots & \varphi_n(q_n) \end{array} \right|$$

is different from zero, then for each $\epsilon > 0$ *there is a* $\delta > 0$ *such that each polynomial* $t(q)$ *in the system* $\{\phi_1, \cdots, \phi_n\}$ *that satisfies the conditions* $|t(q_n)| < \delta$ $(k = 1, \cdots, n)$ *also satisfies the inequality* $|t(q)| < \epsilon$ $(q \in Q)$.

For Čebyšev systems given on an interval, the proofs of these lemmas may be found, for example, in [4]. The proofs are the same in the case of an arbitrary system.

We turn now to the theorem that characterizes almost Čebyšev systems.

Theorem I. *The necessary and sufficient condition that the system* $S_n = \{\phi_1, \cdots, \phi_n\}$ *on the compact metric space* Q *be an almost Čebyšev system is that on each open subset* $G \subset Q$ *at most* $n - N_n(G)$ *linearly independent polynomials in* S_n *can vanish identically.*

Proof. Sufficiency. Let F_m be the set of all those functions f in $G(Q)$ such that from the set $\mathfrak{M}(Q, S_n, f)$ we can choose m points $q_1, \cdots, q_m$ for which the rank of the matrix

$$\left\| \begin{array}{c} \varphi_1, \cdots, \varphi_n \\ \hline q_1, \cdots, q_m \end{array} \right\| = \left(\begin{array}{ccc} \varphi_1(q_1) & \cdots & \varphi_1(q_m) \\ \cdot \cdot \cdot & \cdot & \cdot \cdot \cdot \\ \varphi_n(q_1) & \cdots & \varphi_n(q_m) \end{array} \right) \tag{1}$$

is equal to m. It follows from Lemma I (part c)) that $F_n \subset T(Q, S_n)$. If f is not in the uniqueness class $T(Q, S_n)$, then $f \in F_m$ for $0 \leq m \leq n - 1$, and $f \notin F_n$. The sufficiency of the condition will be shown if we can prove that for each f in F_m $(0 \leq m \leq n - 1)$ and every $\epsilon > 0$ there exists a function $\tilde{f} \in F_{m+1}$ for which $\|f - \tilde{f}\| < \epsilon$. Indeed, in this case the class F_n, and together with it $T(Q, S_n)$, will be dense in $C(Q)$, and consequently, in view of the separability of $C(Q)$, the subspace $L(Q, S_n)$ will be almost Čebyšev.

Let $f \in F_m$ and $f \notin F_{m+1}$. Without loss of generality we may assume that the set $\Delta = \Delta(Q, S_n, f)$ contains the identically zero function. In this case for q in $\mathfrak{M} = \mathfrak{M}(Q, S_n, f)$ we have $|f(q)| = E = E(Q, S_n, f)$ and $t(q) = 0$ $(t \in \Delta)$. Clearly under the hypotheses of the theorem the set $\mathfrak{M}$ for the function f contains at least one limit point q of Q. Indeed, otherwise there would exist $n - m$ linearly independent polynomials, each vanishing identically on an open subset consisting of $m + 1$ isolated points.

For definiteness we shall assume that $f(q_0) = + E$. Consider the function $\Phi(q) = \sup_{t \in \Delta} \{t(q)\}$. It is continuous, nonnegative, and $\Phi(q_0) = 0$. Also, for each point q in Q there exists a polynomial t_q in Δ such that $t_q(q) = \Phi(q)$. Let $q_1, \cdots, q_m$ be points for which the rank of the matrix (1) is equal to m. Taking $\epsilon < E$, choose $\delta > 0$ so small that the sphere $S_\delta(q_0)$ does not contain these points (except possibily for one that coincides with q_0) and such that for $q \in S_\delta(q_0)$ the following inequalities are fulfilled:

$$0 \leqslant \Phi(q) \leqslant \frac{\varepsilon}{2}, \quad 0 \leqslant E - f(q) \leqslant \frac{\varepsilon}{2} \qquad (\varepsilon < E). \qquad (2)$$

A point y can be chosen from the set $S_{\delta/2}(q_0)$ such that the rank of the matrix $\|\phi_1, \cdots, \phi_n/q_1, \cdots, q_m, y\|$ will be equal to $m + 1$. Indeed, if $|\phi_{n_1}, \cdots, \phi_{n_m}/q_1, \cdots, q_m|$ is a nonsingular minor of the matrix (1), then the nontrivial polynomial $p(q) = |\phi_{n_1}, \cdots, \phi_{n_m}, \phi_{n_m+1}/q_1, \cdots, q_m, y|$ by hypothesis does not vanish identically on the set $S_{\delta/2}(q_0)$, since this set is open and contains infinitely many points. Let $t_y(y) = \Phi(y)$. Construct a continuous function $\phi(q)$ such that

$$0 \leqslant \varphi(q) \leqslant 1,$$

$$\varphi(q) = \begin{cases} 1, & \text{if } q \in S_{\frac{\delta}{2}}(q_0), \\ 0, & \text{if } q \overline{\in} S_\delta(q_0). \end{cases}$$

Let R denote that part of the set $S_\delta(q_0)$ on which $[E + t_y(q)]\phi(q) \geq f(q)$, and consider the function

$$\tilde{f}(q) = \begin{cases} [E + t_y(q)]\varphi(q), & \text{if } q \in R, \\ f(q), & \text{if } q \overline{\in} R. \end{cases} \qquad (3)$$

Using (2) it is not difficult to verify that $\tilde{f}(q)$ is continuous and

$$\tilde{f}(q_0) = f(q_0) = E, \qquad (4)$$

$$\tilde{f}(y) = E + t_y(y), \qquad (5)$$

$$\|f - \tilde{f}\| < \varepsilon.$$

Let us show that

$$E(Q, S_n, \tilde{f}) \leqslant E. \qquad (6)$$

For points $q \notin R$ we have $|\tilde{f}(q) - t_y(q)| = |f(q) - t_y(q)| \leq E$. Now if $q \in R$, then $\tilde{f}(q) - t_y(q) = [E + t_y(q)]\phi(q) - t_y(q)$. If this difference is positive, it does not exceed $E + t_y(q) - t_y(q) = E$. If $[E + t_y(q)]\phi(q) \leq t_y(q)$, then again, in view of (2) and (3), we obtain

$$|\tilde{f}(q) - t_y(q)| \leqslant |E + t_y(q)|\varphi(q) + |t_y(q)| \leqslant 2|t_y(q)| \leqslant 2\,\frac{\varepsilon}{2} = \varepsilon < E.$$

Consider an arbitrary polynomial $\tilde{t}(q)$ in $\Delta(Q, S_n, \tilde{f})$. Since $\tilde{f}(q) = f(q)$ outside R, for q not in R we have

$$|f(q) - \tilde{t}(q)| \leqslant E(Q, S_n, \tilde{f}) \leqslant E. \tag{7}$$

For those points of R where $\tilde{t}(q) \leq t(q)$ we have by (2) $\tilde{f}(q) \geq f(q) \geq t(q) \geq \tilde{t}(q)$, and thus for these points

$$|f(q) - \tilde{t}(q)| \leqslant |\tilde{f}(q) - \tilde{t}(q)| \leqslant E. \tag{8}$$

For those points of R where $\tilde{t}(q) \geq t(q)$, for sufficiently small $\lambda > 0$ we have $t_y(q) \leq (1 - \lambda)t_y(q) + \lambda\tilde{t}(q) \leq f(q)$. Therefore, putting $r(q) = (1 - \lambda)t_y(q) + \lambda\tilde{t}(q)$, we have for these points

$$|f(q) - r(q)| \leqslant E. \tag{9}$$

But by (7) and (8), inequality (9) is also valid for all the remaining points of Q; consequently $r(q) \in \Delta(Q, S_n, f)$, and therefore

$$r(q_k) \equiv \lambda\tilde{t}(q_k) + (1 - \lambda)t_y(q_k) = 0 \quad (k = 1, \ldots, m),$$
$$r(y) \equiv \lambda\tilde{t}(y) + (1 - \lambda)t_y(y) \leqslant \Phi(y) = t_y(y). \tag{10}$$

From (10) we have

$$\tilde{t}(q_k) = 0 \quad (k = 1, \ldots, m), \tag{11}$$

$$\tilde{t}(y) \leqslant t_y(y). \tag{12}$$

but by (5), $\tilde{f}(y) = E + t_y(y)$; therefore $\tilde{t}(y) \geq t_y(y)$, since otherwise $|\tilde{f}(y) - \tilde{t}(y)| > E \geq E(Q, S_n, \tilde{f})$. Taking account of (5) and (12), we obtain the equations

$$\tilde{t}(y) = t_y(y) \quad \text{and} \quad |\tilde{f}(y) - \tilde{t}(y)| = E. \tag{13}$$

From (6) we have $E(Q, S_n, \tilde{f}) = E$. And since $q_k \notin R(q_k \neq q_0)$ and $f(q_0) = \tilde{f}(q_0)$, we have $f(q_k) = \tilde{f}(q_k) = \pm E$ $(k = 1, \cdots, m)$. Therefore from (11) and (13) it follows that $q_1, \cdots, q_m, y$ are the points of maximal deviation of the polynomial $\tilde{t}(q)$ from $\tilde{f}(q)$. Since $\tilde{t}(q)$ is an arbitrary polynomial in $\Delta(Q, S_n, \tilde{f})$, these points are in the minimal subset $\mathfrak{M}(Q, S_n, \tilde{f})$. Thus the rank of the matrix $\|\phi_1, \cdots, \phi_n/q_1, \cdots, q_m, y\|$ is equal to $m + 1$, and so $\tilde{f} \in F_{m+1}$.

Necessity. Let the open subset $G_1 \subset Q$ be such that, contrary to what is to be proved, the number of linearly independent polynomials in $L(Q, S_n)$ that vanish identically on G_1 is equal to $n - m_1 \geq n - N_n(G_1) + 1$. This is equivalent to the statement that the number of functions in the system S_n whose restrictions to G_1 are linearly independent is equal to $m_1 \leq N_n(G_1) - 1 < n$. Consider all possible open subsets $\{G_1'\}$ of G_1 that contain at least

m_1 points. Two cases are possible: a) on each subset G_1' the number of linearly independent functions is equal to m_1; b) among the subsets $\{G_1'\}$ there exists a subset G_2 on which the number of linearly independent functions is equal to $m_2 \leq m_1 - 1$. In this latter case we again consider all open subsets $\{G_2'\}$ of G_2 that contain at least m_2 points. As before two cases are possible. In the case analogous to b) we select an open subset G_3 of G_2 on which the number of linearly independent functions is equal to $m_3 \leq m_2 - 1$. Since $n > m_1 > m_2 > \cdots$, in a finite number of steps we shall come to one of the following results.

A) On some open subset G, containing at least $m + 1$ points, the number of linearly independent functions is equal to $m > 0$. These functions remain linearly independent on each open subset G' of G containing at least m points.

B) On the open subset G, containing $m \geq 1$ points, there is no linearly independent function of the system S_n (that is, all the functions vanish identically on G).

Case B) is trivial and we do not discuss it. Let us turn to case A). For convenience we shall assume that on G the functions $\phi_1, \cdots, \phi_m$ are linearly independent, and $\phi_{m+1}, \cdots, \phi_n$ vanish identically. We shall show that in case A) we can choose $m + 1$ points $q_1, \cdots, q_{m+1}$ of G such that all the following determinants are different from zero:

$$D_k \equiv | \varphi_1, \ldots, \varphi_m / q_1, \ldots, q_{k-1}, q_{k+1}, \ldots, q_{m+1} |.$$
$$(k = 1, \ldots, m + 1).$$

If the subset G contains at least $m + 1$ isolated points of Q, then for $\{q_k\}$ we may select any $m + 1$ such points. At the same time the determinants D_k will be different from zero, since any m isolated points constitute an open subset on which, by case A), the system $\{\phi_1, \cdots, \phi_m\}$ is linearly independent.

Suppose now that the number of isolated points in the subset G does not exceed m, and that $q_0 \in G$ is a limit point of Q. Choose $\delta_0 > 0$ so small that the ball $S_{\delta_0}(q_0)$ is entirely contained in G and contains no isolated points of Q. Since $S_{\delta_0}(q_0)$ is an open subset containing infinitely many points, it follows from hypothesis A) that the system S_m is linearly independent on this set. By the same reasoning S_m will be linearly independent on every ball $S_\delta(q)$ for any $\delta > 0$ and any point $q \in S_{\delta_0}(q_0)$. In other words, there is no nontrivial polynomial in the system S_m that vanishes identically on any ball $S_\delta(q)$ $(q \in S_{\delta_0}(q_0))$. Choose m points

$q_1, \cdots, q_m$ in $S_{\delta_0}(q_0)$ such that the determinants

$$D_{m+1} \equiv |\varphi_1, \ldots, \varphi_m / q_1, \ldots, q_m| \tag{14}$$

will be different from zero, and consider the m nontrivial polynomials

$$D_k(q) = |\varphi_1, \ldots, \varphi_m / q_1, \ldots, q_{k-1}, q, q_{k+1}, \ldots, q_m|$$
$$(k = 1, \ldots, m).$$

By the reasoning that has been presented and the continuity of $D_k(q)$ it follows that in any ball $S_\delta(q)$ $(q \in S_{\delta_0}(q_0))$, for each polynomial $D_k(q)$ one can find a ball $S_{\delta_1}(\bar{q}) \subset S_\delta(q)$ at each point of which that polynomial is different from zero. Consequently there exists a point $q_{m+1} \in S_{\delta_0}(q_0)$ at which all the polynomials $D_k(q)$ are different from zero. Thus, taking account of (14), we obtain

$$D_k \equiv |\varphi_1, \ldots, \varphi_m / q_1, \ldots, q_{k-1}, q_{k+1}, \ldots, q_{m+1}| \neq 0$$
$$(k = 1, \ldots, m + 1), \tag{15}$$

which proves the above assertion.

Let us now consider the compact space Q_0 consisting of the $m + 1$ points $q_1, \cdots, q_{m+1}$ that we have chosen. Choose a function $f(q)$ on these $m + 1$ points that does not agree identically on Q_0 with any polynomial of the system S_m. Condition (15) says that the system S_m is a Čebyšev system for the compact set Q_0. Therefore there is a unique polynomial $t_0(q) \in \Delta(Q_0, S_m, f)$ for the function $f(q)$. Without loss of generality we may assume that $t_0(q) = 0$ for all q in Q_0, so that $\max_{q \in Q_0} |f(q)| = E(Q_0, S_m, f) = E$. Construct on Q a function $F(q)$ satisfying the conditions

$$\|F\| \leqslant E,$$
$$F(q) = \begin{cases} f(q), & \text{if } q \in Q_0, \\ 0, & \text{if } q \in Q - G. \end{cases}$$

Since $\phi_{m+1}(q) = \cdots = \phi_n(q) = 0$ if $q \in Q_0$, and $\|F\| \leq E$, it clearly follows that the set $\Delta(Q, S_n, F)$ contains the identically zero function and that $E(Q, S_n, F) = E(Q_0, S_m, f)$. Further, all polynomials $t(q) \in \Delta(Q, S_n, F)$ vanish at the points of Q_0.

Now let $\tilde{F}$ be an arbitrary function in $C(Q)$ that satisfies the condition $\|F - \tilde{F}\| < \delta$, where $\delta > 0$ does not exceed $E/4$ and is so small that $E(Q, S_n, \tilde{F}) \geq E/2$. If $\epsilon > 0$ is given then δ may be so chosen that for the polynomial

$$\tilde{t}(q) = a_1 \varphi_1(q) + \ldots + a_m \varphi_m(q) + \ldots + a_n \varphi_n(q) \in \Delta(Q, S_n, \tilde{F})$$

there exists, by Lemma II, a polynomial $t(q) \in \Delta(Q, S_n, F)$ for which

$$\|t - \tilde{t}\| < \epsilon. \tag{16}$$

It follows from (16) that if $q \in Q_0$, then $|\widehat{t}(q)| = |a_1 \phi_1(q) + \cdots$
$\cdots + a_m \phi_m(q)| < \epsilon$, and therefore, if ϵ is sufficiently small, then by Lemma III on the entire compact space Q we will have

$$|a_1 \varphi_1(q) + \cdots + a_m \varphi_m(q)| \leqslant \frac{1}{4} E. \tag{17}$$

Let us estimate the deviation of the polynomial $p(q) = a_1 \phi_1(q) + \cdots$
$\cdots + a_m \phi_m(q)$ from the function $\widetilde{F}$ on Q. If $q \in G$, then $\phi_{m+1}(q) = \cdots$
$\cdots = \phi_n(q) = 0$, and $|\widetilde{F}(q) - p(q)| = |\widetilde{F}(q) - \widehat{t}(q)| \leq E(Q, S_n, \widetilde{F})$. If $q \in Q - G$, then by (17) we have

$$|\widetilde{F}(q) - p(q)| \leqslant |\widetilde{F}(q)| + |p(q)| \leqslant \epsilon + \frac{1}{4} E \leqslant \frac{1}{4} E + \frac{1}{4} E = \frac{1}{2} E.$$

But then $|\widetilde{F}(q) - p(q) - \lambda \phi_{m+1}(q)| = |\widetilde{F}(q) - p(q)| \leq E(Q, S_n, \widetilde{F})$ if $q \in G$, and

$$|\widetilde{F}(q) - p(q) - \lambda \varphi_{m+1}(q) \leqslant |\widetilde{F}(q) - p(q)|$$
$$+ |\lambda| \, \|\varphi_{m+1}\| \leqslant \frac{1}{2} E + |\lambda| \, \|\varphi_{m+1}\| \leqslant \frac{3}{4} E < E(Q, S_n, \widetilde{F}),$$

if $q \in Q - G$ and $|\lambda| < E/4\|\phi_{m+1}\|$. Thus for sufficiently small λ we have $p(q) + \lambda \phi_{m+1}(q) \in \Delta(Q, S_n, \widetilde{F})$ and, since $\widetilde{F}$ is any function satisfying the condition $\|F - \widetilde{F}\| < \delta$, the subspace $L(Q, S_n)$ is not almost Čebyšev. The theorem is proved.

In the case of a perfect compact set the condition of the theorem can obviously be simplified and it then becomes the "strong" linear independence of the functions $\phi_1, \cdots, \phi_n$.

We now turn to the question of the existence of almost Čebyšev systems. We must first establish

Lemma IV. *On any compact metric space Q one can construct a continuous function that is not identically constant on any open subset containing more than one point.*

Let $X = \{x_k\}$ be a countable dense subset of Q and let M_0 be the set of isolated points of Q. Clearly $M_0 \subset X$. On M_0 we define a function $f_0(q)$ by putting $f_0(x_k) = 1/k$, and extend f_0 by continuity to the closure $\overline{M}_0$ (so that $f(q) = 0$ if q is a limit point of M_0). Now we extend the function $f_0(q)$ to all of Q by defining it outside of $\overline{M}_0$ by means of the equation $f_0(q) = \rho(\overline{M}_0, q)$, where $\rho(\overline{M}_0, q)$ denotes the distance from the point q to the set $\overline{M}_0$. Let $Q^* = Q - \overline{M}_0$. The set Q^* is open and contains no isolated points. Let E_0 denote the union of all open subsets of Q^* on each of which $f_0(q)$ is identically constant; that is $E_0 = \mathbf{U}_{-\infty < c < \infty}(Q^* - \overline{Q^* - P_c})$, where $P_c = \{q \in Q^* : f_0(q) = c\}$. Let x_{k_0} be the first point in the sequence $\{x_k\}$ that belongs to the set E_0. Put $M_1 = (Q - E_0) \cup x_{k_0}$, and construct the function $f_1(q) = f_0(q) + \rho(M_1, q)/1^2$.

Let E_1 be the union of all open subsets of Q^* on each of which $f_1(q) =$ const. Since $f_1(q) = f_0(q)$ outside of E_0, we have $E_1 \subset E_0$. Also, $x_{k_0} \notin E_1$. Indeed, otherwise there would exist a ball $S_\delta(x_{k_0})$ on which $f_0(q) \equiv$ const and $f_1(q) \equiv$ const. But if δ is sufficiently small, then for points $q \in S_\delta(x_{k_0})$ we have $\rho(M_1, q) = \rho(x_{k_0}, q)$, and consequently

$$f_1(q) = f_0(q) + \frac{\rho(M_1, q)}{1^2} = \text{const} + \rho(x_{k_0}, q) \qquad (q \in S(x_{k_0})).$$

But since x_{k_0} is a limit point of Q it follows that $\rho(x_{k_0}, q) \neq$ const on the ball $S_\delta(x_{k_0})$. Thus $x_{k_0} \notin E_1$. Let x_{k_1} be the first point of the sequence $\{x_k\}$ that belongs to E_1. By what has been said, $k_1 \geq k_0 + 1$. Again we put $M_2 = (Q - E_1) \cup x_{k_1}$ and

$$f_2(q) = f_1(q) + \frac{\rho(M_2, q)}{2^2} = f_0(q) + \frac{\rho(M_1, q)}{1^2} + \frac{\rho(M_2, q)}{2^2},$$

and consider the union E_2 of all the open subsets of Q^* on each of which $f_2(q) \equiv$ const. Continuing this process, we see that in a finite or countable number of steps we will reach a situation in which none of the points $\{x_k\}$ lies in the set E_N, which is the union of all the open subsets of Q^* on each of which the continuous function

$$f_N(q) = f_0(q) + \sum_{i=1}^{N} \frac{\rho(M_i, q)}{i^2} \qquad (N \leqslant \infty)$$

is constant. Therefore E_N is empty. Recalling the way in which $f_0(q)$ was defined on $\overline{M}_0 = Q - Q^*$, we see that $f_N(q)$ has the desired property.

Theorem II. *Let $f(q)$ be a continuous function on the compact metric space Q that is not constant on any open subset of Q containing more than one point. Then the system of functions $\{1, f(q), [f(q)]^2, \cdots\}$ is an almost Markov system on Q.*

We must show that the system $S_n = \{1, f, \cdots, f^{n-1}\}$ is almost Čebyšev. Let G be an open subset of Q. Consider the two possible cases.

1) The subset G consists of a finite number of isolated points. Since for any n isolated points $x_1, \cdots, x_n$ the values $f(x_1), \cdots, f(x_n)$ are all distinct, it follows that the determinant $|1, f, \cdots, f^{n-1}/x_1, \cdots, x_n|$ is different from zero and therefore the system S_n is a Čebyšev system on the set M_0 of isolated points of Q. Therefore the number of linearly independent polynomials in the system S_n that vanish identically on the subset $G \subset M_0$ does not exceed $n - N_n(G)$.

2) The subset G contains infinitely many points. Then $N_n(G) = n$ and we must show that on G we cannot have the identity

$$a_0 + a_1 f(q) + \ldots + a_{n-1}[f(q)]^{n-1} \equiv 0. \tag{18}$$

But the left side of (18) can vanish at most for $n - 1$ values of the function $f(q)$. And in this case, as is easily seen, the function $f(q)$, being continuous, must be constant on some infinite open subset of G, which is impossible.

As is known, Čebyšev systems S_n are characterized by the fact that for any n points of Q every Lagrange interpolation problem has a solution in the system S_n. If Q is not homeomorphic to a subset of the circle, then it does not have any Čebyšev systems. In this connection it is intersesting to consider interpolation properties of almost Čebyšev systems.

Let Q^n be the space of all systems of n points $\omega = \{q_1, \cdots, q_n\}$ $(q_k \in Q)$ with the metric $p(\omega_1, \omega_2) = \max_{1 \le k \le n} \rho(q_k^1, q_k^2)$, where $\omega_i = \{q_k^i\}$. From the characterization of almost Čebyšev systems given in Theorem 1 it is not difficult to deduce that the system $S_n = \{\phi_1, \cdots, \phi_n\}$ is almost Čebyšev if and only if the determinant

$$F(\omega) \equiv |\varphi_1, \ldots, \varphi_n / q_1, \ldots, q_n|$$

is different from zero on a dense subset Λ of the space Q^n. Clearly every Lagrange interpolation problem on the set Λ can be solved in the system S_n. Bearing this in mind and using the continuity of $F(\omega)$ on Q^n, we obtain the following proposition.

Theorem III. *Let S_n be an almost Čebyšev system of functions on the compact metric space Q. Then the systems of points $\{q_1, \cdots, q_n\}$ at which Lagrange interpolation may not always be possible in the system S_n form only a closed nowhere dense subset of the space Q^n.*

We mention that the theorem of P. P. Korovkin [4] on the completeness of a Markov system on a certain perfect subset $P \subset Q$ may be carried over to almost Markov systems.

In conclusion let us note that Theorem 1 remains valid for the space $C(S)$ of all continuous functions on a compact Hausdorff space S. However in this case the space $C(S)$ may not contain any almost Čebyšev systems $S_n (n > 1)$. It is easy to verify, for example, that if the cardinality of the set of isolated points of S is greater than the continuum, then the space $C(S)$ will not have any almost Čebyšev systems containing more than one function.

BIBLIOGRAPHY

[1]   A. L. Garkavi, *On Čebyšev and almost-Čebyšev subspaces,* Izv. Akad.

Nauk SSSR Ser. Mat. **28** (1964), 799–818; English transl., Amer. Math. Soc. Transl. (2) **96** (1970), 153–175   MR **29** #2635.

[2] ————, *On Čebyšev and almost-Čebyšev subspaces*, Dokl. Akad. Nauk SSSR **149** (1963), 1250–1252 = Soviet Math. Dokl. **4** (1963), 532–534. MR **26** #6737.

[3] J. C. Mairhuber, *On Haar's theorem concerning Čebyšev approximation problems having unique solutions*, Proc. Amer. Math. Soc. **7** (1956), 609–615.   MR **18**, 125.

[4] P. P. Korovkin, *On the closure of Čebyšev functions*, Dokl. Akad. Nauk SSSR **78** (1951), 853–855. (Russian)   MR **13**, 126.

Translated by:
A. Shields

Amer. Math. Soc. Transl.
(2) Vol. 96, 1970

ON MULTIPLICATIVE REPRESENTATIONS OF
J-NONEXPANSIVE OPERATOR-FUNCTIONS. I[*]

Ju. P. GINZBURG

It is known that a function $f(\zeta)$ $(|f(\zeta)| < 1)$ that is analytic for $|\zeta| < 1$ admits the representation

$$f(\zeta) = e^{i\alpha}\prod_k \frac{\zeta_k - \zeta}{1 - \overline{\zeta}_k\zeta} \cdot \frac{|\zeta_k|}{\zeta_k} \exp\left\{\int_0^{2\pi} \frac{\zeta + e^{i\theta}}{\zeta - e^{i\theta}} d\sigma(\theta)\right\} \tag{1}$$

in which $\operatorname{Im}\alpha = 0$, $|\zeta_k| < 1$, $\Sigma_k(1 - |\zeta_k|) < \infty$ and $\sigma(\theta)$ $(\sigma(\theta - 0) = \sigma(\theta))$ is a nondecreasing function. This classical result served as the starting point for the deep investigations undertaken by V. P. Potapov [1,2] on the multiplicative structure of analytic J-nonexpansive (finite-dimensional) matrix functions $Y(\zeta)$ $(|\zeta| < 1, Y^*(\zeta)JY(\zeta) \leq J, J^2 = I, J^* = J, \det Y(\zeta) \neq 0)$. Potapov proved that every such matrix function can be represented in the form

$$Y(\zeta) = U\overset{\frown}{\int}_0^l \exp\left[\frac{\zeta + e^{i\theta(t)}}{\zeta - e^{i\theta(t)}}\right]dE(t) \overset{\frown}{\prod}_k \left[\left[\frac{\zeta_k - \zeta}{1 - \overline{\zeta}_k\zeta} \cdot \frac{|\zeta_k|}{\zeta_k}\right]^{\pm 1} P_k + Q_k\right] \tag{2}$$

Here $U^*JU = J$, $P_k^2 = P_k$, $\pm JP_k \geq 0$, $Q_k = I - P_k$, $\theta(t)$ is a nondecreasing scalar function, $JE(t)$ is a Hermitian-increasing matrix function and $\operatorname{tr} JE(t) \equiv t$.

One of the main results of the present article is a generalization of a theorem of V. P. Potapov on the class $K_J^{\mathfrak{S}}$ defined in §2 of (infinite-dimensional) operator functions (the theorems of §§5 and 8; these propositions were obtained in a less finished form in [3] and cited without proof in the note [4]).

It is known that the parameters of the representation (1) are uniquely determined by the function $f(\zeta)$. In the article we study the uniqueness problem for representations of form (2) (§§9 and 10); but a complete solution of it is obtained only in the case $J = I$ (these results were announced in [5,6]).

---

[*] Translation of Mat. Issled. 2 (1967), no. 2, 52–83.

190   Ju. P. GINZBURG

$\S 6$ is devoted to a study of the convergence of a product of Blaschke type for the more general (than $K_J^{\mathfrak{G}}$) class M_J.

The theorems of $\S\S 1-4$ (obtained in the finite-dimensional case by Potapov [2]) and of $\S 7$ are auxiliary theorems for proving the main results of the paper. In $\S 1$ we give only the formulations of the theorems; detailed proofs are presented in [7].

The connections existing between the theory of multiplicative representations and the theory of nonselfadjoint operators has permitted a number of authors to, on the one hand, obtain important information on the properties of such operators ([8-12] and others), and, on the other hand, study the multiplicative structure of the characteristic functions of operators that are similar in a well defined sense to selfadjoint and unitary operators [8,9,13-16]. The totality of operator functions studied by this method is very extensive, but it is not contained in, nor does it contain, the class $K_J^{\mathfrak{G}}$.

In contrast to the above-mentioned papers [8,9,13-16] our arguments do not make use of facts from the theory of characteristic functions of nonself-adjoint operators. In a number of places our presentation is similar to the purely analytic constructions of Potapov.

The first part of the paper, which is published here, contains $\S\S 1-6$. The second part ($\S\S 7-10$) will be published in the next issue of "Matematičeskie Issledovanija."

$\S 1$. J-nonexpansive operator functions

1. Let $\mathfrak{H}$ be a Hilbert space, P_+ an orthogonal projection in $\mathfrak{H}$ and $P_- = I - P_+$. We put $J = P_+ - P_-$. A linear operator U mapping $\mathfrak{H}$ onto itself is called J-unitary if

$$U^* J U = J. \tag{1.1}$$

It is shown in [17] that a J-unitary operator is necessarily bounded. It is obvious that the adjoint U^* of a J-unitary operator U is also J-unitary.

A bounded linear operator A is called J-Hermitian if $JA = A^* J$, J-non-expansive if

$$A^* J A \leq J \tag{1.2}$$

and J-binonexpansive when both A and A^* are J-nonexpansive. If $A = I - T$, where T is a compact operator (such an operator A is called a <u>Fredholm</u> operator), then (1.2) implies that A is J-binonexpansive [18].[1]

1) A detailed study of J-nonexpansive and more general operators has recently been carried out in [19-23].

2. Let $Y(\zeta)$ be an operator function that is holomorphic interior to the unit circle with the possible exception of a set of isolated points and that is equal to a *J*-binonexpansive operator at each point of holomorphy. The set of such operator functions is denoted by S_J. The following five theorems are proved in [7].

Theorem 1.1. *If* $Y(\zeta) \in S_J$ *then the following inequality holds for any* λ, μ $(|\lambda| < 1, |\mu| < 1)$ *in the domain of holomorphy of* $Y(\zeta)$ *and any* $f, g \in \mathfrak{H}$:

$$\frac{|([Y(\lambda) - Y(\mu)]f, g)|^2}{|\lambda - \mu|^2} \leq \frac{([J - Y^*(\lambda)JY(\lambda)]f, f)\, ([J - Y(\mu)JY^*(\mu)]g, g)}{(1 - |\lambda|^2)\, (1 - |\mu|^2)}. \tag{1.3}$$

The main object of our attention will be the class K_J consisting of those operator functions $Y(\zeta) \in S_J$ for each of which there exists a point $\zeta_0,\ |\zeta_0| < 1$, such that $Y(\zeta_0)$ has a bounded inverse and the operator $J - Y^*(\zeta_0)JY(\zeta_0)$ is compact.

Theorem 1.2. *If* $Y(\zeta) \in K_J$ *then*

1) $Y(\zeta) = U\tilde{Y}(\zeta)$,

where U *is a constant J-unitary operator and* $\tilde{Y}(\zeta)$ *is equal to a Fredholm operator at each point of holomorphy of* $Y(\zeta)$, *while the operator* $\tilde{Y}(\zeta_0)$ *is J-Hermitian and has a positive spectrum, and*

2) $Y(\zeta)$ *and* $Y^{-1}(\zeta)$ *do not have any singularities interior to the unit circle except poles. The leading coefficient of the Laurent series in a neighborhood of such a pole is a finite-dimensional operator.*

Theorem 1.2 implies in particular that a study of the functions of the class K_J essentially reduces to a study of *J*-nonexpansive functions $Y(\zeta)$ that take Fredholm values and are such that $Y(0)$ is a *J*-Hermitian operator with a positive spectrum.

Theorem 1.3. *Suppose* $Y(\zeta)$ $(\in K_J)$ *has a pole of order* $k > 0$ *at the point* ζ_0.[1] *Then the following representation holds:*

$$Y(\zeta) = Y_1(\zeta)\left[\frac{1 - \bar{\zeta}_0\zeta}{\zeta_0 - \zeta}\, P + Q\right],$$

where P *is a finite-dimensional idempotent* $(P^2 = P)$ *J-Hermitian-negative* $(JP \leq 0)$ *operator,* $Q = I - P$ *and* $Y_1(\zeta) \in K_J$ *and has a pole of order* $k - 1$

---

1) I.e. $Y(\zeta) = A_{-k}(\zeta - \zeta_0)^{-k} + A_{-k+1}(\zeta - \zeta_0)^{-k+1} + \cdots + A_0 + \cdots$, where $A_{-k} \neq 0$.

at ζ_0. If ζ_0 is a pole of order $m \geq 0$ of the operator function $Y^{-1}(\zeta)$ then the order of the pole of $Y_1^{-1}(\zeta)$ at this point is equal to m.

Theorem 1.4. *If* $Y(\zeta) \in K_J$ *and* $Y^{-1}(\zeta)$ *has a pole of order* $k > 0$ *at the point* ζ_0, *then, for* $|\zeta| < 1$, $Y(\zeta)$ *admits the representation*

$$Y(\zeta) = Y_1(\zeta)\left[\frac{\zeta_0 - \zeta}{1 - \bar{\zeta}_0\zeta} \; P + Q\right],$$

where P *is a finite-dimensional idempotent* J-*Hermitian-positive* $(JP \geq 0)$ *operator,* $Y_1(\zeta) \in K_J$ *and* $Y_1^{-1}(\zeta)$ *has a pole of order* $k - 1$ *at* ζ_0. *If* ζ_0 *is a pole of order* $m \geq 0$ *of the operator function* $Y(\zeta)$ *then* $Y_1(\zeta)$ *also has a pole of order* m *at this point.*

Theorem 1.5.[1] *Suppose* $Y(\zeta)$ $(\in K_J)$ *is holomorphic and* J-*unitary on an arc* Γ *of the unit circle except at the point* ζ_0, *where it has a pole of order* k. *Then*

$$Y(\zeta) = Y_1(\zeta)\left[I + \frac{\zeta + \zeta_0}{\zeta - \zeta_0} \; G\right],$$

where $Y_1(\zeta) \in K_J$ *and has a pole of order* $k - 1$ *at* ζ_0 *while* G *is a compact* J-*Hermitian-positive nilpotent* $(G^2 = 0)$ *operator. In addition,* $\overline{Q\mathfrak{H}} = \overline{A\mathfrak{H}}$, *where* A *is the leading Laurent coefficient of* $Y(\zeta)$ *at* ζ_0.

It is easily seen that under the conditions of the theorem $Y^{-1}(\zeta)$ also has a pole of order k at ζ_0 while the order of the pole of $Y_1^{-1}(\zeta)$ at this point is equal to $k - 1$. This is implied by the following symmetry principle.

Theorem 1.6. *If an operator function* $F(\zeta)$ *is holomorphic in a domain* D *that is symmetric with respect to the unit circle and takes* J-*unitary values on an arc* Γ *of this circle, then the operator* $F^{-1}(\zeta)$ *exists and is bounded for each* $\zeta \in D$ *and the following formula holds:*

$$F^{-1}(\zeta) = JF^*(\bar{\zeta}^{-1})J.$$

In particular, if ζ_0 *is a pole of* $F(\zeta)$ *of order* k *then* $F^{-1}(\zeta)$ *has a pole of the same order at* $\bar{\zeta}_0^{-1}$.

The proof follows directly from the fact that the holomorphic (in D)

---

1) We note that Theorems 1.3–1.5 are proved in [7] for operator functions of the class M_J, which is wider than K_J. In the same place are indicated formulas which connect the operators P and G appearing in these theorems with the coefficients in the Laurent expansions of $Y(\zeta)$ and $Y^{-1}(\zeta)$.

operator functions

$$\Phi_1(\zeta) = F^*(\bar{\zeta}^{-1})JF(\zeta) - J, \quad \Phi_2(\zeta) = F(\zeta)JF^*(\bar{\zeta}^{-1}) - J$$

vanish on Γ and hence are identically equal to zero.

3. We consider the class R_J consisting of those operator functions $Y(\zeta) \in K_J$ which satisfy the following conditions: 1) $Y(\zeta)$ is holomorphic in the extended plane with the exception of a finite number of points at which it has poles; 2) $Y(\zeta)$ takes J-unitary values at the points of holomorphy lying on the unit circle; and 3) the operator $I - Y(\zeta_0)$ is finite dimensional for some point of holomorphy ζ_0 ($|\zeta_0| < 1$) of $Y(\zeta)$. It follows from Theorem 1.6 that if $Y(\zeta) \in R_J$ then $Y^{-1}(\zeta)$ also has only a finite number of poles in the complex plane.

It is not difficult to see that when $Y(\zeta) \in R_J$ the operator $J - Y(\zeta_0)JY^*(\zeta_0) = T$ is also finite dimensional. Let $\mathfrak{L}$ denote the range of T and $\mathfrak{L}^\perp = \mathfrak{H} \ominus \mathfrak{L}$. Then $([J - Y(\zeta_0)JY^*(\zeta_0)]g, g) = 0$ for any $g \in \mathfrak{L}^\perp$ and it follows from inequality (1.3) that $[Y(\zeta) - Y(\zeta_0)]f \in \mathfrak{L}$ for any $f \in \mathfrak{H}$ if ζ is a point of holomorphy of $Y(\zeta)$. Thus *if* $Y(\zeta) \in R_J$ *then the ranges of all of the operators* $I - Y(\zeta)$ *belong to one and the same finite-dimensional subspace.*

We consider the operator functions (below U is a J-unitary operator such that $I - U$ is finite dimensional)

$$U\left[\frac{1 - \bar{\zeta}_0\zeta}{\zeta_0 - \zeta} P + Q\right] \quad (\dim P\mathfrak{H} < \infty, \ P^2 = P, \ JP \leq 0, \ Q = I - P, \ |\zeta_0| < 1),$$

$$U\left[\frac{\zeta_0 - \zeta}{1 - \bar{\zeta}_0\zeta} P + Q\right] \quad (\dim P\mathfrak{H} < \infty, \ P^2 = P, \ JP \geq 0, \ Q = I - P, \ |\zeta_0| < 1),$$

$$U\left[I + \frac{\zeta + \zeta_0}{\zeta - \zeta_0} G\right] \quad (\dim G\mathfrak{H} < \infty, \ G^2 = 0, \ JG \geq 0, \ |\zeta_0| = 1),$$

which we will call elementary factors of the first, second and third kind respectively. It is easily verified that

an elementary factor $B(\zeta)$ belongs to the class K_J, has a unique singularity in the complex plane, viz. a pole of first order (respectively interior to, outside and on the unit circle) and takes J-unitary values for $|\zeta| = 1$.

In addition, $I - B(\zeta)$ is a finite-dimensional operator.

Thus $B(\zeta) \in R_J$. It follows from Theorem 1.7 that the above properties are characteristic for an elementary factor.

Theorem 1.7. *Each operator function* $Y(\zeta)$ *belonging to* R_J *admits the representation*

$$Y(\zeta) = U B_n(\zeta) \cdots B_1(\zeta),$$

where U *is a constant* J*-unitary operator and the* $B_k(\zeta)$ $(k = 1, \cdots, n)$ *are elementary factors* [2,7].

§ 2. The class $K_J^{\mathfrak{S}}$ and approximation by finite products of elementary factors

1. We will give here a definition of the class of operator functions $K_J^{\mathfrak{S}}$, which is the main object of study in the present paper. As a preliminary we recall that an operator A acting in a separable Hilbert space $\mathfrak{H}$ is said to be *nuclear* [24] if its modulus $(A^*A)^{\frac{1}{2}}$ has a finite trace: $\mathrm{tr}\,(A^*A)^{\frac{1}{2}} < \infty$. We denote the class of all nuclear operators acting in $\mathfrak{H}$ by $\mathfrak{S}$[1]) The quantity $|A| = \mathrm{tr}\,(A^*A)^{\frac{1}{2}}$, which satisfies all of the axioms for a norm, is called the nuclear (or trace) norm of A. It will be assumed below that the reader is familiar with the properties of the operators of the class $\mathfrak{S}$ [24,25]. We will say that a sequence of (not necessarily nuclear) operators A_n $\mathfrak{S}$-converges to an operator A $(A_n \to A)$ if $A_n - A \in \mathfrak{S}$ $(n = 1, 2, \cdots)$ and $|A_n - A| \to 0$. We will also use without further explanation the terms $\mathfrak{S}$-compactness, $\mathfrak{S}$-continuity, etc.

We will say that an operator function $Y(\zeta)$ $(\in K_J)$ belongs to the class $K_J^{\mathfrak{S}}$ if there exists a point ζ_0 $(|\zeta_0| < 1)$ such that

$$J - Y^*(\zeta_0) J Y(\zeta_0) \in \mathfrak{S}.$$

Suppose $Y(\zeta) \in K_J^{\mathfrak{S}}$. Since $Y(\zeta) \in K_J$, the equality $Y(\zeta) = U \widetilde{Y}(\zeta)$ holds (Theorem 1.2), where U is a J-unitary operator and $\widetilde{Y}(\zeta)$ takes Fredholm values at each point of holomorphy of $Y(\zeta)$. In particular $\widetilde{Y}(\zeta_0)$ is a Fredholm J-nonexpansive operator and hence [18,23] admits the polar representation

$$\widetilde{Y}(\zeta_0) = VR,$$

where V is a J-unitary operator and R is a J-Hermitian operator with a

---

1) This set is usually denoted by $\mathfrak{S}_1$ (see, for example, [24]).

nonnegative spectrum $(R^2 = J\widetilde{Y}^*(\zeta_0)J\widetilde{Y}(\zeta_0))$. Let $\widetilde{\widetilde{Y}}(\zeta) = V^{-1}\widetilde{Y}(\zeta)$. Then $\widetilde{\widetilde{Y}}(\zeta_0) = R$ and consequently

$$I - \widetilde{\widetilde{Y}}(\zeta_0) = I - R = (I + R)^{-1}(I - R^2) = (I + R)^{-1}J[J - Y^*(\zeta_0)JY(\zeta_0)] \in \mathfrak{S}.$$

Below, setting $\zeta_0 = 0$, we will assume that $Y(\zeta) \in K_J^{\mathfrak{S}}$ already has these properties, viz. $Y(0)$ *is a J-Hermitian operator with a nonnegative spectrum and* $I - Y(0) \in \mathfrak{S}$.

Such an operator function $Y(\zeta)$ $(\in K_J^{\mathfrak{S}})$ will be said to be *normalized*. If $Y(\zeta)$ is normalized then the operator $[I + Y(0)]^{-1}$ exists and is bounded, and hence [2,7] the operator function

$$\Omega(\zeta) = i[I - Y(\zeta)][I + Y(\zeta)]^{-1} \tag{2.1}$$

is holomorphic for $|\zeta| < 1$. In addition,

$$\mathrm{Im}\, J\Omega(\zeta) = \tfrac{1}{2}i[J\Omega(\zeta) - \Omega^*(\zeta)J] = [I + Y^*(\zeta)]^{-1}[J - Y^*(\zeta)JY(\zeta)][I + Y(\zeta)]^{-1} \geq 0,$$

$$\mathrm{Re}\, J\Omega(0) = \tfrac{1}{2}[J\Omega(0) + \Omega^*(0)J] = i[I + Y^*(0)]^{-1}[Y^*(0)J - JY(0)][I + Y(0)]^{-1} = 0.$$

It follows that, as in the scalar case, the following representation holds:

$$J\Omega(\zeta) = i \int_0^{2\pi} \frac{e^{i\theta} + \zeta}{e^{i\theta} - \zeta} d\Sigma(\theta) \quad (|\zeta| < 1), \tag{2.2}$$

where $\Sigma(\theta)$ $(\Sigma(0) = 0)$ is a Hermitian-increasing operator function.

Clearly,

$$0 \leq \Sigma(\theta) \leq \Sigma(2\pi) = \tfrac{1}{2}i[J\Omega(0) - \Omega^*(0)J] \quad (0 \leq \theta \leq 2\pi).$$

Since $\Omega(0) \in \mathfrak{S}$, we have $\Sigma(\theta) \in \mathfrak{S}$ $(0 \leq \theta \leq 2\pi)$. Consequently $\Sigma(\theta)$ is a function of bounded $\mathfrak{S}$-variation on $[0, 2\pi]$:

$$\sum_{j=0}^{n-1} |\Sigma(\theta_{j+1}) - \Sigma(\theta_j)| = \sum_{j=0}^{n-1} \mathrm{tr}\,[\Sigma(\theta_{j+1}) - \Sigma(\theta_j)]$$

$$= \mathrm{tr}\,\Sigma(2\pi) = \mathrm{tr}(1/2i)[J\Omega(0) - \Omega^*(0)J] < \infty.$$

Thus the integral sums $\mathfrak{S}$-converge to $-iJ\Omega(\zeta)$. Consequently $\Omega(\zeta)$ takes nuclear values for $|\zeta| < 1$. Since this convergence is uniform interior to the unit disk, $\Omega(\zeta)$ is an $\mathfrak{S}$-holomorphic operator function.

Suppose now ζ_1 is a point of (ordinary) holomorphy of an operator function $Y(\zeta) \in K_J^{\mathfrak{S}}$. Then $I - Y(\zeta_1) = -i\Omega(\zeta_1)[I + Y(\zeta_1)] \in \mathfrak{S}$ and the equality

$$(i/h)[Y(\zeta_1 + h) - Y(\zeta_1)]$$

$$= (i/h)[\Omega(\zeta_1 + h) - \Omega(\zeta_1)][I + Y(\zeta_1 + h)] + i\Omega(\zeta_1)\,\frac{Y(\zeta_1 + h) - Y(\zeta_1)}{h}$$

implies[1] the $\mathfrak{S}$-differentiability of $Y(\zeta)$ at ζ_1. Analogous assertions hold for $Y^{-1}(\zeta)$. We have proved

Theorem 2.1. *If $Y(\zeta)$ is a normalized function in $K_J^{\mathfrak{S}}$ while $D(Y)$ and $D(Y^{-1})$ are the domains, lying interior to the unit disk, of (ordinary) holomorphy of the operator functions $Y(\zeta)$ and $Y^{-1}(\zeta)$ respectively, then $I - Y(\zeta) \in \mathfrak{S}$ for $\zeta \in D(Y)$ and $I - Y^{-1}(\zeta) \in \mathfrak{S}$ for $\zeta \in D(Y^{-1})$. The operator function $Y(\zeta)$ is $\mathfrak{S}$-holomorphic in $D(Y)$ while $Y^{-1}(\zeta)$ is $\mathfrak{S}$-holomorphic in $D(Y^{-1})$.*

2. We will make use of the apparatus of the Cayley transform to prove a theorem which plays an essential role in §8.

Theorem 2.2. *If $Y(\zeta) \in K_J^{\mathfrak{S}}$ then there exists a sequence*

$$Y_n(\zeta) = V_n B_{k_n}^{(n)}(\zeta) \cdots B_1^{(n)}(\zeta) \tag{2.3}$$

(the V_n are J-unitary operators and the $B_j^{(n)}(\zeta)$ are elementary factors of the first, second and third kind) such that

$$Y_n(\zeta) \to Y(\zeta), \; Y_n^{-1}(\zeta) \to Y^{-1}(\zeta)$$

uniformly in ζ interior to $D(Y)$ and $D(Y^{-1})$ respectively.

The proof is carried out under the assumption, which is sufficient for the subsequent applications, that both $D(Y)$ and $D(Y^{-1})$ coincide with the open unit disk. We will also assume that the function $Y(\zeta)$ is normalized.

We form the operator function $\Omega(\zeta)$ by means of formula (2.1). By means of the representation (2.2) we construct the operator function

$$\Omega_k(\zeta) = iJ \sum_{j=1}^{n_k} \frac{e^{i\theta_j^{(k)}} + \zeta}{e^{i\theta_j^{(k)}} - \zeta} \Delta_j^{(k)}$$

$$(\Delta_j^{(k)} = \Sigma(\theta_j^{(k)}) - \Sigma(\theta_{j-1}^{(k)}); \; 0 = \theta_0^{(k)} < \cdots < \theta_n^{(k)} = 2\pi)$$

----

1) Here, as above, we make use of the facts that from $A \in \mathfrak{S}$ and $\|B\| < \infty$ it follows that $AB \in \mathfrak{S}$, $BA \in \mathfrak{S}$ and that $A_n \to A$, $A_n \in \mathfrak{S}$, $B_n \Rightarrow B$ (the symbol $\Rightarrow$ denotes $\mathfrak{R}$-convergence, i.e. convergence in the uniform operator topology) implies $A_n B_n \to AB$, $B_n A_n \to BA$.

so that $|\Omega(\zeta) - \Omega_k(\zeta)| \to 0$ uniformly interior to the unit disk. Taking advantage of the nuclearity of the operators $\Delta_j^{(k)}$ ($k = 1, 2, \cdots; j = 1, \cdots, n_k$), we find finite-dimensional operators $\widetilde{\Delta}_j^{(k)}$ such that

$$|\Delta_j^{(k)} - \widetilde{\Delta}_j^{(k)}| < \frac{1}{2 n_k k}, \quad 0 \leq \widetilde{\Delta}_j^{(k)} \leq \Delta_j^{(k)}.$$

Putting

$$\widetilde{\Omega}_k(\zeta) = iJ \sum_{j=1}^{n_k} \frac{e^{i\theta_j^{(k)}} + \zeta}{e^{i\theta_j^{(k)}} - \zeta} \widetilde{\Delta}_j^{(k)},$$

we see that for $|\zeta| < \rho < 1$

$$|\Omega_k(\zeta) - \widetilde{\Omega}_k(\zeta)| \leq \frac{2}{1-\rho} \sum_{j=1}^{n_k} |\Delta_j^{(k)} - \widetilde{\Delta}_j^{(k)}| < \frac{1}{k(1-\rho)}.$$

Thus $\widetilde{\Omega}_k(\zeta) \to \Omega(\zeta)$ uniformly interior to the unit disk, and, since the function $[iJ + \Omega(\zeta)]^{-1} = (1/2i)[Y(\zeta) + I]$ is holomorphic for $|\zeta| < 1$, it follows that for $k \geq k\rho$ the functions

$$Y_k(\zeta) = [iJ + \widetilde{\Omega}_k(\zeta)]^{-1} [iJ - \widetilde{\Omega}_k(\zeta)] = 2i[iJ + \widetilde{\Omega}_k(\zeta)]^{-1} - I$$

are holomorphic and uniformly bounded for $|\zeta| \leq \rho$ and

$$|Y(\zeta) - Y_k(\zeta)| = 2|[iJ + \widetilde{\Omega}_k(\zeta)]^{-1} - [iJ + \Omega(\zeta)]^{-1}|$$

$$\leq 2 \|[iJ + \widetilde{\Omega}_k(\zeta)]^{-1}\| \, \|[iJ + \Omega(\zeta)]^{-1}\| \, |\widetilde{\Omega}_k(\zeta) - \Omega(\zeta)|$$

and hence $Y_k(\zeta) \to Y(\zeta)$ for $|\zeta| \leq \rho$. The assertion concerning $Y_k^{-1}(\zeta)$ is proved analogously.

Starting from the fact that $\widetilde{\Omega}_k(\zeta)$ is a finite-dimensional rational operator function taking Hermitian values for $|\zeta| = 1$, it is not difficult to show that $Y_k(\zeta) \in R_J$ and hence by virtue of Theorem 1.7 can be represented in the form (2.3).

We note that a proposition analogous to Theorem 2.2 (with the symbol $\to$ replaced by $\Rightarrow$) holds for an operator function $Y(\zeta) \in K_J$. The proof carries over to this case almost without change.

§ 3. The infinite product and the multiplicative integral in the nuclear metric

In this section we cite a simple theorem on infinite products of operators. The theorems on multiplicative integrals which are usually formulated [2,9] for the case of the uniform operator metric are extended here to the nuclear metric (the proofs carry over to our case with only slight changes).[1]

1. We first cite some inequalities which will be needed in the sequel (below $A_1, \cdots, A_n, B_1, \cdots, B_n$ are nuclear operators):

$$\|(I + A_n) \cdots (I + A_1)\| \leq \exp(|A_1| + \cdots + |A_n|) \tag{3.1}$$

$$|e^{A_n} \cdots e^{A_n}| \leq \exp(|A_1| + \cdots + |A_n|) \tag{3.2}$$

$$|(I + A_n) \cdots (I + A_1) - (I + B_n) \cdots (I + B_1)| \leq e^{\Sigma(|A_j| + |B_j|)} \Sigma|A_j - B_j| \tag{3.3}$$

$$|(I + A_n) \cdots (I + A_1) - I| \leq e^{\Sigma|A_j|} \Sigma|A_j| \tag{3.4}$$

$$|e^{A_n} \cdots e^{A_1} - I| \leq e^{\Sigma|A_j|} \Sigma|A_j| \tag{3.5}$$

$$|e^{A_n} \cdots e^{A_1} - [I + A_1 + \cdots + A_n]| \leq e^{\Sigma|A_j|} (\Sigma|A_j|)^2. \tag{3.6}$$

The proof of (3.1) follows from the fact that $\|I + A\| \leq 1 + \|A\| \leq 1 + |A| \leq e^{|A|}$, while (3.2) follows from the inequality $\|e^A\| \leq e^{\|A\|} \leq e^{|A|}$. The validity of (3.3) follows from the identity

$$(I + A_n) \cdots (I + A_1) - (I + B_n) \cdots (I + B_1)$$

$$= \sum_{j=1}^{n} (I + A_n) \cdots (I + A_{j+1})(A_j - B_j)(I + B_{j-1}) \cdots (I + B_1)$$

and inequality (3.1). Inequality (3.4) is obtained from (3.3) for $B_1 = \cdots \cdots = B_n = 0$. As to (3.5) and (3.6), these inequalities easily follow from the identity

$$e^{A_n} \cdots e^{A_1} = \sum_{k=0}^{\infty} \frac{1}{k!} \sum_{m_1 + \cdots + m_n = k} \frac{k!}{m_1! \cdots m_n!} A_n^{m_n} \cdots A_1^{m_1}.$$

---

1) We note that in the present section $\mathfrak{S}$ can be regarded as an arbitrary symmetrically normed ideal [24] of the ring $\mathfrak{R}$ of bounded operators, with $|\cdot|$ being the corresponding symmetric norm. It can also be supposed that $\mathfrak{S} = \mathfrak{R}$ and $|\cdot| = \|\cdot\|$, so that the theorems on multiplicative integrals in the $\mathfrak{R}$ metric are a special case of the ones cited below.

2. Theorem 3.1. *Suppose the operator functions* $A_k(\zeta)$ *take nuclear values for* $\zeta \in \mathfrak{M}$. *Let the following inequalities hold:*

$$|A_k(\zeta)| \leq \alpha_k \quad (\zeta \in \mathfrak{M}),$$

where

$$\sigma = \sum_{k=1}^{\infty} \alpha_k < \infty, \qquad (3.7)$$

Then the following assertions are valid.

1) *The sequence*

$$\Pi_n(\zeta) = \overset{n}{\underset{k=1}{\overleftarrow{\Pi}}} [I + A_k(\zeta)] \quad (n = 1, 2, \cdots)$$

uniformly $\mathfrak{S}$-*converges for* $\zeta \in \mathfrak{M}$.

2) *There exists a* k_0 *such that for* $k \geq k_0$ *and* $\zeta \in \mathfrak{M}$ *the operators* $I + A_k(\zeta)$ *have bounded inverses and the sequence*

$$\overset{n}{\underset{k=k_0}{\overrightarrow{\Pi}}} [I + A_k(\zeta)]^{-1} \quad (n = k_0, k_0 + 1, \cdots)$$

also uniformly $\mathfrak{S}$-*converges on* $\mathfrak{M}$.

Proof. Taking advantage of (3.1) and (3.4), we get

$$|\Pi_{n+p}(\zeta) - \Pi_n(\zeta)| \leq \left| \overset{n+p}{\underset{k=n+1}{\overleftarrow{\Pi}}} [I + A_k(\zeta)] - I \right| \, \|\Pi_n(\zeta)\|$$

$$\leq e^{\sum_{k=1}^{n+p} |A_k(\zeta)|} \overset{n+p}{\underset{k=n+1}{\Sigma}} |A_k(\zeta)|$$

·or

$$|\Pi_{n+p}(\zeta) - \Pi_n(\zeta)| \leq e^{\sigma} \overset{n+p}{\underset{k=n+1}{\Sigma}} \alpha_k \quad (\zeta \in \mathfrak{M}),$$

which by virtue of (3.7) proves assertion 1). Further, (3.7) implies the existence of a k_0 such that for $k \geq k_0$ and $\zeta \in \mathfrak{M}$

$$\|A_k(\zeta)\| \leq |A_k(\zeta)| < \tfrac{1}{2}.$$

Consequently the operators $I + A_k(\zeta)$ have bounded inverses, and $\|[I + A_k(\zeta)]^{-1}\| \leq 2 \quad (k \geq k_0, \zeta \in \mathfrak{M})$. Since

$$C_k(\zeta) \equiv [I + A_k(\zeta)]^{-1} - I = -[I + A_k(\zeta)]^{-1} A_k(\zeta),$$

we get $|C_k(\zeta)| \leq 2|A_k(\zeta)| \leq 2\alpha_k$ $(k \geq k_0, \zeta \in \mathfrak{M})$, and the validity of 2) is established by virtue of (3.7) and the already proved assertion 1).

3. Suppose $H(t)$ $(a \leq t \leq b)$ is an operator function of bounded $\mathfrak{S}$-variation, i.e.

$$v = \mathop{\mathfrak{S}\text{- var}}_{[a,b]} H = \sup \sum_{j=1}^{n} |H(t_j) - H(t_j) - H(t_{j-1})| < \infty.$$

Here the supremum is taken over all possible partitionings

$$a = t_0 < t_1 < \cdots < t_{n-1} < t_n = b \tag{3.8}$$

of the segment $[a, b]$.

If $f(t)$ is a bounded $(|f(t)| \leq m)$ scalar function on $[a, b]$ (other restrictions will be imposed on $f(t)$ later), we consider for an arbitrary partitioning (3.8) of the segment $[a, b]$ the "integral product"

$$\Pi = \overset{\curvearrowleft}{\prod_{j=1}^{n}} e^{f(\tau_j)\Delta_j} \quad (\Delta_j = H(t_j) - H(t_{j-1}), \ t_{j-1} \leq \tau_j \leq t_j; \ j = 1, \cdots, n). \tag{3.9}$$

We compare Π with the integral product

$$\Pi' = \overset{\curvearrowleft}{\prod_{j=1}^{n}} e^{f(\tau_{k_j}^{(j)})\Delta_{k_j}^{(j)}} \cdots e^{f(\tau_1^{(j)})\Delta_1^{(j)}}$$

$$(\Delta_s^{(j)} = H(t_s^{(j)}) - H(t_{s-1}^{(j)}), \ t_{s-1}^{(j)} \leq \tau_s^{(j)} \leq t_s^{(j)} \ (j = 1, \cdots, n; \ s = 1, \cdots, k_j)),$$

which corresponds to a refinement of the partitioning (3.8). Taking advantage of inequality (3.3), we get

$$|\Pi - \Pi'| \leq e^{2mv} \sum_{j=1}^{n} |e^{f(\tau_j)\Delta_j} - e^{f(\tau_{k_j}^{(j)})\Delta_{k_j}^{(j)}} \cdots e^{f(\tau_1^{(j)})\Delta_1^{(j)}}|.$$

Estimating each summand with the use of (3.6), we obtain the inequality

$$|\Pi - \Pi'| \leq e^{2mv} \sum_{j=1}^{n} \{|f(\tau_j) - f(\tau_{k_j}^{(j)})||\Delta_{k_j}^{(j)}| + \cdots + |f(\tau_j) - f(\tau_1^{(j)})||\Delta_1^{(j)}| +$$

$$+ (|f(\tau_j)||\Delta_j|)^2 e^{|f(\tau_j)||\Delta_j|} + (|f(\tau_{k_j}^{(j)})||\Delta_{k_j}^{(j)}| + \cdots$$

$$\cdots + |f(\tau_1^{(j)})||\Delta_1^{(j)}|)^2 e^{|f(\tau_{k_j}^{(j)})||\Delta_{k_j}^{(j)}|+\cdots+|f(\tau_1^{(j)})||\Delta_1^{(j)}|} \}.$$

Let $\omega_j(f)$ denote the oscillation of $f(t)$ on $[t_{j-1},\, t_j]$, and let $v(t) = \mathfrak{S}\text{-var}_{[a,t]}H$. Then

$$|\Pi - \Pi'| \le e^{2mv} \sum_{j=1}^{n} \{\omega_j(f)[v(t_j) - v(t_{j-1})]$$

$$+ 2m^2[v(t_j) - v(t_{j-1})]^2 e^{mv}\}.$$

Thus

$$|\Pi - \Pi'| \le e^{2mv} \sum_{j=1}^{n} \omega_j(f)[v(t_j) - v(t_{j-1})]$$

$$+ 2m^2 e^{3mv} v \max_{j} [v(t_j) - v(t_{j-1})]. \tag{3.10}$$

We now prove the following proposition.

Theorem 3.2. *Under unlimited refinements of the partitioning* (3.8) *the integral products* (3.9) $\mathfrak{S}$-*converge to the multiplicative integral*

$$\overset{\leftarrow}{\underset{a}{\overset{b}{\int}}} e^{f(t)dH(t)}, \tag{3.11}$$

if one of the following two conditions is satisfied:

A) $f(t)$ *is Riemann integrable over* $[a, b]$ *and the inequality*

$$|H(t'') - H(t')| \le l|t'' - t'|$$

holds for some $l > 0$ *and any* $t',\, t'' \in [a, b]$.

B) $f(t)$ *is continuous while* $H(t)$ *is of bounded* $\mathfrak{S}$-*variation and* $\mathfrak{S}$-*continuous on* $[a, b]$.

For suppose condition A) is satisfied. Then $|v(t'') - v(t')| \le l|t'' - t'|$ $(t',\, t'' \in [a, b])$ and from (3.10) we get

$$|\Pi - \Pi'| \le e^{2mv} l \sum_{j=1}^{n} \omega_j(f)(t_j - t_{j-1})$$

$$+ 2m^2 e^{3mv} vl \max_{j} (t_j - t_{j-1}). \tag{3.12}$$

202   Ju. P. GINZBURG

Since $f(t)$ is Riemann integrable, $|\Pi - \Pi'| \to 0$ for $\max_j(t_j - t_{j-1}) \to 0$.

Suppose now that B) is satisfied. In this case it is easily seen that the function $v(t)$ is continuous. The inequality

$$|\Pi - \Pi'| \le e^{2mv}\, v \max_j \omega_j(f) + 2m^2 e^{3mv}\, v \max_j \omega_j(v),$$

which follows from (3.10), then completes the proof of the theorem.

Below are cited estimates for the multiplicative integral which are used in the sequel and which directly follow from inequalities (3.2), (3.5) and (3.6) respectively:

$$\left\| \overset{\leftarrow}{\int_a^b} e^{f(t)dH(t)} \right\| \le e^{\int_a^b |f(t)|\,dv(t)} \tag{3.13}$$

$$\left| \overset{\leftarrow}{\int_a^b} e^{f(t)dH(t)} - I \right| \le \left[\int_a^b |f(t)|\,dv(t) \right]^2 e^{\int_a^b |f(t)|\,dv(t)} \tag{3.14}$$

$$\left| \overset{\leftarrow}{\int_a^b} e^{f(t)dH(t)} - \left[I + \int_a^b f(t)dH(t) \right] \right|$$
$$\le \left[\int_a^b |f(t)|\,dv(t) \right]^2 e^{\int_a^b |f(t)|\,dv(t)} \tag{3.15}$$

4. We will cite here the operator, including "multiplicative," analogs of the classical theorems of Helly (cf. [2,9]) which are needed for the sequel.

Theorem 3.3. Suppose a family of nuclear operator functions $\{H_n(t)\}$ ($a \le t \le b$) satisfies the following conditions:

1) there exists a number $\mu > 0$ not depending on n such that

$$\underset{[a,b]}{\mathfrak{S}\text{-var}}\, H_n \le \mu;$$

2) the family of operators $\{H_n(t)\}$ is $\mathfrak{S}$-compact for each $t \in [a, b]$.

Then a subsequence $\{H_{n_j}(t)\}$ can be chosen from $\{H_n(t)\}$ which is $\mathfrak{S}$-convergent for each $t \in [a, b]$; the limit function $H(t)$ satisfies the inequality

$$\mathfrak{S}\text{-var}\, H \le \mu$$

For the proof one should choose a subsequence from $\{H_n(t)\}$, by taking advantage of the separability of $\mathfrak{H}$ and the classical first theorem of Helly, which is weakly convergent for each $t \in [a, b]$ and then show that it is $\mathfrak{S}$-convergent by making use of 2).

Theorem 3.4. *If a sequence which is uniformly bounded on* $[a, b]$ *of Riemann integrable functions* $f_n(t)$ $(|f_n(t)| \leq m)$ *(pointwise) converges to a Riemann integrable function* $f(t)$, *and if a sequence of operator functions* $H_n(t)$ *satisfying the condition*

$$|H_n(t'') - H_n(t')| \leq l|t'' - t'| \quad (t', t'' \in [a, b]),$$

$\mathfrak{S}$-*converges to* $H(t)$, *then*

$$\int_a^b e^{f_n(t)dH_n(t)} \longrightarrow \int_a^b e^{f(t)dH(t)}.$$

The proof of this theorem is an almost word-for-word repetition of the proof of the corresponding proposition for the case of the uniform metric [2,9].

§ 4. Theorems on the modulus of a multiplicative integral

1. We consider an operator function $H(t)$ $(a \leq t \leq b)$ of bounded uniform variation:

$$\operatorname*{var}_{[a,b]} H = \sup \sum_{j=1}^n \|H(t_j) - H(t_{j-1})\| < \infty \quad (a = t_0 < t_1 < \cdots < t_n = b).$$

If the space $\mathfrak{H}$ is separable (which will be assumed in the sequel) then $H(t)$ has almost everywhere on $[a, b]$ a weak derivative $M(t) = H'(t)$ with the following properties:

1) $M(t)$ is weakly measurable,[1] i.e. the scalar functions $(M(t)x, y)$ $(x, y \in \mathfrak{H})$ are measurable;

2) $\int_a^b \|M(t)\| dt < \infty$.

The first property is obvious. And as to the second, its proof follows directly from 1) and the inequality $\|M(t)\| \leq h'(t)$, where $h(t) = \operatorname{var}_{[a,t]} H$.

Following D. L. Kučer [26], we will say that an operator function having properties 1) and 2) is *normal* on $[a, b]$. With each such operator

---

1) The separability of $\mathfrak{H}$ implies that $M(t)$ is strongly measurable [27]. But we will not make use of this fact.

function $M(t)$ we can associate a bounded operator S defined by the equality

$$(Sx, y) = \int_a^b (M(t)x, y)dt \quad (x, y \in \mathfrak{H})$$

and called the integral of $M(t)$ over the segment $[a, b]$:

$$S = \int_a^b M(t)dt.$$

In this connection we have the inequality

$$\left\| \int_a^b M(t)dt \right\| \leq \int_a^b \|M(t)\| dt.$$

Suppose now $H(t)$ is an operator function that is absolutely continuous (in the uniform operator metric) on $[a, b]$. Then the following equality clearly holds:

$$H(t) = H(a) + \int_a^t M(\tau)d\tau, \tag{4.1}$$

where $M(t) = H'(t)$. Conversely, every operator function $H(t)$ defined by equality (4.1) with a normal function $M(t)$ is absolutely continuous and $H'(t) = M(t)$ almost everywhere on $[a, b]$.

If $H(t)$ is absolutely continuous on $[a, b]$, we consider the operator function

$$W(t) = \overset{\leftarrow}{\int_a^t} e^{dH(\tau)} \equiv \overset{\leftarrow}{\int_a^t} e^{M(\tau)d\tau} \quad (a \leq t \leq b). \tag{4.2}$$

It is known $[2,9]$ (cf. (3.15)) that the following formula holds:

$$\overset{\leftarrow}{\int_\alpha^\beta} e^{M(t)dt} = I + \int_\alpha^\beta M(t)dt + A,$$

where the operator A satisfies the inequality

$$\|A\| \leq \left[\int_\alpha^\beta \|M(t)\| dt \right]^2 e^{\int_\alpha^\beta \|M(t)\| dt}$$

It follows that $W(t)$ is an absolutely continuous operator function which has a weak derivative $W'(t)$ almost everywhere on $[a, b]$. In this connection

$$W'(t) = M(t)W(t), \quad W(a) = I, \tag{4.3}$$

which implies that $W'(t)$ is normal. The fact that the differential problem (4.3) has a unique solution (4.2) in the class of absolutely continuous operator functions follows from the equivalence of (4.3) in this class to the integral equation

$$W(t) = I + \int_a^t M(\tau)W(\tau)d\tau$$

which can be solved by an iterative method (see [26]).

2. We will assume below that in (4.2) the absolutely continuous function $H(t)$ $(H(a) = 0)$ takes *J*-Hermitian values on $[a, b]$ while $W(t)$ admits for each t the polar representation

$$W(t) = U(t)R(t), \tag{4.4}$$

in which $U(t)$ is a *J*-unitary operator while $R(t)$ is a *J*-Hermitian operator with a positive spectrum. Since $W(t)$ is an absolutely continuous weakly differentiable operator function, the same properties are also possessed by $B(t) = JW^*(t)JW(t)$ as well as $B^{-1}(t)$ —the latter by virtue of the fact that $W^{-1}(t)$ admits a representation analogous to (4.2):

$$W^{-1}(t) = \int_a^t \overrightarrow{e^{-M(\tau)d\tau}}.$$

If $\max_{a \leq t \leq b} \|B(t)\| = \rho_1$, and $\max_{a \leq t \leq b} \|B^{-1}(t)\| = 1/\rho_2$, then the spectra of the operators $B(t)$ are positive by virtue of the equality $B(t) = R^2(t)$ and lie on the segment $[\rho_2, \rho_1]$. If Γ is any rectifiable contour enclosing this segment and not containing the origin, for example, the circle with center at the point $(\rho_1 + \rho_2)/2$ and radius $\rho_1/2$, then for all $t \in [a, b]$

$$R(t) = (1/2\pi i)\int_\Gamma \sqrt{\zeta}[\zeta I + B(t)]^{-1}d\zeta.$$

This implies the equality

$$R(t_2) - R(t_1) = (1/2\pi i)\int_\Gamma \sqrt{\zeta}[\zeta I - B(t_2)]^{-1}[B(t_2) - B(t_1)][\zeta I - B(t_1)]^{-1}d\zeta. \tag{4.5}$$

If $\mu = \max_{\zeta \in \Gamma, a \leq t_1, t_2 \leq b} \sqrt{|\zeta|} \, \|[\zeta I - B(t_1)]^{-1}\| \, \|[\zeta I - B(t_2)]^{-1}\|$, then

$$\|R(t_2) - R(t_1)\| \leq \tfrac{1}{2}\mu\rho_1\|B(t_2) - B(t_1)\|$$

and the absolute continuity of $B(t)$ on $[a, b]$ implies the same property for $R(t)$. Equality (4.5) also implies the existence of a weak derivative $R'(t)$ almost everywhere on $[a, b]$ and the formula

$$R'(t) = (1/2\pi i)\int_\Gamma \sqrt{\zeta}\,[\zeta I - B(t)]^{-1} B'(t)[\zeta I - B(t)]^{-1} d\zeta. \qquad (4.6)$$

Since (4.3) and the fact that $M(t)$ is J-Hermitian imply the inequalities

$$W^*(t)JW'(t) = W^*(t)JM(t)W(t), \quad W^{*\prime}(t)JW(t) = W^*(t)JM(t)W(t),$$

$$B'(t) = [JW^*(t)JW(t)]' = 2JW^*(t)JM(t)W(t), \qquad (4.7)$$

$B'(t)$ is a normal operator function together with $M(t)$. In this case (4.6) implies that $R'(t)$ is normal on $[a, b]$. It follows from (4.4) that $U(t) = W(t)R^{-1}(t)$ and hence is an absolutely continuous weakly differentiable operator function. In addition,

$$U'(t) = W'(t)R^{-1}(t) - W(t)R^{-1}(t)R'(t)R^{-1}(t),$$

and by virtue of (4.3) and (4.4)

$$U'(t) = M(t)U(t) - U(t)R'(t)R^{-1}(t) \qquad (4.8)$$

On the other hand, (4.7) can be rewritten as follows:

$$(R^2(t))' = 2JW^*(t)JM(t)W(t),$$

$$R'(t)R(t) + R(t)R'(t) = 2JR^*(t)U^*(t)JM(t)U(t)R(t),$$

$$M(t) = \tfrac{1}{2}U(t)[R'(t)R^{-1}(t) + R^{-1}(t)R'(t)]U^{-1}(t). \qquad (4.9)$$

From (4.8) and (4.9) it follows that almost everywhere on $[a, b]$

$$- U'(t) = U(t)N(t), \qquad (4.10)$$

where

$$N(t) = \tfrac{1}{2}[R^{-1}(t)R'(t) - R'(t)R^{-1}(t)] \qquad (4.11)$$

is a normal operator function. Since $U(t)$ is absolutely continuous and satisfies, in addition to the differential equation (4.10), the initial condition $U(0) = I$, we have

$$U(t) = \int_a^t \overrightarrow{e^{N(\tau)d\tau}}. \qquad (4.12)$$

We note further that (4.9) implies

$$H(t) = \int\limits_a^t M(\tau)d\tau = \tfrac{1}{2} \int\limits_a^t U(\tau)[R'(\tau)R^{-1}(\tau) + R^{-1}(\tau)R'(\tau)]U^{-1}(\tau)d\tau. \qquad (4.13)$$

We have thus proved

Theorem 4.1. *Suppose* $H(t)$ $(H(a) = 0)$ *is an absolutely continuous operator function that takes J-Hermitian values on* $[a, b]$, *and suppose*

$$W(t) = \int\limits_a^{\overset{\leftarrow}{t}} e^{dH(\tau)}$$

admits the polar representation $W(t) = U(t)R(t)$ *for each* $t \in [a, b]$. *Then* $U(t)$ *and* $R(t)$ *are absolutely continuous and* $H(t)$ *is uniquely determined by* $R(t)$ *for* $t \in [a, b]$ *by formulas* (4.13), (4.12) *and* (4.11).

3. Theorem 4.2 formulated below together with Theorem 4.1 will play an important role in §§ 5, 6, 8 and 9.

Theorem 4.2. *Suppose* $H(t)$ *is a J-Hermitian-nonincreasing absolutely continuous*[1] *operator function on* $[a, b]$. *Then the following assertions are valid.*

1) *For each* $t \in [a, b]$ *the operator*

$$W(t) = \int\limits_a^{\overset{\leftarrow}{t}} e^{dH(\tau)}$$

has a bounded inverse, is J-binonexpansive, and hence [18,23] *admits the polar representation*

$$W(t) = U(t)R(t).$$

2) *The operator functions* $J - W^*(t)JW(t)$, $W^{-1}(t)JW^{-1*}(t) - J$, $J - JR(t)$ *and* $R^{-1}(t)J - J$ *are Hermitian-nondecreasing.*

3) *The inequalities*

$$0 \le J - JR(t) \le J - W^*(t)JW(t)(= J - R^*(t)JR(t)), \qquad (4.14)$$

$$0 \le R^{-1}(t)J - J \le W^{-1}(t)JW^{-1*}(t) - J \ (= R^{-1}(t)JR^{-1*}(t) - J) \qquad (4.15)$$

are valid.

---

1) It can be shown that the requirement of absolute continuity here can be replaced by the requirement that $H(t)$ be of bounded variation on $[a, b]$.

Proof. We write the identity (4.7) in the form

$$[J - W^*(t)JW(t)]' = -2W^*(t)JH'(t)W(t).$$

Then the inequality $JH'(t) \leq 0$ and the absolute continuity of $J - W^*(t)JW(t)$ implies that this operator function increases on $[a, b]$ and hence

$$J - W^*(t)JW(t) \geq J - W^*(0)JW(0) = 0.$$

Since

$$[W^{-1}(t)JW^{-1\,*}(t) - J]' = W^{-1}(t)JW^{-1\,*}(t)[J - W^*(t)JW(t)]'W^{-1}(t)JW^{-1\,*}(t),$$

the operator function $W^{-1}(t)JW^{-1\,*}(t) - J$ increases on $[a, b]$ and

$$J - W(t)JW^*(t) = W(t)[W^{-1}(t)JW^{-1\,*}(t) - J]W^*(t) \geq 0.$$

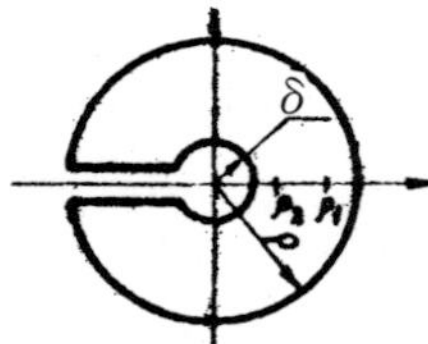

For a proof of the increase of $J - JR(t)$ we make use of (4.6), where the contour depicted in the figure can be taken as Γ. Passing to the limit for $\delta \to 0$ and $\rho \to \infty$, we get

$$R'(t) = (1/\pi) \int_{-\infty}^{0} \sqrt{|\zeta|}\,[\zeta I - B(t)]^{-1} B'(t)[\zeta I - B(t)]^{-1}d\zeta \quad (B(t) = JW^*(t)JW(t)).$$

It follows that

$$JR'(t) = (1/\pi) \int_{0}^{\infty} \sqrt{\lambda}\,[B^*(t) + \lambda I]^{-1}[W^*(t)JW(t)]'[B(t) + \lambda I]^{-1}d\lambda \leq 0$$

and hence $J - JR(t)$ increases. As to the assertion concerning $R^{-1}(t)J - J$, it follows from the identity

$$[R^{-1}(t)J]' = -[R^{-1}(t)J]JR'(t)[R^{-1}(t)J]^*.$$

Thus assertions 1) and 2) of the theorem are proved.

For a proof of 3) we note that 2) implies the inequality $J - JR(t) \geq 0$, which can be rewritten in the form

$$R^{\frac{1}{2}\,*}(t)JR^{\frac{1}{2}}(t) \leq J. \tag{4.16}$$

Here

$$R^{\frac{1}{2}} = (1/2\pi i) \int_{\Gamma} \sqrt{\zeta}\,(\zeta I - R)^{-1}d\zeta,$$

and the fact that R is J-Hermitian and has a positive spectrum implies [18] that $R^{\frac{1}{2}}$ has the same properties. It follows from (4.16) that

$$W^*(t)JW(t) = R^*(t)JR(t) \leq R^{\frac{1}{2}*}(t)JR^{\frac{1}{2}}(t) = JR(t)$$

and hence $J - JR(t) \leq J - W^*(t)JW(t)$. The assertion concerning $R^{-1}(t)J - J$ is proved analogously. Theorem 4.2 is proved.

$$\S\,5.\ \textbf{Convergence of the product of elementary factors for}$$
$$\textbf{operator functions of the class } K_J^{\mathfrak{S}}$$

1. Suppose $Y(\zeta) \in K_J^{\mathfrak{S}}$. Since the set of singularities of $Y(\zeta)$ and $Y^{-1}(\zeta)$ interior to the unit disk consists of isolated points, it can be assumed without loss of generality that $Y(\zeta)$ and $Y^{-1}(\zeta)$ are holomorphic at $\zeta = 0$.

We consider the elementary factors $B_k(\zeta)$ of $Y(\zeta)$ corresponding to the singularities lying interior to the unit disk. As we know ($\S\,1.3$),

$$B_k^{(\mathrm{I})}(\zeta) = U_k^{(\mathrm{I})}\left[\frac{1 - \bar{\zeta}_k\zeta}{\zeta_k - \zeta}\, P_k^{(\mathrm{I})} + Q_k^{(\mathrm{I})}\right]$$

$$(\dim P_k^{(\mathrm{I})} < \infty,\ P_k^{(\mathrm{I})^2} = P_k^{(\mathrm{I})},\ JP_k^{(\mathrm{I})} \leq 0,\ Q_k^{(\mathrm{I})} = I - P_k^{(\mathrm{I})})$$

if ζ_k is a pole of $Y(\zeta)$, and

$$B_k^{(\mathrm{II})}(\zeta) = U_k^{(\mathrm{II})}\left[\frac{\zeta_k - \zeta}{1 - \bar{\zeta}_k\zeta}\, P_k^{(\mathrm{II})} + Q_k^{(\mathrm{II})}\right]$$

$$(\dim P_k^{(\mathrm{II})} < \infty,\ P_k^{(\mathrm{II})^2} = P_k^{(\mathrm{II})},\ JP_k^{(\mathrm{II})} \geq 0,\ Q_k^{(\mathrm{II})} = I - P_k^{(\mathrm{II})}),$$

if ζ_k is a pole of $Y^{-1}(\zeta)$. The J-unitary operators $U_k^{(\mathrm{I})}$ and $U_k^{(\mathrm{II})}$ are defined here by the formulas

$$U_k^{(\mathrm{I})} = \frac{\zeta_k}{|\zeta_k|}\, P_k^{(\mathrm{I})} + Q_k^{(\mathrm{I})},\quad U_k^{(\mathrm{II})} = \frac{|\zeta_k|}{\zeta_k}\, P_k^{(\mathrm{II})} + Q_k^{(\mathrm{II})}$$

and hence

$$B_k^{(\mathrm{I})}(\zeta) = \frac{1 - \bar{\zeta}_k\zeta}{\zeta_k - \zeta}\cdot\frac{\zeta_k}{|\zeta_k|}P_k^{(\mathrm{I})} + Q_k^{(\mathrm{I})},$$

$$B_k^{(\mathrm{II})}(\zeta) = \frac{\zeta_k - \zeta}{1 - \bar{\zeta}_k\zeta}\cdot\frac{|\zeta_k|}{\zeta_k}\, P_k^{(\mathrm{II})} + Q_k^{(\mathrm{II})}. \tag{5.1}$$

We put $\zeta_k = \rho_k e^{i\theta_k}$. Simple calculations show that

$$B_k^{(I)}(\zeta) = I + \frac{\zeta + e^{i\theta_k}}{\zeta - \rho_k e^{i\theta_k}}\,(\rho_k - 1)P_k^{(I)},$$

$$B_k^{(II)}(\zeta) = I + \frac{\zeta + e^{i\theta_k}}{\zeta - \rho_k^{-1} e^{i\theta_k}}\left[\frac{1}{\rho_k} - 1\right]P_k^{(II)}.$$

It follows that the elementary factors of both the first and second kind are given by the formula

$$B_k(\zeta) = I + \frac{\zeta + e^{i\theta_k}}{\zeta - r_k e^{i\theta_k}}\,(r_k - 1)P_k \tag{5.2}$$

$$(r_k = |\zeta_k|^{\pm 1},\ \theta_k = \arg \zeta_k,\ P_k^2 = P_k,\ \dim P_k \mathfrak{H} < \infty,\ \pm JP_k \leq 0).$$

In order to apply Theorem 3.1 to an investigation of the convergence of an infinite product of elementary factors we must show that

$$\sum_{k=1}^{\infty} |r_k - 1|\,|P_k| < \infty. \tag{5.3}$$

In view of the fact that $r_k \to 1$ for $k \to \infty$, (5.3) is equivalent to the convergence of the series $\sum |(\ln r_k)P_k|$. For a proof of this convergence we first note that

$$B_k(0) = e^{-(\ln r_k)P_k}.$$

We introduce on $[0, \infty)$ the operator function

$$M(t) = -(\ln r_k)P_k \quad (k - 1 \leq t < k,\ k = 1, 2, \cdots). \tag{5.4}$$

Clearly, $M(t) \leq 0$, and hence the operator function

$$H(t) = \int_a^t M(\tau)d\tau \quad (0 \leq t < \infty)$$

is J-Hermitian-nonincreasing. If

$$\Pi_n(\zeta) = \overset{n}{\underset{k=1}{\overleftarrow{\Pi}}} B_k(\zeta),$$

then

$$\Pi_n(0) = \overset{n}{\underset{0}{\overleftarrow{\int}}} e^{dH(\tau)}.$$

Let

$$W'(t) = \overset{t}{\underset{0}{\overleftarrow{\int}}} e^{dH(\tau)} = \overset{t}{\underset{0}{\overleftarrow{\int}}} e^{M(\tau)d\tau} .$$

The operator function $W(t)$ satisfies the conditions of Theorems 4.1 and 4.2. It follows in particular that

$$0 \leq J - W^*(t)JW(t) \leq J - W^*([t] + 1)JW([t] + 1) = J - \Pi^*_{[t]+1}(0)J\Pi_{[t]+1}(0),$$

$$0 \leq W^{-1}(t)JW^{-1\,*}(t) - J \leq W^{-1}([t] + 1)JW^{-1\,*}([t] + 1) - J \leq \Pi^{-1}_{[t]+1}(0)J\Pi^{-1\,*}_{[t]+1}(0) - J$$

In view of the fact that $Y(0)\Pi^{-1}_{[t]+1}(0)$ is a J-binonexpansive operator the following inequalities hold:

$$J - \Pi^*_{[t]+1}(0)J\Pi_{[t]+1}(0) \leq J - Y^*(0)JY(0),$$

$$\Pi^{-1}_{[t]+1}(0)J\Pi^{-1\,*}_{[t]+1}(0) - J \leq Y^{-1}(0)JY^{-1\,*}(0) - J,$$

and hence for any $t \in [0, \infty)$

$$0 \leq J - W^*(t)JW(t) \leq J - Y^*(0)JY(0),$$

$$0 \leq W^{-1}(t)JW^{-1\,*}(t) - J \leq Y^{-1}(0)JY^{-1\,*}(0) - J. \tag{5.5}$$

Let $W(t) = U(t)R(t)$ be the polar representation of $W(t)$. Then (5.5), (4.14) and (4.15) imply

$$0 \leq J - JR(t) \leq J - Y^*(0)JY(0),$$

$$0 \leq R^{-1}(t)J - J \leq Y^{-1}(0)JY^{-1\,*}(0) - J. \tag{5.6}$$

With the intention of taking advantage of Theorem 4.1, we carry out the following estimates:

$$\int\limits_{0}^{t} |R^{-1}(\tau)R'(\tau)|\,d\tau \le \int\limits_{0}^{t} \|R^{-1}(\tau)\| |[J - JR(\tau)]'|\,d\tau$$

$$\le (1 + \|Y^{-1}(0)JY^{-1\,*}(0) - J\|) \int\limits_{0}^{t} \mathrm{tr}\,[J - JR(\tau)]'\,d\tau$$

$$= (1 + \|Y^{-1}(0)JY^{-1\,*}(0) - J\|)\,\mathrm{tr}\,[J - JR(t)]$$

$$\le (1 + \|Y^{-1}(0)JY^{-1\,*}(0) - J\|)\,\mathrm{tr}\,[J - Y^{*}(0)JY(0)]. \qquad (5.7)$$

Here we have made use of inequality (5.6) as well as the monotone increase of the function $J - JR(t)$ (Theorem 4.2). The legitimacy of transposing the integral and trace signs readily follows from the finite dimensionality of the operators P_k.

From (5.7) we obtain the convergence of the improper integral $\int_0^\infty |N(t)|\,dt$, where $N(t) = \frac{1}{2}[R^{-1}(t)R'(t) - R'(t)R^{-1}(t)]$ (see (4.11)), and hence, on the basis of (4.12) and inequality (3.13),

$$\|U(t)\| \le e^{\int\limits_{0}^{\infty} |N(\tau)|\,d\tau}, \quad \|U^{-1}(t)\| \le e^{\int\limits_{0}^{\infty} |N(\tau)|\,d\tau}$$

Taking advantage of (4.9), we get

$$\int\limits_{0}^{\infty} |M(\tau)|\,d\tau \le e^{2\int\limits_{0}^{\infty} |N(\tau)|\,d\tau} \int\limits_{0}^{\infty} |R^{-1}(\tau)R'(\tau)|\,d\tau < \infty.$$

Since, on the basis of (5.4),

$$\int\limits_{0}^{\infty} |M(\tau)|\,d\tau = \sum\limits_{k=1}^{\infty} |(\ln r_k)P_k|,$$

we have proved

Lemma 5.1. *If* $Y(\zeta) \in K_J^{\mathfrak{S}}$ *and the* $B_k(\zeta)$ $(k = 1, 2, \cdots)$ *are the elementary factors of form (5.1) (or (5.2)) of* $Y(\zeta)$*, then the series*

$$\sum\limits_{k=1}^{\infty} |(\ln r_k)P_k| = \sum\limits_{k=1}^{\infty} \ln \frac{1}{|\zeta_k|}\,|P_k|$$

converges, which is equivalent to the convergence of the series

$$\sum_{k=1}^{\infty} (1 - |\zeta_k|)|P_k|.$$

2. We agree to say that an infinite product $\overleftarrow{\prod}_{k=1}^{\infty} B_k(\zeta)$ of elementary factors uniformly $\mathfrak{G}$- ($\mathfrak{R}$-, strongly) converges interior to the unit disk if for each $\rho \in (0, 1)$ there exists a number k_ρ such that the functions $B_k(\zeta)$ and $B_k^{-1}(\zeta)$ are holomorphic in the disk $|\zeta| \leq \rho$ for $k \geq k_\rho$ and the sequences of products

$$\overleftarrow{\prod_{k=k_\rho}^{k_\rho+n}} B_k(\zeta), \quad \overleftarrow{\prod_{k=k_\rho}^{k_\rho+n}} B_k^{-1}(\zeta) \quad (n = 1, 2, \cdots)$$

uniformly $\mathfrak{G}$- ($\mathfrak{R}$-, strongly) converge interior to the same disk for $n \to \infty$.

It is not difficult to see that a product of elementary factors that is uniformly $\mathfrak{G}$-convergent in the unit disk defines a function

$$\Pi(\zeta) = \overleftarrow{\prod_{k=1}^{\infty}} B_k(\zeta),$$

given for $|\zeta| \leq \rho$ by the equality

$$\Pi(\zeta) = \left[\lim_{n \to \infty} \overleftarrow{\prod_{k=k_\rho}^{k_\rho+n}} B_k(\zeta) \right] \overleftarrow{\prod_{k=1}^{k_\rho-1}} B_k(\zeta),$$

which belongs to the class $K_J^{\mathfrak{G}}$. This result and Lemma 5.1 imply the necessity part of the following assertion.

Theorem 5.1. *In order for an infinite product of elementary factors of form* (5.1) *to be uniformly $\mathfrak{G}$-convergent for $|\zeta| < 1$ it is necessary and sufficient that*

$$\sum_{k=1}^{\infty} (1 - |\zeta_k|)|P_k| < \infty. \tag{5.8}$$

Let us prove the sufficiency part. We select an arbitrary $\rho \in (0, 1)$. Taking advantage of (5.8), we find a k_ρ such that $|\zeta_k| > (1 + \rho)/2$ for $k \geq k_\rho$. Then the operator functions $B_k(\zeta)$ and $B_k^{-1}(\zeta)$ are holomorphic in the disk $|\zeta| \leq \rho$ for $k \geq k_\rho$. Further, $|r_k| > (1 + \rho)/2$ for $k \geq k_\rho$ and hence the following inequalities hold for such k and $|\zeta| \leq \rho$:

$$\left| \frac{\zeta + e^{i\theta_k}}{\zeta - r_k e^{i\theta_k}} (r_k - 1) P_k \right| \leq \frac{4}{1-\rho} (1 - |\zeta_k|) |P_k|.$$

We complete the proof by making use of (5.8) and Theorem 3.1.

Theorem 5.2. *Each operator function* $Y(\zeta) \in K_J^{\mathfrak{S}}$ *can be represented in the form*

$$Y(\zeta) = Y_0(\zeta) \Pi(\zeta),$$

where $\Pi(\zeta)$ *is a uniformly* $\mathfrak{S}$*-convergent product of elementary factors* (5.1) *and* $Y_0(\zeta)$ *is a function in* $K_J^{\mathfrak{S}}$ *that is holomorphic together with* $Y_0^{-1}(\zeta)$ *interior to the unit disk.*

Proof. Let $\Pi(\zeta)$ be the infinite product of elementary factors $B_k(\zeta)$ corresponding to all of the singularities of $Y(\zeta)$ and $Y^{-1}(\zeta)$ interior to the unit disk, which is uniformly $\mathfrak{S}$-convergent on the basis of Lemma 5.1 and Theorem 5.1. We choose a $\rho \in (0, 1)$ and suppose $|\zeta_k| > \rho$ for $k \geq k_0$. Then the operator functions $Y(\zeta)(\prod_{k=1}^{k_0} B_k(\zeta))^{-1}$ and $(\prod_{k=1}^{k_0} B_k(\zeta)) Y^{-1}(\zeta)$ are holomorphic for $|\zeta| \leq \rho$. The sequences

$$Y_n(\zeta) = Y(\zeta) \left[\overset{n}{\underset{k=1}{\overleftarrow{\Pi}}} B_k(\zeta) \right]^{-1} = \left[Y(\zeta) \left[\overset{k_0}{\underset{k=1}{\overleftarrow{\Pi}}} B_k(\zeta) \right]^{-1} \right] \left[\overset{n}{\underset{k=k_0+1}{\overleftarrow{\Pi}}} B_k(\zeta) \right]^{-1},$$

$$Y_n^{-1}(\zeta) = \left[\overset{n}{\underset{k=1}{\overleftarrow{\Pi}}} B_k(\zeta) \right] Y^{-1}(\zeta) \quad (n \geq k_0)$$

uniformly $\mathfrak{S}$-converge in the disk $|\zeta| \leq \rho$ for $n \to \infty$. Let $Y_0(\zeta) = \lim_{n \to \infty} Y_n(\zeta)$. Since $Y_n(\zeta) \in K_J^{\mathfrak{S}}$, we have $Y_0(\zeta) \in K_J^{\mathfrak{S}}$. In addition, the holomorphicity of $Y_n(\zeta)$ and $Y_n^{-1}(\zeta)$ for $|\zeta| \leq \rho$ implies the holomorphicity of $Y_0(\zeta)$ and $Y_0^{-1}(\zeta)$ interior to the unit disk. The theorem is proved.

§ 6. The product of elementary factors for operator functions that are meromorphic in the unit disk

1. Let the symbol M_J denote the set of all operator functions $Y(\zeta)$ contained in S_J (see §1) and such that the only singularities of $Y(\zeta)$ and $Y^{-1}(\zeta)$ interior to the unit disk are poles. Clearly, $K_J \subset M_J$ (Theorem 1.2). As before, the symbol $D(Y)$ $(D(Y^{-1}))$ denotes the set of points of holomorphy of $Y(\zeta)$ $(Y^{-1}(\zeta))$ lying in the disk $|\zeta| < 1$.

We write inequality (1.3) for $Y(\zeta) \in M_J$:

$$\frac{|([Y(\lambda) - Y(\mu)]f, g)|^2}{|\lambda - \mu|^2} \leq \frac{([J - Y^*(\lambda)JY(\lambda))]f, f)([J - Y(\mu)JY^*(\mu)]g, g)}{(1 - |\lambda|^2)(1 - |\mu|^2)}$$

$$(\lambda, \mu \in D(Y); f, g \in \mathfrak{H}). \tag{6.1}$$

Suppose now $\mu \in D(Y) \cap D(Y^{-1})$ and $h \in \mathfrak{H}$. Putting $g = Y^{*-1}(\mu)h$ in (6.1), we prove the inequality

$$\frac{|([Y^{-1}(\mu)Y(\lambda) - I]f, h)|^2}{|\lambda - \mu|^2} \leq \frac{([J - Y^*(\lambda)JY(\lambda)]f, f)([Y^{-1}(\mu)JY^{-1*}(\mu) - J]h, h)}{(1 - |\lambda|^2)(1 - |\mu|^2)}$$

$$(\lambda \in D(Y), \mu \in D(Y^{-1}), f, h \in \mathfrak{H}). \tag{6.2}$$

2. Suppose $Y(\zeta) \in M_J$. We take advantage of the fact that Theorems 1.3 and 1.4 (excluding the assertion on the finite dimensionality of the operator P; see [7], Theorems 4 and 5) hold for such an operator function. We will successively detach from $Y(\zeta)$ (for example, in the order of increasing moduli of the singularities) the elementary factors of the form

$$B_j(\zeta) = U_j \left[\left[\frac{1 - \bar{\zeta}_j \zeta}{\bar{\zeta}_j - \zeta} \right]^{\pm 1} P_j + Q_j \right].$$

Here U_j is a J-unitary operator, $P_j^2 = P_j$, $\pm JP_j \leq 0$, $Q_j = I - P_j$ and ζ_j is a pole of the operator function $Y^{\pm 1}(\zeta)$ such that $|\zeta_j| < 1$.

Let

$$\Pi_n(\zeta) = B_n(\zeta) \cdots B_1(\zeta).$$

Then $Y_n(\zeta) \equiv Y(\zeta)\Pi_n^{-1}(\zeta) \in M_J$ and $\Pi_n(\zeta) \in M_J$. The first inclusion is established by an n-fold application of the analogs of Theorems 1.3 and 1.4. As to the second, it follows from the fact that the $B_j(\zeta)$ $(j = 1, 2, \cdots, n)$ belong to M_J.

The equalities

$$Y(\zeta) = Y_n(\zeta)\Pi_n(\zeta), \quad \Pi_{n+1}(\zeta) = B_{n+1}(\zeta)\Pi_n(\zeta)$$

imply the inequalities

$$0 \leq J - \Pi_n^*(\zeta)J\Pi_n(\zeta) \leq J - \Pi_{n+1}^*(\zeta)J\Pi_{n+1}(\zeta) \leq J - Y^*(\zeta)JY(\zeta), \tag{6.3}$$

216   Ju. P. GINZBURG

$$0 \leq \Pi_n^{-1}(\bar{\zeta}) J \Pi_n^{-1*}(\zeta) - J \leq \Pi_{n+1}^{-1}(\zeta) J \Pi_{n+1}^{-1*}(\zeta) - J \leq Y^{-1}(\zeta) J Y^{-1*}(\zeta) - J$$

$$(6.4)$$

and hence

     1°. *The sequences*

$$\{J - \Pi_n^*(\zeta) J \Pi_n(\zeta)\} \quad (\zeta \in D(Y)),$$

$$\{\Pi_n^{-1}(\zeta) J \Pi_n^{-1*}(\zeta) - J\} \quad (\zeta \in D(Y^{-1}))$$

strongly converge.

     We now write (6.2) for the operator function $\Pi_n(\zeta) \Pi_m^{-1}(\zeta) = B_n(\zeta) \cdots$
$\cdots B_{m+1}(\zeta) \in M_J$ $(m < n)$. Replacing f by $\Pi_m(\lambda) f$ and h by $\Pi_m^{-1*}(\mu) h$ in
the resultant inequality, we prove that

$$\frac{|([\Pi_m^{-1}(\mu) \Pi_m(\lambda) - \Pi_n^{-1}(\mu) \Pi_n(\lambda)] f, h)|^2}{|\lambda - \mu|^2}$$

$$\leq \frac{([\Pi_n^*(\lambda) J \Pi_n(\lambda) \Pi_m^*(\lambda) J \Pi_m(\lambda)] f, f)([\Pi_m^{-1}(\mu) J \Pi_m^{-1*}(\mu) - \Pi_n^{-1}(\mu) J \Pi_n^{-1*}(\mu)] h, h)}{(1 - |\lambda|^2)(1 - |\mu|^2)}$$

$$(m < n, \lambda \in D(Y), \mu \in D(Y^{-1}); f, h \in \mathfrak{H}). \qquad (6.5)$$

It follows from this result and (6.4) that

$$\|[\Pi_m^{-1}(\mu) \Pi_m(\lambda) - \Pi_n^{-1}(\mu) \Pi_n(\lambda)] f\|^2$$

$$\leq \frac{|\lambda - \mu|^2}{(1 - |\lambda|^2)(1 - |\mu|^2)} \|[\Pi_n^*(\lambda) J \Pi_n(\lambda) - \Pi_m^*(\lambda) J \Pi_m(\lambda)] f\| \, \|f\| \, \|Y^{-1}(\mu) J Y^{-1*}(\mu) - J\|.$$

By now making use of 1°, we prove the first part of the following proposition.

     2°. *The sequence* $\{\Pi_n^{-1}(\mu) \Pi_n(\lambda)\}$ *strongly converges for* $\lambda \in D(Y)$ *and*
$\mu \in D(Y^{-1})$. *This convergence is uniform in* λ (μ) *interior to* $D(Y)$ ($D(Y^{-1})$)
for any fixed $\mu \in D(Y^{-1})$ ($\lambda \in D(Y)$).

     As to the second assertion, in order to prove it we note that inequality
(6.2), when written for $\Pi_n(\zeta)$, implies

$$\|\Pi_n^{-1}(\mu) \Pi_n(\lambda) - I\| \leq \frac{|\lambda - \mu|^2}{(1 - |\lambda|^2)(1 - |\mu|^2)} \|J - Y^*(\lambda) J Y(\lambda)\| \, \|Y^{-1}(\mu) J Y^{-1*}(\mu) - J\|$$

$$(\lambda \in D(Y), \mu \in D(Y^{-1})).$$

It follows that for any compact sets $Q_1 \subset D(Y)$ and $Q_2 \subset D(Y^{-1})$ there exists a number $\alpha(Q_1, Q_2)$ such that $\|\Pi_n^{-1}(\mu)\Pi_n(\lambda)\| \leq \alpha(Q_1, Q_2)$ for $\lambda \in Q_1$, $\mu \in Q_2$ and $n = 1, 2, \cdots$. Applying the generalized Vitali theorem ([27], Theorem 3.14.1), we establish the required uniform convergence.

In order to prove the convergence of the sequence $\Pi_n(\zeta)$ we must choose the *J*-unitary operators U_j in the expression for the elementary factors $B_j(\zeta)$ in a special way. Namely, let the U_j be such that the operators $\Pi_n(0) = B_n(0) \cdots B_1(0)$ $(n = 1, 2, \cdots)$ are *J*-moduli,[1] i.e. *J*-Hermitian operators with a positive spectrum (the existence of such U_j follows from the possibility of a polar representation of an invertible *J*-binonexpansive operator [18,23]).

It is not difficult to show (cf. §5.1) that $\Pi_n(0)$ can be represented in the form

$$\Pi_n(0) = V_n \overleftarrow{\int_0^n} e^{dH(t)} \quad (n = 1, 2, \cdots), \tag{6.6}$$

where the V_n are *J*-unitary operators and $H(t)$ is an absolutely continuous *J*-Hermitian-nonincreasing operator function. Thus $\Pi_n(0) \equiv R(n)$ is the *J*-modulus of the operator $W(n) \equiv \overleftarrow{\int_0^n} e^{dH(t)}$. Taking advantage of Theorem 4.2 and inequalities (6.3) and (6.4), we get

$$0 \leq J - J\Pi_n(0) \leq J - J\Pi_{n+1}(0) \leq J - \Pi_{n+1}^*(0)J\Pi_{n+1}(0) \leq J - Y^*(0)JY(0), \tag{6.7}$$

$$0 \leq \Pi_n^{-1}(0)J - J \leq \Pi_{n+1}^{-1}(0)J - J \leq \Pi_{n+1}^{-1}(0)J\Pi_{n+1}^{-1*}(0) - J \leq Y^{-1}(0)JY^{-1*}(0) - J. \tag{6.8}$$

This implies the strong convergence of the sequences $\{\Pi_n(0)\}$ and $\{\Pi_n^{-1}(0)\}$. Since on the basis of $2°$ the sequences $\{\Pi_n^{-1}(0)\Pi_n(\zeta)\}$ and $\{\Pi_n^{-1}(\zeta)\Pi_n(0)\}$ converge strongly and uniformly in ζ interior to the domains $D(Y)$ and $D(Y^{-1})$ respectively, the same assertion is true for the sequences $\{\Pi_n(\zeta)\}$ and $\{\Pi_n^{-1}(\zeta)\}$. Recalling that $Y(\zeta) = Y_n(\zeta)\Pi_n(\zeta)$, where $Y_n(\zeta)$ and $Y_n^{-1}(\zeta)$ are holomorphic for $|\zeta| < |\zeta_n|$, we complete the proof of the following assertion.

Theorem 6.1. *Each operator function* $Y(\zeta) \in M_J$ *can be represented in the form*

---

1) We assume that $0 \in D(Y)$, $0 \in D(Y^{-1})$.

$$Y(\zeta) = Y_0(\zeta)\Pi(\zeta),$$

where $\Pi(\zeta)$ *is a strongly convergent (for* $|\zeta| < 1$*) product of elementary factors and* $Y_0(\zeta) \in M_J$*, with* $Y_0(\zeta)$ *and* $Y_0^{-1}(\zeta)$ *being holomorphic interior to the unit disk.*

Remark. *If* $Y(\zeta) \in K_J$ *then the Blaschke product* $\Pi(\zeta)$ *is* $\Re$*-convergent.* In fact, inequalities (6.3), (6.4), (6.7) and (6.8) imply the $\Re$-convergence of the sequences

$$\{J - \Pi_n^*(\zeta)J\Pi_n(\zeta)\}, \quad \{\Pi_n^{-1}(\zeta)J\Pi_n^{-1*}(\zeta) - J\}, \quad \{J - J\Pi_n(0)\},$$

and $\{\Pi_n^{-1}(0)J - J\}$. Inequality (6.5) shows that the sequence $\{\Pi_n^{-1}(\mu)\Pi_n(\lambda)\}$ also $\Re$-converges. This result implies our assertion.

3. As can be seen from what has been stated in the present section and in §5, the elementary factors in an infinite Blaschke-Potapov product can be normalized in various ways (through the choice of the left unitary factor). Also, the order in which the factors are detached can be different. At any rate, we have

Theorem 6.2. *Suppose* $Y(\zeta) \in M_J$ *and*

$$Y(\zeta) = Y_0^{(1)}(\zeta)\Pi^{(1)}(\zeta) = Y_0^{(2)}(\zeta)\Pi^{(2)}(\zeta),$$

where $\Pi^{(1)}(\zeta)$ *and* $\Pi^{(2)}(\zeta)$ *are strongly convergent products of elementary factors while* $Y_0^{(j)}(\zeta) \in M_J$ *and* $Y_0^{(j)}(\zeta)$ *and* $Y_0^{(j)-1}(\zeta)$ *are holomorphic for* $|\zeta| < 1$ $(j = 1, 2)$. *Then* $\Pi^{(2)}(\zeta) = U\Pi^{(1)}(\zeta)$*, where* U *is a constant* J*-unitary operator.*

This theorem was established by V. P. Potapov [2] for a finite-dimensional $\mathfrak{H}$. In the general case the arguments follow the same idea, although they become significantly more cumbersome. Theorem 6.1 will be proved in §9 for the case $J = I$.

BIBLIOGRAPHY

[1] V. P. Potapov, *On holomorphic matrix functions bounded in the unit circle*, Dokl. Akad. Nauk SSSR 72 (1950), 849–852. (Russian) MR 13, 736.

[2] ——————, *The multiplicative structure of J-contractive matrix functions*, Trudy Moskov. Mat. Obšč. 4 (1955), 125–236; English transl., Amer. Math. Soc. Transl. (2) 15 (1960), 131–243.   MR 17, 958; 22 #5733.

[3] Ju. P. Ginzburg, *J-nonexpansive analytic operator functions,* Dissertation, Odessa, 1958. (Russian)

[4] ———, *On J-contractive operator functions,* Dokl. Akad. Nauk SSSR 117 (1957), 171–173. (Russian)   MR 20 #1203.

[5] ———, *The factorization of analytic matrix functions,* Dokl. Akad. Nauk SSSR 159 (1964), 489–492 = Soviet Math. Dokl. 5 (1964), 1510–1514.   MR 30 #3228.

[6] ———, *Multiplicative representations of bounded analytic operator-functions,* Dokl. Akad. Nauk SSSR 170 (1966), 23–26 = Soviet Math. Dokl. 7 (1966), 1125–1128.   MR 34 #611.

[7] ———, *A maximum principle for J-contracting operator functions and some of its consequences,* Izv. Vysš. Učebn. Zaved. Matematika 1963, no. 1 (32), 42–53. (Russian)   MR 26 #6793.

[8] M. S. Livšic, *On spectral resolution of non-selfadjoint linear operators,* Mat. Sb. 34 (76) (1954), 145–199; English transl., Amer. Math. Soc. Transl. (2) 5 (1957), 67–114.   MR 16, 48; 18, 748.

[9] M. S. Brodskiĭ and M. S. Livšic, *Spectral analysis of non-selfadjoint operators and intermediate systems,* Uspehi Mat. Nauk 13 (1958), no. 1 (79), 3–85; English transl., Amer. Math. Soc. Transl. (2) 13 (1960), 265–346.   MR 20 #7221; 22 #3982.

[10] V. T. Poljackiĭ, *On the reduction of quasi-unitary operators to a triangular form,* Dokl. Akad. Nauk SSSR 113 (1957), 756–759. (Russian) MR 19, 873.

[11] A. V. Kužel', *Spectral analysis of unbounded non-selfadjoint operators,* Dokl. Akad. Nauk SSSR 125 (1959), 35–37. (Russian)   MR 22 #2905.

[12] L. A. Sahnovič, *Dissipative operators with an absolutely continuous spectrum,* Dokl. Akad. Nauk SSSR 167 (1966), 760–763 = Soviet Math. Dokl. 7 (1966), 483–486.   MR 34 #3334.

[13] M. S. Brodskiĭ, *A multiplicative representation of certain analytic operator-functions,* Dokl. Akad. Nauk SSSR 138 (1961), 751–754 = Soviet Math. Dokl. 2 (1961), 695–698.   MR 24 #A1031.

[14] I. C. Gohberg and M. G. Kreĭn, *On the multiplicative representation of the characteristic functions of operators close to the unitary ones,* Dokl. Akad. Nauk SSSR 164 (1965), 732–735 = Soviet Math. Dokl. 6 (1965), 1279–1283.   MR 33 #571.

[15] V. M. Brodskiĭ, *Multiplicative representation of the characteristic functions of contraction operators*, Dokl. Akad. Nauk SSSR 173 (1967), 256–259 = Soviet Math. Dokl. 8 (1967), 362–366.　MR 36 #3154.

[16] ――――, *Some theorems on operator integrals over chains of orthogonal projections and their applications to the theory of characteristic functions*, Izv. Akad. Nauk SSSR Ser. Mat. (to appear)

[17] I. S. Iohvidov, *Boundedness of J-isometric operators*, Uspehi Mat. Nauk 16 (1961), no. 4 (100), 167–170; English transl., Amer. Math. Soc. Transl. (2) 47 (1965), 67–71.　MR 24 #A1628.

[18] Ju. P. Ginzburg, *On J-nonexpansive operators in Hilbert space*, Naučn. Zap. Fiz.-Mat. Fak. Odessk. Gos. Ped. Inst. 22 (1958), no. 1, 13–20. (Russian)

[19] I. S. Iohvidov, *Linear-fractional transformations of J-non-dilating operators*, Akad. Nauk Armjan. SSR Dokl. 42 (1966), no. 1, 3–8. (Russian)　MR 33 #4678.

[20] ――――, *J-nondilating operators in a Banach space*, Dokl. Akad. Nauk SSSR 169 (1966), 519–522 = Soviet Math. Dokl. 7 (1966, 962–965. MR 33 #6399.

[21] M. G. Kreĭn and Ju. L. Šmul'jan, *A class of operators in a space with an indefinite metric*, Dokl. Akad. Nauk SSSR 170 (1966), 34–37 = Soviet Math. Dokl. 7 (1966), 1137–1141.　MR 34 #615.

[22] ――――, *On the plus-operators in a space with an indefinite metric*, Mat. Issled. 1 (1966), no. 1, 131–161; English transl., Amer. Math. Soc. Transl. (2) 85 (1969), 93–113.　MR 34 #4923.

[23] ――――, *J-polar representations of plus-operators*, Mat. Issled. 1 (1966), no. 2, 172–210; English transl., Amer. Math. Soc. Transl. (2) 85 (1969), 115–143.　MR 34 #8183.

[24] I. C. Gohberg and M. G. Kreĭn, *Introduction to the theory of linear non-selfadjoint operators in Hilbert space*, "Nauka", Moscow, 1965; English transl., Transl. Math. Monogrpahs, vol. 18, Amer. Math. Soc., Providence, R.I., 1969.　MR 36 #3137.

[25] I. M. Gel'fand and N. Ja. Vilenkin, *Generalized functions. Vol. 4: Applications of harmonic analysis*, Fizmatgiz, Moscow, 1961; English transl., Academic Press, New York, 1964.　MR 26 #4173; 30 #4152.

[26] D. L. Kučer, *Some questions on the integration of normal differential equations in Banach spaces*, Naučn. Zap. Kafedr. Mat.-Fiz. Estest. Odessk. Gos. Ped. Inst. 25 (1961), no. 2, 25–37. (Russian)

[27] E. Hille and R. S. Phillips, *Functional analysis and semi-groups*, rev. ed., Amer. Math. Soc. Colloq. Publ., vol. 31, Amer. Math. Soc., Providence, R.I., 1957; Russian transl., IL, Moscow, 1962. MR **19**, 664.

Translated by:
S. Smith

Amer. Math. Soc. Transl.
(2) Vol. 96, 1970

ON MULTIPLICATIVE REPRESENTATIONS OF
J-NONEXPANSIVE OPERATOR-FUNCTIONS. II[*]

Ju. P. GINZBURG

§ 7. Tests for compactness of operator families

1. In the following sections we need certain tests for compactness of families of completely continuous and nuclear operators.

Theorem 7.1. *If the sequences* $\{A_n\}$ *and* $\{B_n\}$ *of completely continuous nonnegative operators* $\Re$-*converge, i.e. converge in the uniform operator topology, and if the operators* C_n, $n = 1, 2, \cdots$ *satisfy the inequality*

$$|(C_n f, g)| \leq \|A_n f\| \, \|B_n g\| \tag{7.1}$$

for all $f, g \in \mathfrak{H}$, *then the operators* C_n *are completely continuous and the family* $\{C_n\}$ *is* $\Re$-*compact.*

Proof. Because the sequences $\{A_n\}$ and $\{B_n\}$ converge it follows that there exist constants μ_A and μ_B such that $\|A_n\| \leq \mu_A$ and $\|B_n\| \leq \mu_B$. This means that $\|C_n\| \leq \mu_A \mu_B$, so that the sequence $\{C_n\}$ is weakly compact. Without loss of generality we may suppose that this sequence weakly converges to an operator C. We shall show that indeed $C_n \Rightarrow C$.

To this end we first note that (7.1) implies the inequalities

$$\|C_n f\| \leq \mu_B \, \|A_n f\|, \quad \|C_n^* f\| \leq \mu_A \, \|B_n f\| \quad (f \in \mathfrak{H}, \, n = 1, 2, \cdots),$$

which means that the C_n are completely continuous. If $A_n \Rightarrow A$ and $B_n \Rightarrow B$, then

$$|(C f, g)| \leq \|A f\| \, \|B g\|$$

and

$$\|C f\| \leq \mu_B \, \|A f\|, \quad \|C^* f\| \leq \mu_A \, \|B f\| \quad (f \in \mathfrak{H}).$$

Then for any $f \in \mathfrak{H}$

$$\|C f - C_n f\|^2 = ((C^* - C_n^*)(C - C_n) f, f)$$
$$\leq |((C - C_n) f, C f)| + \|C_n^* (C - C_n) f\| \, \|f\| \leq$$

---

*Translation of Mat. Issled. 2 (1967), no 3, 20–51.

$$\leq |((C - C_n)f,\, Cf)| + \mu_A \|B_n (C - C_n)f\| \,\|f\|$$

$$\leq |((C - C_n)f,\, Cg)| + \mu_A \|B_n - B\| \,(\|C\| + \|C_n\|)\, \|f\|$$

$$+ \mu_A \|B(C - C_n)f\|.$$

The first term on the right tends to zero because of the weak convergence of of C_n to C, the second because $B_n \Rightarrow B$, and the third because the operator B is completely continuous, which means that it transforms the weakly converging sequence $C_n f$ into a strongly converging one. Thus we have proved that $\|Cf - C_n f\| \longrightarrow 0$ $(f \in \mathfrak{H})$.

Suppose we are given $\epsilon > 0$. Let $\mathfrak{K}$ be the unit sphere in $\mathfrak{H}$. Then the set $A\mathfrak{K}$ is compact. Now we choose $g_i \in \mathfrak{K}$ $(i = 1, \cdots, k)$, so that the set $\{Ag_1, \cdots, Ag_k\}$ is an ϵ-net in $A\mathfrak{K}$. If f is an arbitrary vector of $\mathfrak{K}$, then there exists a $j = j(f)$ such that $\|Af - Ag_j\| < \epsilon$. Then

$$\|Cf - C_n f\| \leq \| C(f - g_j)\| + \|Cg_j - C_n g_j\| + \|C_n(g_j - f)\|$$

$$\leq \mu_B \|Af - Ag_j\| + \|Cg_j - C_n g_j\| + \mu_B \|A_n(f - g_j)\|$$

$$\leq \mu_B \{\|Af - Ag_j\| + \|A_n - A\| \,\|f - g_j\| + \|Af - Ag_j\|\} + \|Cg_j - C_n g_j|.$$

Now we choose an n_0 such that if $n \geq n_0$ the inequalities $\|A_n - A\| < \epsilon$ and $\|Cg_i - C_n g_i\| < \epsilon$ $(i = 1, \cdots, k)$ are satisfied. Then

$$\|Cf - C_n f\| < (4\mu_B + 1)\epsilon$$

for all $f \in \mathfrak{K}$, so that $C_n \Rightarrow C$. The theorem is proved.

Corollary 7.1. (M. S. Livšic [2]). *If $0 \leq C_n \leq A_n$ and the sequence $\{A_n\}$ of completely continuous operators $\mathfrak{K}$-converges, then the family $\{C_n\}$ is $\mathfrak{K}$-compact.*

The proof follows immediately from the fact that

$$|(C_n f,\, g)| \leq \sqrt{(C_n f,\, f)(C_n g,\, g)} \leq \sqrt{(A_n f,\, f)(A_n g,\, g)} = \|A_n^{1/2}f\| \,\|A_n^{1/2}g\|.$$

Corollary 7.2. *If $0 \leq C_1 \leq \cdots \leq C_n \leq A$, where A is a completely continuous operator, then the sequence $\{C_n\}$ $\mathfrak{K}$-converges.*

For the proof we note that because of Corollary 7.1 the sequence $\{C_n\}$ is $\mathfrak{K}$-compact, and that because of the monotonicity it is not possible for it to have several partial limits.

2. Now we shall establish an analog of Theorem 7.1 for nuclear operators. First we shall prove the following auxiliary assertion.

Lemma 7.1. *If for all $f,\, g \in \mathfrak{H}$ the inequality*

$$|(Cf,\, g)|^2 \leq (Af,\, f)(Bg,\, g) \tag{7.2}$$

holds, where $0 \leq A \in \mathfrak{S}$, $0 \leq B \in \mathfrak{S}$, then the operator C is in $\mathfrak{S}$ and

$$|C|^2 \le |A|\,|B|. \tag{7.3}$$

Proof. Suppose that $\{f_n\}$ and $\{g_n\}$ are two arbitrary bases in $\mathfrak{H}$. It follows from (2.2) that

$$\left[\sum_{n=1}^{\infty} |(Cf_n, g_n)|\right]^2 \le \left[\sum_{n=1}^{\infty} \sqrt{(Af_n, f_n)(Bg_n, g_n)}\right]^2$$

$$\le \sum_{n=1}^{\infty}(Af_n, f_n)\sum_{n=1}^{\infty}(Bg_n, g_n) = |A|\,|B| < \infty.$$

This means $[^3]$ that $C \in \mathfrak{S}$, and since

$$|C| = \sup \sum_{n=1}^{\infty}|(Cf_n, g_n)|,$$

where the upper limit is taken over all possible orthonormal bases in $\mathfrak{H}$, then $|C|^2 \le |A|\,|B|$.

Theorem 7.2. *If the sequences $\{A_n\}$ and $\{B_n\}$ of nonnegative nuclear operators $\mathfrak{S}$-converge, and if the operators C_n satisfy the inequalities*

$$|(C_n f, g)|^2 \le (A_n f, f)(B_n g, g) \quad (n = 1, 2, \cdots) \tag{7.4}$$

for all $f, g \in \mathfrak{H}$, then the sequence $\{C_n\}$ is $\mathfrak{S}$-compact.

Proof. Suppose that $A_n \longrightarrow A$ and $B_n \longrightarrow B$. Then $A_n \Rightarrow A$ and $B_n \Rightarrow B$, and Theorem 7.1 implies that the sequence $\{C_n\}$ is $\mathfrak{R}$-compact. If C is one of its partial limits, then we will take $C_n \Rightarrow C$. We shall prove that indeed $C_n \longrightarrow C$.

Suppose that $\{f^{(j)}\}$ is an orthonormal basis in $\mathfrak{H}$, $P^{(k)}$ an orthoprojector onto the linear hull of the vectors $f^{(1)}, \cdots, f^{(k)}$, and $Q^{(k)} = I - P^{(k)}$. Now suppose that $\epsilon > 0$. Since

$$\sum_{j=1}^{\infty}(Af^{(j)}, f^{(j)}) = |A| < \infty, \quad \sum_{j=1}^{\infty}(Bf^{(j)}, f^{(j)}) = |B| < \infty,$$

there exists a k_0 such that

$$\sum_{j=k_0}^{\infty}(Af^{(j)}, f^{(j)}) < \frac{\epsilon}{4}, \quad \sum_{j=k_0}^{\infty}(Bf^{(j)}, f^{(j)}) < \frac{\epsilon}{4},$$

i.e.

$$|Q^{(k_0)}AQ^{(k_0)}| < \frac{\epsilon}{4}, \quad |Q^{(k_0)}BQ^{(k_0)}| < \frac{\epsilon}{4}. \tag{7.5}$$

Now we choose an n_0 such that for $n \ge n_0$ the inequalities $|A_n - A| < \epsilon/4$ and $|B_n - B| < \epsilon/4$ are satisfied. Then for $n \ge n_0$

$$|Q^{(k0)}A_n Q^{(k0)}| \le |Q^{(k0)}AQ^{(k0)}| + |Q^{(k0)}(A_n - A)Q^{(k0)}|$$

$$\le |Q^{(k0)}AQ^{(k0)}| + |A_n - A|.$$

Thus for $n \ge n_0$ we have proved the first of the inequalities

$$|Q^{(k0)}A_n Q^{(k0)}| < \frac{\epsilon}{2}, \quad |Q^{(k0)}B_n Q^{(k0)}| < \frac{\epsilon}{2}. \tag{7.6}$$

The second is proved analogously. From (7.4) it follows that

$$|(Q^{(k0)}C_n Q^{(k0)}f,\, g)|^2 \le (Q^{(k0)}A_n Q^{(k0)}f,\, f)\, (Q^{(k0)}B_n Q^{(k0)}g,\, g),$$

$$|(Q^{(k0)}CQ^{(k0)}f,\, g)|^2 \le (Q^{(k0)}f,\, f)\, (Q^{(k0)}g,\, g),$$

so that, using Lemma 7.1 and (7.5) and (7.6), we get

$$|Q^{(k0)}C_n Q^{(k0)}| \le \sqrt{|Q^{(k0)}A_n Q^{(k0)}|\ |Q^{(k0)}B_n Q^{(k0)}|} < \frac{\epsilon}{2},$$

$$|Q^{(k0)}CQ^{(k0)}| \le \sqrt{|Q^{(k0)}AQ^{(k0)}|\ |Q^{(k0)}BQ^{(k0)}|} < \frac{\epsilon}{4}.$$

Since

$$|C_n - C| \le |P^{(k0)}(C_n - C)P^{(k0)}| + |P^{(k0)}(C_n - C)Q^{(k0)}|$$

$$+ |Q^{(k0)}(C_n - C)P^{(k0)}| + |Q^{(k0)}(C_n - C)Q^{(k0)}| \le 3k_0 \|C_n - C\|$$

$$+ |Q^{(k0)}C_n Q^{(k0)}| + |Q^{(k0)}CQ^{(k0)}| < 3k_0 \|C_n - C\| + \frac{3\epsilon}{4},$$

on taking $n_1 \ge n_0$ and such that $\|C_n - C\| < \epsilon/12k_0$ if $n \ge n_1$ we get $|C_n - C| < \epsilon$ for $n \ge n_1$. The theorem is proved.

The propositions formulated below are proved in just the same way as Corollaries 7.1 and 7.2.

Corollary 7.3. *If* $0 \le C_n \le A_n$ $(n = 1, 2, \cdots)$ *and the sequence* $\{A_n\}$ *of nuclear operators* $\mathfrak{S}$-*converges, then the sequence* $\{C_n\}$ *is* $\mathfrak{S}$-*compact.*

Corollary 7.4. *If*

$$0 \le C_1 \le C_2 \le \cdots \le C_n \le \cdots \le A,$$

where $A \in \mathfrak{S}$, *then the sequence* $\{C_n\}$ $\mathfrak{S}$-*converges.*

§8. Proof of the fundamental theorem and representations of operator-functions of class $K_J^{\mathfrak{S}}$

1. Suppose that the operator-function $Y(\zeta)$ lies in the class $K_J^{\mathfrak{S}}$. We suppose moreover that $Y(\zeta)$ and $Y^{-1}(\zeta)$ are holomorphic within the unit

disk. The set of all such operator-functions is denoted by $\tilde{K}_J^G$. Using Theorem 2.2, we find that for each function $Y(\zeta) \in \tilde{K}_J^G$ there exists a sequence

$$Y_n(\zeta) = V_n B_{k_n}^{(n)}(\zeta) \cdots B_1^{(n)}(\zeta), \qquad (8.1)$$

where V_n is a J-unitary operator and the $B_j^{(n)}(\zeta)$ are elementary factors, such that

$$Y_n(\zeta) \longrightarrow Y(\zeta), \ Y_n^{-1}(\zeta) \longrightarrow Y^{-1}(\zeta) \qquad (8.2)$$

uniformly inside the unit disk.

We enumerate the factors of the first and second kind in the same way as in §5. We take the J-unitary operator in the expression for a factor of the third kind to be equal to I. Thus the factors of the first and second kind have the form

$$B_j^{(n)}(\zeta) = I + \frac{\zeta + e^{i\theta_j^{(n)}}}{\zeta - r_j^{(n)} e^{i\theta_j^{(n)}}} (r_j^{(n)} - 1) P_j^{(n)} \qquad (8.3)$$

$$(r_j^{(n)} = |\zeta_j^{(*)}|^{\pm 1}, \ \theta_j^{(n)} = \arg \zeta_j^{(n)}, \ P_j^{(n)2} = P_j^{(n)}, \ \pm J P_j^{(n)} \le 0, \ \dim P_j^{(n)} \mathfrak{H} < \infty),$$

and factors of the third kind

$$B_j^{(n)}(\zeta) = I + \frac{\zeta + e^{i\theta_j^{(n)}}}{\zeta - e^{i\theta_j^{(n)}}}$$

$$G_j^{(n)} \ (\theta_j^{(n)} = \arg \zeta_j^{(n)}, \ G_j^{(n)2} = 0, \ J G_j^{(n)} \ge 0, \ \dim G_j^{(n)} \mathfrak{H} < \infty). \qquad (8.4)$$

Accordingly, any elementary factor may be written in the form

$$B_j^{(n)}(\zeta) = I + \frac{\zeta + e^{i\theta_j^{(n)}}}{\zeta - r_j^{(n)} e^{i\theta_j^{(n)}}} G_j^{(n)}, \qquad (8.5)$$

where $G_j^{(n)}$ is a J-Hermitian-nonegative finite-dimensional operator, and $r_j^{(n)} > 0$.

It follows from (8.2) that

$$\lim_{n \to \infty} \max_{1 \le j \le k_n} |1 - r_j^{(n)}| = 0. \qquad (8.6)$$

It is not hard to see that the following relation holds:

$$G_j^{(n)2} = (r_j^{(n)} - 1) G_j^{(n)}. \qquad (8.7)$$

2.　Just as in §5, a fundamental role is played here by the study of the values of the operator-function

$$\Pi_n(\zeta) = B_{k_n}^{(n)}(\zeta) \cdots B_1^{(n)}(\zeta)$$

for $\zeta = 0$.　Here we essentially use the results of §§4 and 7.　Put

$$M_j^{(n)} = \frac{\ln r_j^{(n)}}{1 - r_j^{(n)}} G_j^{(n)} (r_j^{(n)} \neq 1),\quad M_j^{(n)} = -G_j^{(n)} (r_j^{(n)} = 1). \tag{8.8}$$

Obviously

$$JM_j^{(n)} \leq 0,\ \dim M_j^{(n)} \mathfrak{H} < \infty,\ B_j^{(n)}(0) = I - \frac{1}{r_j^{(n)}} G_j^{(n)} = e^{M_j^{(n)}}.$$

Thus

$$\Pi_n(0) = e^{M_{k_n}^{(n)}} \cdots e^{M_1^{(n)}}.$$

Suppose that

$$M_n(t) = M_j^{(n)} (j - 1 \leq t < j;\ j = 1, 2, \cdots, k_n),\ H_n(t) = \int_0^t M_n(\tau)d\tau.$$

Then

$$\Pi_n(0) = \overset{k_n}{\underset{0}{\overleftarrow{\int}}} e^{dH_n(\tau)} = \overset{k_n}{\underset{0}{\overleftarrow{\int}}} e^{M_n(\tau)d\tau}.$$

Consider the vector-functions

$$W_n(t) = \overset{t}{\underset{0}{\overleftarrow{\int}}} e^{dH_n(\tau)}.$$

On the basis of Theorem 4.1 the components of the polar representation $W_n(t) = U_n(t)R_n(t)$, which exists on the basis of Theorem 4.2, and of the operator-function $H_n(t)$ are connected by the formulas

$$H_n(t) = \tfrac{1}{2} \int_0^t U_n(\tau)[R_n'(\tau) R_n^{-1}(\tau) + R_n^{-1}(\tau) R_n'(\tau)]U_n^{-1}(\tau)\, d\tau, \tag{8.9}$$

$$U_n(t) = \overset{t}{\underset{0}{\overleftarrow{\int}}} e^{N_n(\tau)d\tau},\ N_n(t) = \tfrac{1}{2}[R_n^{-1}(t)R_n'(t) = R_n'(t)R_n^{-1}(t)]. \tag{8.10}$$

We study first of all the operator-functions

$$F_n(t) = \int_0^t R_n^{-1}(t)R_n'(t)dt\ (0 \leq t \leq k_n).$$

The following estimate holds (cf. §5, the proof of inequality (5.7)):

$$\mathfrak{S}\text{-}\underset{[0,k_n]}{\text{var}}\ F_n = \int_0^{k_n} |R_n^{-1}(\tau)R_n'(\tau)|\,d\tau \le \underset{[0,k_n]}{\max}\ \|R_n^{-1}(\tau)\|\int_0^{k_n}\text{tr}\,[-JR_n'(\tau)]\,d\tau$$

$$\le [1 + \|Y_n^{-1}(0)\,JY^{-1*}(0) - J\|]\,\text{tr}\,[J - Y_n^*(0)\,JY_n(0)].$$

Then it follows from (8.2) that there exists a constant $\alpha > 0$ such that

$$\alpha_n \equiv \mathfrak{S}\text{-}\underset{[0,k_n]}{\text{var}}\ F_n \le \alpha \quad (n = 1, 2, \cdots). \tag{8.11}$$

Taking into account the fact that

$$\mathfrak{S}\text{-}\text{var}\,F_n = \int_0^{k_n} |M_n(\tau)|\,d\tau = \sum_{j=1}^{k_n} |M_j^{(n)}|,$$

we verify the validity of the following proposition:

1) *The sums* $\sigma_n = \sum_{j=1}^{k_n} |M_j^{(n)}|$ *are uniformly bounded.*

In order to study the operator sums $S_n = \sum_{j=1}^{k_n} M_j^{(n)}$, we proceed as follows. Fixing n, we introduce the new variable

$$s = \phi_n(t) = \text{arc tg}\,t + \mathfrak{S}\text{-}\underset{[0,t]}{\text{var}}\ F_n \quad (0 \le t \le k_n).$$

Since $\phi_n(t)$ increases strictly on $[0, k_n]$, we can define an inverse function $t = \psi_n(s)$ on the segment $[0, \mu_n]\,(\mu_n = \text{arc tg}\,k_n + \alpha_n)$, and then extend it to $(\mu_n, \mu]\,(\mu = \pi/2 + \alpha)$ as a constant: $\psi_n(s) = k_n, \ \mu_n < s \le \mu.$

Put $\widetilde{F}_n(s) = F_n[\psi_n(s)] \ (0 \le s \le \mu).$ Then for $0 \le s' \le s'' \le \mu$ the following inequalities hold:

$$|\widetilde{F}_n(s'') - \widetilde{F}_n(s')| = |F_n[\psi_n(s'')] - F_n[\psi_n(s')]| \le \mathfrak{S}\text{-}\underset{[t',t'']}{\text{var}}\ F_n$$

$$< \phi_n(t'') - \phi_n(t') \le s'' - s' \quad (t' = \psi_n(s'),\ t'' = \psi_n(s'')).$$

Thus

$$|\widetilde{F}_n(s'') - \widetilde{F}_n(s')| \le |s'' - s'| \quad (s', s'' \in [0, \mu], n = 1, 2, \cdots). \tag{8.12}$$

Now we prove the compactness of the family $\{\widetilde{F}_n(s)\}$ of operator-functions. To this end we note that for fixed n, for any $f, g \in \mathfrak{H}$ and $s \in [0, \mu]$ the following inequalities are valid:

$$|(\widetilde{F}_n(s)f, g)|^2 = |(F_n(t)f, g)|^2 = \left|\int_0^t (R_n^{-1}(\tau)\,R_n'(\tau)f, g)\,d\tau\right|^2$$

$$\le \left\{\int_0^t |([-JR_n'(\tau)]^{\frac12}f, [-JR_n'(\tau)]^{\frac12}R_n^{-1}(\tau)\,Jg)|\,d\tau\right\}^2 \le$$

$$\leq \left[\int_0^t \sqrt{(-JR_n'(\tau)f,\, f)}\,\sqrt{(-R_n^{-1}(\tau)J \cdot JR_n'(\tau)R_n^{-1}(\tau)Jg,\, g)}\;d\tau\right]^2$$

$$\leq \int_0^t (-JR_n'(\tau)f,\, f)\,d\tau \int_0^t ([R_n^{-1}(\tau)J]'\,g,\, g)\,d\tau$$

$$= ([J - JR_n(t)]f,\, f)([R_n^{-1}(t)J - J]g,\, g).$$

In the above calculations we used the monotone increasing property and the
absolute continuity of the operator-functions $J - JR_n(t)$ and $R_n^{-1}(t)J - J$
(Theorems 4.1 and 4.2). Taking account of the inequalities

$$0 \leq J - JR_n(t) \leq J - Y_n^*(0)JY_n(0),\; 0 \leq R_n^{-1}(t)J - J \leq Y_n^{-1}(0)JY_n^{-1*}(0) - J$$

(cf. (5.6)), we obtain

$$|(\widetilde{F}_n(s)f,\, g)|^2 \leq ([J - Y_n^*(0)JY_n(0)]f,\, f)\,([Y_n^{-1}(0)JY_n^{-1*}(0) - J]g,\, g).$$

Now use Theorem 7.2. Since

$$J - Y_n^*(0)JY_n(0) \twoheadrightarrow J - Y^*(0)JY(0),\; Y_n^{-1}(0)JY_n^{-1*}(0) - J \twoheadrightarrow Y^{-1}(0)JY^{-1*}(0) - J,$$

for each $s \in [0,\, \mu]$ the family $\{\widetilde{F}_n(s)\}$ is $\mathfrak{S}$-compact. On the basis of Theorem
3.3 and inequality (8.12) one may select from the sequence $\{\widetilde{F}_n(s)\}$ of operator-
functions a sequence which is $\mathfrak{S}$-convergent for each $s \in [0,\, \mu]$, for which we
use the same notation:

$$\widetilde{F}_n(s) \twoheadrightarrow \widetilde{F}(s)\;\;(0 \leq s \leq \mu). \tag{8.13}$$

Consider the operator-functions

$$\widetilde{U}_n(s) = U_n[\psi_n(s)]\;\;(0 \leq s \leq \mu).$$

Since on the basis of (8.10)

$$U_n(t) = \int_0^t e^{\frac{1}{2}d[F_n(\tau) - JF_n^*(\alpha)J]},$$

we have

$$\widetilde{U}_n(s) = \int_0^s e^{\frac{1}{2}d[\widetilde{F}_n(\sigma) - J\widetilde{F}_n^*(\sigma)J]}.$$

We use Theorem 3.4. Taking account of (8.12) and (8.13), we prove that for
each $s \in [0,\, \mu]$

$$\widetilde{U}_n(s) \twoheadrightarrow \widetilde{U}(s) = \int_0^s e^{\frac{1}{2}d[\widetilde{F}(\sigma) - J\widetilde{F}^*(\sigma)J]}. \tag{8.14}$$

We now pass directly to the study of the operators

$$S_n = \sum_{j=1}^{k_n} M_j^{(n)} = \int_0^{k_n} M_n(t)\,dt = H_n(k_n) =$$

$$= \tfrac{1}{2} \int_0^{k_n} U_n(\tau) \frac{d}{d\tau} [F_n(\tau) + J F_n^*(\tau)J] \, U_n^{-1}(\tau) \, d\tau.$$

Using the strict monotonicity and the absolute continuity of the function $s = \phi_n(t)$ on $[0, k_n]$, we make a change of variable in the integral

$$J_n = \int_0^{\mu_n} \widetilde{U}_n(\sigma) \frac{d}{d\sigma} [\widetilde{F}_n(\sigma) + J \widetilde{F}_n^*(\sigma)J] \widetilde{U}_n^{-1}(\sigma) \, d\sigma$$

which exists because of the absolute continuity of $\widetilde{F}_n(s)$ and the continuity of $\widetilde{U}_n(s)$ and $\widetilde{U}_n^{-1}(s)$. We get

$$J_n = \int_0^{k_n} U_n(\tau) \frac{d}{d\tau} [F_n(\tau) + J F_n^*(\tau)J] \, \frac{1}{\phi_n'(\tau)} \, U_n^{-1}(\tau) \, \phi_n'(\tau) \, d\tau,$$

and thus

$$S_n = \int_0^{\mu} \widetilde{U}_n(\sigma) \Phi_n'(\sigma) \widetilde{U}_n^{-1}(\sigma) \, d\sigma,$$

where

$$\Phi_n(s) \equiv \tfrac{1}{2}[\widetilde{F}_n(s) + J \widetilde{F}_n^*(s)J] \longrightarrow \tfrac{1}{2}[\widetilde{F}(s) + J \widetilde{F}^*(s)J] \equiv \Phi(s). \qquad (8.15)$$

Here we have taken account of the fact that $\Phi_n'(s) = 0$ for $s \in (\mu_n, \mu]$.

We shall show that the operators S_n can be uniformly $\mathfrak{S}$-approximated, relative to n, by the sums

$$S_n^{(m)} = \sum_{k=1}^{m} \widetilde{U}_n\!\left(\frac{k\mu}{m}\right)\!\left[\Phi_n\!\left(\frac{k\mu}{m}\right) - \Phi_n\!\left(\frac{(k-1)\mu}{m}\right)\right] \widetilde{U}_n^{-1}\!\left(\frac{k\mu}{m}\right),$$

$$\qquad (8.16)$$

$$S_n^{(m)} = \sum_{k=1}^{m} \int_{\frac{(k-1)\mu}{m}}^{\frac{k\mu}{m}} \widetilde{U}_n\!\left(\frac{k\mu}{m}\right)\!\Phi_n'(\sigma)\,\widetilde{U}_n^{-1}\!\left(\frac{k\mu}{m}\right) d\sigma.$$

The following estimate holds:

$$|S_n - S_n^{(m)}| \leq \sum_{k=1}^{m} \int_{\frac{(k-1)\mu}{m}}^{\frac{k\mu}{m}} \left|\widetilde{U}_n(\sigma) - \widetilde{U}_n\!\left(\frac{k\mu}{m}\right)\right| \, |\Phi_n'(\sigma)| \, \|\widetilde{U}_n^{-1}(\sigma)\| \, d\sigma$$

$$+ \sum_{k=1}^{m} \int_{\frac{(k-1)\mu}{m}}^{\frac{k\mu}{m}} \left\|\widetilde{U}_n\!\left(\frac{k\mu}{m}\right)\right\| \, |\Phi_n'(\sigma)| \, \left|\widetilde{U}_n^{-1}(\sigma) - \widetilde{U}_n^{-1}\!\left(\frac{k\mu}{m}\right)\right| d\sigma. \qquad (8.17)$$

Because of (8.12) $|\Phi_n(s'') - \Phi_n(s')| \leq |s'' - s'| \; (s', s'' \in [0, \mu])$, which means that $|\Phi'_n(s)| \le 1$ almost everywhere on $[0, \mu]$.[1] Further, because of (3.13),

$$\|\tilde{U}_n(s)\| \le e^{\mathfrak{S} \underset{[0,\mu]}{\mathrm{var}} \tilde{F}_n} \le e^{\mu}.$$

Use inequality (3.14). For $s \in [(k-1)\mu/m, k\mu/m]$ we get

$$\left| \tilde{U}_n(s) - \tilde{U}_n\left[\frac{k\mu}{m}\right] \right| \le \|\tilde{U}_n(s)\| \left| I - \int_0^{\frac{k\mu}{m}} e^{\frac{1}{2}d[\tilde{F}_n(\sigma) - J\tilde{F}_n^*(\sigma)J]} \right|$$

$$\le e^{2\mu} \quad \underset{\left[\frac{(k-1)\mu}{m}, \frac{k\mu}{m}\right]}{\mathfrak{S}\text{-var}} \tilde{F}_n \le \frac{\mu e^{2\mu}}{m}.$$

An analogous inequality is valid for $\tilde{U}_n^{-1}$. Then from (8.17) there follows the estimate

$$|S_n - S_n^{(m)}| \le \frac{2\mu e^{3\mu}}{m},$$

which does not depend on n. This proves the possibility of uniform (relative to n) $\mathfrak{S}$-approximation of the operators S_n by the operators $S_n^{(m)}$.

In order to prove the $\mathfrak{S}$-convergence of the sequence $\{S_n\}$, we find for any $\epsilon > 0$ an index m such that

$$|S_n - S_n^{(m)}| < \frac{\epsilon}{3}.$$

Fixing m and using (8.14) and (8.15), and also the expression (8.16) for $S_n^{(m)}$, we find an n_0 such that for n', $n'' \ge n_0$

$$|S_{n''}^{(m)} - S_{n'}^{(m)}| < \frac{\epsilon}{3}.$$

Then for n', $n'' \ge n_0$

$$|S_{n''} - S_{n'}| \le |S_{n''} - S_{n''}^{(m)}| + |S_{n''}^{(m)} - S_{n'}^{(m)}| + |S_{n'}^{(m)} - S_{n'}| < \epsilon,$$

which proves what was required. Thus:

II) *The sequence* $S_n = \sum_{j=1}^{k_n} M_j^{(n)} (n = 1, 2, \cdots)$ *is* $\mathfrak{S}$-*compact.*

3. Now we study the sequence of sums

$$\tau_n = \sum_{j=1}^{k_n} |G_j^{(n)}|, \quad T_n = \sum_{j=1}^{k_n} G_j^{(n)}.$$

It follows from (8.8) and (8.7) that $G_j^{(n)} = -\gamma_j^{(n)} M_j^{(n)}$, where $0 < \gamma_j^{(n)} \to 1$ as

---

1) We recall that the values of $R'_n(t)$ and thus of $\tilde{F}'_n(s)$ and $\Phi'_n(s)$ lie in some finite-dimensional subspace $\mathfrak{L}_n$ (cf. §5).

$n \longrightarrow \infty$, uniformly relative to j. We write

$$\gamma = \sup_{j,n} \gamma_j^{(n)} \, (n = 1, 2, \cdots ; j = 1, \cdots, k_n).$$

Then

$$\tau_n \leq \gamma \sum_{j=1}^{k_n} |M_j^{(n)}| = \gamma \sigma_n$$

so that, on the basis of I), we have established the validity of the following assertion:

III) *The sequence* $\tau_n = \sum_{j=1}^{k_n} |G_j^{(n)}|$ *is bounded* $(\sup \tau_n = l < \infty)$.

Since

$$0 \leq JT_n = \sum_{j=1}^{k_n} JG_j^{(n)} \leq -\gamma \sum_{j=1}^{k_n} JM_j^{(n)} = -\gamma JS_n,$$

in view of proposition II) and Corollary 7.3 we have

IV) *The sequence* $T_n = \sum_{j=1}^{k_n} G_j^{(n)}$ *is* $\mathfrak{S}$-*compact*.

We need still a further proposition:

V) *The sequence* $\tau_n^{(2)} = \sum_{j=1}^{k_n} |G_j^{(n)\,2}|$ *tends to zero*.

In fact, in view of (8.7)

$$\tau_n^{(2)} = \sum_{j=1}^{k_n} |r_j^{(n)} - 1| \, |G_j^{(n)}| \leq \max_{1 \leq j \leq k_n} |r_j^{(n)} - 1| \sum_{j=1}^{k_n} |G_j^{(n)}|,$$

which, on the basis of (8.6) and proposition III), proves V).

4. Consider along with the sequence

$$\Pi_n(\zeta) = \prod_{j=1}^{k_n} B_j^{(n)}(\zeta) = \prod_{j=1}^{k_n} \left[I + \frac{\zeta + e^{i\theta_j^{(n)}}}{\zeta - r_j^{(n)} e^{i\theta_j^{(n)}}} \, G_j^{(n)} \right]$$

of operator-functions the sequences

$$\widetilde{\Pi}_n(\zeta) = \sum_{j=1}^{k_n} \left[I + \frac{\zeta + e^{i\theta_j^{(n)}}}{\zeta - e^{i\theta_j^{(n)}}} \, G_j^{(n)} \right],$$

$$\widetilde{\widetilde{\Pi}}_n(\zeta) = \sum_{j=1}^{k_n} \exp \left\{ \frac{\zeta + e^{i\theta_j^{(n)}}}{\zeta - e^{i\theta_j^{(n)}}} \, G_j^{(n)} \right\}.$$

VI) $|\Pi_n(\zeta) - \widetilde{\widetilde{\Pi}}_n(\zeta)| \longrightarrow 0$ *uniformly within the unit disk*.

To prove this we fix a ρ, $0 < \rho < 1$, and take n_ρ so that for $n \geq n_\rho$ the inequality $r_j^{(n)} - \rho > (1 - \rho)/2$ holds. The choice of such an n_ρ is possible in view of (8.6). Suppose that $|\zeta| \leq \rho$. Then, using inequality (3.3),

$$|\Pi_n(\zeta) - \widetilde{\Pi}_n(\zeta)| \; e^{\frac{4}{1-\rho} \Sigma |G_j^{(n)}|} \sum_{j=1}^{k_n} \left| \frac{\zeta + e^{i\theta_j^{(n)}}}{\zeta - e^{i\theta_j^{(n)}}} - \frac{\zeta + e^{i\theta_j^{(n)}}}{\zeta - r_j^{(n)} e^{i\theta_j^{(n)}}} \right| \; |G_j^{(n)}|$$

$$\leq \frac{8}{(1 - \rho)^2} \; e^{\frac{4}{1-\rho}\tau_n} \cdot \tau_n \cdot \max_{1 \leq j \leq k_n} |1 - r_j^{(n)}|$$

so that in view of equation (8.6) and proposition III) $|\Pi_n(\zeta) - \widetilde{\Pi}_n(\zeta)| \to 0$ uniformly for $|\zeta| \leq \rho$. Further,

$$|\widetilde{\Pi}_n(\zeta) - \widetilde{\widetilde{\Pi}}_n(\zeta)| \leq e^{\frac{2}{1-\rho}\tau_n} \sum_{j=1}^{k_n} \left| \exp\left\{ \frac{\zeta + e^{i\theta_j^{(n)}}}{\zeta - e^{i\theta_j^{(n)}}} G_j^{(n)} \right\} - \left[I + \frac{\zeta + e^{i\theta_j^{(n)}}}{\zeta - e^{i\theta_j^{(n)}}} G_j^{(n)} \right] \right|.$$

Using inequality (3.6), we obtain

$$|\widetilde{\Pi}_n(\zeta) - \widetilde{\widetilde{\Pi}}_n(\zeta)| \leq e^{\frac{3}{1-\rho}\tau_n} \cdot \frac{4}{(1 - \rho)^2} \sum_{j=1}^{k_n} |G_j^{(n)2}|$$

and in view of propositions III) and V) $|\widetilde{\Pi}_n(\zeta) - \widetilde{\widetilde{\Pi}}_n(\zeta)| \to 0$ uniformly for $|\zeta| \leq \rho$. Since

$$|\Pi_n(\zeta) - \widetilde{\widetilde{\Pi}}_n(\zeta)| \leq |\Pi_n(\zeta) - \widetilde{\Pi}_n(\zeta)| + |\widetilde{\Pi}_n(\zeta) - \widetilde{\widetilde{\Pi}}_n(\zeta)|,$$

proposition VI) is proved.

Now we represent the operator-function $\widetilde{\Pi}_n(\zeta)\,(n = 1, 2, \cdots)$ in the form of a multiplicative integral. To this end recall the number l defined in III). We define on the segment $[0, l]$ an operator-function $G_n(t)$ by the equations

$$G_n(t) = \begin{cases} |G_1^{(n)}|^{-1} G_1^{(n)} & (0 \leq t < |G_1^{(n)}|) \\ |G_m^{(n)}|^{-1} G_m^{(n)} & \left(\sum_{j=1}^{m-1} |G_j^{(n)}| \leq t < \sum_{j=1}^{m} |G_j^{(n)}|, \; m = 2, \cdots, k_n \right) \\ 0 & (\tau_n \leq t \leq l). \end{cases}$$

Since $JG_j^{(n)} \geq 0$, we have $JG_n(t) \geq 0$. Moreover, $\operatorname{tr} JG_n(t) = |G_n(t)| = 1$ for $0 \leq t < \tau_n$. Introduce the function

$$E_n(t) = \int_0^t G_n(\tau)\, d\tau.$$

Obviously, $JE_n(t) \geq 0$ and it is a Hermitian-increasing operator-function. As a consequence of this

$$0 \leq JE_n(t) \leq JE_n(l) = J\int_0^l G_n(\tau)\, d\tau = J\sum_{j=1}^{k_n} G_j^{(n)} = JT_n. \tag{8.18}$$

Moreover, for $0 \leq t \leq \tau_n$

$$\operatorname{tr} JE_n(t) = t \tag{8.19}$$

so that

$$\big|E_n(t'') - E_n(t')\big| \leq |t'' - t'| \;\; (t', \, t'' \in [0, \, l]). \tag{8.20}$$

Now consider also the scalar functions

$$\theta_n(t) = \begin{cases} \theta_1^{(n)} & (0 \leq t < |G_1^{(n)}|) \\[2mm] \theta_m^{(n)} & \left[\sum_{j=1}^{m-1} |G_j^{(n)}| \leq t < \sum_{j=1}^{m} |G_j^{(n)}|; \, m = 2, \cdots, k_n\right] \\[2mm] \theta_{k_n}^{(n)} & (\tau_n \leq t \leq l). \end{cases}$$

Obviously we may suppose that the elementary factors of the functions $Y_n(\zeta)$ are split off in the order of nondecreasing arguments of the singularities, so that the functions $\theta_n(t)$ are decreasing.

It is not hard to see that the function $\overset{\approx}{\Pi}_n(\zeta)$ may be expressed using the functions $E_n(t)$ and $\theta_n(t)$ as follows:

$$\overset{\approx}{\Pi}_n(\zeta) = \int_0^l \exp\left\{ \frac{\zeta + e^{i\theta_n(t)}}{\zeta - e^{i\theta_n(t)}}\, dE_n(t)\right\}.$$

Since on the basis of IV) the sequence $\{T_n\}$ is $\mathfrak{S}$-compact, it follows from (8.18) and Corollary 7.3 that for each $t \in [0, \, l]$ the family of operators $E_n(t)$ is $\mathfrak{S}$-compact. Then it follows from (8.20) and Theorem 3.3 that there exists a sequence $\{E_{n_j}(t)\}$ which $\mathfrak{S}$-converges for each $t \in [0, \, l]$. If $E(t)$ is its limit, then (8.19) implies the equation $\operatorname{tr} JE(t) \equiv t \; (0 \leq t \leq l)$. Obviously, in view of the classical Helly's theorem, we may suppose that the sequence of functions $\theta_{n_j}(t)$ is convergent. Then $\theta(t) = \lim \theta_{n_j}(t)$ is a nondecreasing function satisfying the inequality $0 \leq \theta(t) \leq 2\pi$.

Using Theorem 3.4, we prove that for $|\zeta| < 1$

$$\overset{\approx}{\Pi}_{n_j}(\zeta) \longrightarrow \overset{\curvearrowleft}{\int_0^l} \exp\left\{ \frac{\zeta + e^{i\theta(t)}}{\zeta - e^{i\theta(t)}} dE(t) \right\}, \quad \overset{\approx}{\Pi}_{n_j}^{-1}(\zeta) \longrightarrow \overset{\curvearrowright}{\int_0^l} \exp\left\{ \frac{\zeta + e^{i\theta(t)}}{\zeta - e^{i\theta(t)}} dE(t) \right\}.$$

It follows from proposition VI) that $\Pi_{n_j}(\zeta)$ and $\Pi_{n_j}^{-1}(\zeta)$ have the same $\mathfrak{G}$-limits. Since $Y_n(\zeta) = V_n \Pi_n(\zeta)$ (see (8.1)), (8.2) implies the $\mathfrak{G}$-convergence of the sequence V_{n_j}. If $V_{n_j} \longrightarrow V$, then V is a J-unitary operator. From the arguments just presented, and from (8.2), it follows that the following representation for the operator-function $Y(\zeta) \in \widetilde{K}_j^{\mathfrak{G}}$ holds:

$$Y(\zeta) = V \overset{\curvearrowleft}{\int_0^l} \exp\left\{ \frac{\zeta + e^{i\theta(t)}}{\zeta - e^{i\theta(t)}} dE(t) \right\}.$$

Calling on Theorem 5.2, we verify that we have proved the necessary part of the following theorem.

Theorem 8.1[1]. *For the operator-function $Y(\zeta)$ to lie in the class $K_j^{\mathfrak{G}}$ it is necessary and sufficient that for $|\zeta| < 1$ it admit the representation[2]*

$$Y(\zeta) = V \overset{\curvearrowleft}{\int_0^l} \exp\{k[\zeta, \theta(t)] dE(t)\} \cdot \overset{\curvearrowright}{\prod_{j=1}^{\infty}} \left[\left(\frac{\zeta_j - \zeta}{\zeta - \bar{\zeta}_j \zeta} \cdot \frac{|\zeta_j|}{\zeta_j} \right)^{\epsilon_j} P_j + Q_j \right],$$

(8.21)

in which V is a J-unitary operator, $l \geq 0$, $\theta(t)$ is a nondecreasing scalar function on $[0, l]$ with $0 \leq \theta(t) \leq 2\pi$, and $JE(t)$ is a Hermitian-nondecreasing operator-function on $[0, l]$, satisfying the condition

$$\operatorname{tr} JE(t) = t \quad (0 \leq t \leq l), \tag{8.22}$$

$$|\zeta_j| < 1, \quad \epsilon_j = \pm 1, \quad \dim P_j \mathfrak{H} < \infty, \quad P_j^2 = P_j, \quad \epsilon_j J P_j \geq 0, \quad Q_j = I - P_j,$$

and

$$\sum_{j=1}^{\infty} (1 - |\zeta_j|) |P_j| < \infty. \tag{8.23}$$

For the proof of sufficiency we need the following proposition.

Lemma 8.1. *If A is a bounded operator in $\mathfrak{H}$ and $JA + A^*J \leq 0$, then $W \equiv e^A$ is a J-binonexpansive operator.*

Indeed, suppose that $W(t) = e^{tA}$ $(0 \leq t \leq 1)$. Since $W'(t) = AW(t) = W(t)A$,

---

1) This theorem, partially formalated in $[^4]$, generalizes a basic theorem of V. P. Potapov $[^5]$.

2) Here and in what follows we adopt the notation

$$k(\zeta, \theta) = \frac{\zeta + e^{i\theta}}{\zeta - e^{i\theta}}.$$

we have

$$\frac{d}{dt}[W^*(t)JW(t)] = W^*(t)[JA + A^*J]W(t) \leq 0 \ (0 \leq t \leq 1)$$

so that $W^*(1)JW(1) \leq W^*(0)JW(0)$ or $W^*JW \leq J$. Analogously $WJW^* \leq J$, which proves the lemma.

Now consider the operator-function

$$Y_0(\zeta) = \overset{l}{\underset{0}{\overset{\curvearrowleft}{\int}}} \exp\{k[\zeta, \ \theta(t)] \ dE(t)\}$$

and suppose that $\theta(t)$ and $E(t)$ satisfy the hypotheses of the theorem. Since $f_\zeta(t) \equiv k[\zeta, \ \theta(t)]$ is Riemann-integrable in t, and since (8.22) implies the equation $|E(t'') - E(t')| = |t'' - t'|$, the integral product

$$\overset{n}{\underset{j=1}{\overset{\curvearrowleft}{\Pi}}} \exp\{f_\zeta(\tau_j)[E(t_j) - E(t_{j=1})]\} \tag{8.24}$$

$$(0 = t_0 < t_1 < \cdots < t_n = l, \ t_{j-1} \leq \tau_j \leq t_j)$$

$\mathfrak{S}$-converges to a multiplicative integral $Y_0(\zeta)$ for $(\zeta) < 1$ (Theorem 3.2, A). Suppose that $|\zeta| \leq \rho < 1$. Then

$$m_\zeta \equiv \underset{[0, l]}{\sup} |f_\zeta(t)| \leq \frac{2}{1 - \rho},$$

$$\omega_j(f_\zeta) \equiv \underset{t_{j-1} \leq t' < t'' \leq t_j}{\sup} |f_\zeta(t'') - f_\zeta(t')| \leq \frac{2\rho\omega_j(e^{i\theta(t)})}{(1 - \rho)^2}.$$

Hence, on the basis of inequality (3.12) it follows that the products (8.24) $\mathfrak{S}$-converge to $Y_0(\zeta)$ uniformly in the disk $|\zeta| \leq \rho$. Since

$$J\{f_\zeta(\tau_j)[E(t_j) - E(t_{j-1})]\} + \{f_\zeta(\tau_j)[E(t_j) - E(t_{j-1})]\}^*J$$

$$= 2 \operatorname{Re} k[\zeta, \ \theta(\tau_j)] J[E(t_j) - E(t_{j-1})] \leq 0,$$

by Lemma 8.1 each integral product (8.24) lies in the class $\widetilde{K}_J^{\mathfrak{S}}$. Accordingly, taking into account what was said above, $Y_0(\zeta) \in \widetilde{K}_J^{\mathfrak{S}}$.

That the condition (8.23) guarantees the $\mathfrak{S}$-convergence of the infinite product of elementary factors

$$\Pi(\zeta) = \overset{\infty}{\underset{j=1}{\overset{\curvearrowleft}{\Pi}}} \left[\left[\frac{\zeta_j - \zeta}{1 - \bar{\zeta}_j\zeta} \cdot \frac{|\zeta_j|}{\zeta_j}\right]^{\epsilon_j} P_j + Q_j \right],$$

follows from Theorem 5.1. Accordingly $\Pi(\zeta) \in K_J^{\mathfrak{S}}$, which means that

$$Y(\zeta) = VY_0(\zeta)\Pi(\zeta) \in K_J^{\mathfrak{S}}.$$

Theorem 8.1 is completely proved.

§ 9. Multiplicative representation of nonexpansive
operator-functions and their determinants.
Uniqueness theorems

1. In the present section the object of our investigation will be operator-functions analytic in the unit disk and having the following properties:
1) $\|Y(\zeta)\| \leq 1$ for $|\zeta| < 1$; 2) the operator $Y(0)$ is boundedly invertible;
3) $Y(0) - I \in \mathfrak{S}$. The collection of such operator-functions will be denoted by $C^{\mathfrak{S}}$. Since $I - Y^*(0) Y(0) \in \mathfrak{S}$ and $Y^*(\zeta) Y(\zeta) \leq I$ for $|\zeta| < 1$ for $Y(\zeta) \in C^{\mathfrak{S}}$, from the considerations of § 2, in particular from Theorem 2.1, it follows that $C^{\mathfrak{S}} \subset K_I^{\mathfrak{S}}$ (the class $K_J^{\mathfrak{S}}$ for $J = I$), $Y(\zeta) - I \in \mathfrak{S}$ for $|\zeta| < 1$ and $Y^{-1}(\zeta)$ is meromorphic inside the unit disk.

We shall formulate Theorem 8.1 for the class $C^{\mathfrak{S}}$.

Theorem 9.1. *For the operator-function $Y(\zeta)$ to lie in the class $C^{\mathfrak{S}}$ it is necessary and sufficient that it should admit for $|\zeta| < 1$ the representation*[1]

$$Y(\zeta) = \overset{l}{\underset{0}{\int}} \exp\{k[\zeta, \, \theta(t)]dE(t)\}V \prod_{j=1}^{\infty} \left[\frac{\zeta_j - \zeta}{1 - \overline{\zeta}_j\zeta} \cdot \frac{|\zeta_j|}{\zeta_j} P_j + Q_j \right] \tag{9.1}$$

in which V is a unitary operator such that $V - I \in \mathfrak{S}$, $l \geq 0$, $\theta(t)$ is a nondecreasing scalar function defined on $[0, \, l]$ with $0 \leq \theta(t) \leq 2\pi$, $E(t)$ is a Hermitian-nondecreasing operator-function defined on the same interval, satisfying the condition

$$\operatorname{tr} E(t) = t \ (0 \leq t \leq l), \tag{9.2}$$

$|\zeta_j| < 1$, P_j *are orthoprojectors,* $\dim P_j\mathfrak{H} = m_j < \infty$, $Q_j = I - P_j$ *and*

$$\sum_{j=1}^{\infty} m_j(1 - |\zeta_j|) < \infty. \tag{9.3}$$

Now we use the following facts from the theory of determinants of infinite order [6]: 1) If $A \in \mathfrak{S}$ then the determinant $\det(I - A) = \prod_j[1 - \lambda_j(A)]$ exists, where $\lambda_j(A)$, $j = 1, 2, \cdots$, are the eigenvalues of A; 2) if $A, B \in \mathfrak{S}$, then

---

1) In order to obtain (9.1), it suffices to replace $E(t)$ by $V^{-1}E(t)V$ in the representation (8.21), written out for the case $J = I$.

$\det[(I - A)(I - B)] = \det(I - A)\det(I - B)$; 3) $\det(I - A)$ is an $\mathfrak{S}$-continuous functional of A; 4) $\det e^{A} = r^{\operatorname{tr} A}$ for $A \in \mathfrak{S}$.

Since in (9.1) the Blaschke-Potapov infinite product $\mathfrak{S}$-converges and the integral products also $\mathfrak{S}$-converge to a multiplicative integral, (9.1) implies the following representation:

$$\det Y(\zeta) = e^{i\alpha} \prod_{j=1}^{\infty} \left[\frac{\zeta_j - \zeta}{1 - \bar{\zeta}_j\zeta} \cdot \frac{|\zeta_j|}{\zeta_j} \right]^{m_j} \exp\left\{ \int_0^l k\,[\zeta,\,\theta(t)]\,dt \right\}. \qquad (9.4)$$

2. Now we shall prove Theorem 7.2 formulated earlier for the case $J = I$.

Theorem 9.2. *Suppose that* $Y(\zeta) \in M_I$ *(in particular,* $Y(\zeta) \in C^{\mathfrak{S}}$*) and*

$$Y(\zeta) = Y_0^{(1)}(\zeta)\,\Pi^{(1)}(\zeta) = Y_0^{(2)}(\zeta)\,\Pi^{(2)}(\zeta), \qquad (9.5)$$

where $\Pi^{(1)}(\zeta)$ *and* $\Pi^{(2)}(\zeta)$ *are strongly converging products of elementary factors, and* $Y_0^{(j)}(\zeta) \in M_I$ *and* $Y_0^{(j)^{-1}}(\zeta)$ $(j = 1, 2)$ *are holomorphic for* $|\zeta| < 1$. *Then there exists a constant unitary operator* U *such that*

$$\Pi^{(2)}(\zeta) = U\,\Pi^{(1)}(\zeta)\ (|\zeta| < 1). \qquad (9.6)$$

First we prove two lemmas. In order to formulate them, we agree to say that the elementary factor

$$B(\zeta) = U\left[\frac{\zeta_0 - \zeta}{1 - \bar{\zeta}_0\zeta}\,P + Q \right]$$

$(|\zeta_0| < 1,\ P$ is an orthoprojector, $Q = I - P$ and U is a unitary operator) splits from the operator-function $V(\zeta) \in M_I$ if $Y(\zeta)\,B^{-1}(\zeta) \in M_I$.

Lemma 9.1. *For the elementary factor* $B(\zeta)$ *to split from* $Y(\zeta) \in M_1$, *it is necessary and sufficient that the equation* $Y(\zeta_0)\,P = 0$ *is satisfied.*

For the proof of the lemma we suppose, without loss of generality, that $\zeta_0 = 0$. Here

$$B(\zeta) = U(\zeta P + Q),\quad B^{-1}(\zeta) = (\zeta^{-1}P + Q)\,U^{-1}.$$

If

$$Y(\zeta) = A_0 + \zeta A_1 + \cdots\ (A_0 = Y(0),\ |\zeta| < 1),$$

then the decomposition

$$Y(\zeta)B^{-1}(\zeta) = \zeta^{-1}A_0 P U^{-1} + (A_1 P + A_0 Q)U^{-1} + \cdots$$

is warranted.

Suppose that $B(\zeta)$ splits from $Y(\zeta)$. Then $Y(\zeta)B^{-1}(\zeta) \in M_I$. Accordingly, the function $Y(\zeta)B^{-1}(\zeta)$ is holomorphic for $|\zeta| < 1$, so that $A_0 P U^{-1} = 0$, i.e. $Y(0)P = 0$. The necessity is proved.

For the proof of sufficiency we note that it follows from the equation $Y(0)P = 0$ that the function $Y(\zeta)B^{-1}(\zeta)(|\zeta| < 1)$ is holomorphic. Since $\|Y(\zeta)B^{-1}(\zeta)\| \leq 1/\rho$ for $|\zeta| = \rho$, by the maximum principle this inequality is satisfied for $|\zeta| \leq \rho$. Fixing on any ζ with $|\zeta| < 1$ and passing to the limit as $\rho \to 1$, we find that $\|Y(\zeta)B^{-1}(\zeta)\| \leq 1$, i.e. $Y(\zeta)B^{-1}(\zeta) \in M_I$. The lemma is proved.

The following assertion follows immediately from Lemma 9.1.

Lemma 9.2. *Suppose that* $Y_1(\zeta)Y_2(\zeta) \in M_I$ *and that the elementary factor*

$$B(\zeta) = U\left[\frac{\zeta_0 - \zeta}{1 - \bar{\zeta}_0 \zeta}P + Q\right]$$

splits from $Y(\zeta) \equiv Y_2(\zeta)Y_1(\zeta)$. *Suppose that the operator* $Y_2(\zeta_0)$ *has a bounded inverse. Then* $B(\zeta)$ *splits from* $Y_1(\zeta)$.

We pass to the proof of Theorem 9.2. Suppose that (9.5) holds and $\Pi^{(j)}(\zeta) = \lim_{n \to \infty} \Pi_n^{(j)}(\zeta)$, where $\Pi_n^{(j)}(\zeta) = B_n^{(j)}(\zeta) \cdot \ldots \cdot B_1^{(j)}(\zeta)$ $(j = 1, 2; B_k^{(j)}(\zeta)$ are the elementary factors). From the equation $Y(\zeta) = Y_0^{(1)}(\zeta)\Pi^{(1)}(\zeta)$ it follows that $B_1^{(1)}(\zeta)$ splits from $Y(\zeta)$, and since $Y(\zeta) = Y_0^{(2)}(\zeta)\Pi^{(2)}(\zeta)$, on the basis of Lemma 9.2 $B_1^{(1)}(\zeta)$ splits from $\Pi^{(2)}(\zeta)$, i.e. $\Pi^{(2)}(\zeta)B_1^{(1)-1}(\zeta) \in M_I$. Analogously $\Pi^{(1)}(\zeta)B_1^{(2)-1}(\zeta) \in M_I$. Arguing by induction, we establish that $\Pi^{(2)}(\zeta)\Pi_n^{(1)-1}(\zeta) \in M_I$ and $\Pi^{(1)}(\zeta)\Pi_n^{(2)-1}(\zeta) \in M_I$ $(n = 1, 2, \cdots)$. Passing to the limit as $n \to \infty$, we prove the validity of the inclusion

$$\Pi^{(2)}(\zeta)\Pi^{(1)-1}(\zeta) \in M_I, \quad \Pi^{(1)}(\zeta)\Pi^{(2)-1}(\zeta) \in M_I. \tag{9.7}$$

(9.7) implies the identity

$$[\Pi^{(2)}(\zeta)\Pi^{(1)-1}(\zeta)]^* [\Pi^{(2)}(\zeta)\Pi^{(1)-1}(\zeta)] \equiv I \ (|\zeta| < 1),$$

so that, on the basis of the maximum principle (see for example [7], Theorem 2), (9.6) follows. Theorem 9.2 is proved.

3. The scalar function $\theta(t)$ $(0 \leq t \leq l)$ is called *canonical* if is nondecreasing, continuous on the left, and on the semisegment $[0, l)$ takes on all the values of the semisegment $[0, 2\pi)$.

Theorem 9.3 ([8]). *The operator-function* $Y(\zeta) \in C^{\mathfrak{S}}$ *admits the representation* (9.1) *in which* $\theta(t)$ *is a canonical function. The number* l *and the function* $\theta(t)$ *are determined by* $Y(\zeta)$ *uniquely.*

In the proof of this theorem we shall use the following auxiliary proposition:[1]

Lemma 9.3. *If* $X^{(1)}(\zeta)$, $X^{(2)}(\zeta) \in C^{\mathfrak{S}}$, *then there exist* $Y^{(1)}(\zeta)$, $Y^{(2)}(\zeta) \in C^{\mathfrak{S}}$, *such that*

$$X^{(1)}(\zeta) X^{(2)}(\zeta) = Y^{(2)}(\zeta) Y^{(1)}(\zeta) \tag{9.8}$$

and

$$\det X^{(1)}(\zeta) = \det Y^{(1)}(\zeta), \ \det X^{(2)}(\zeta) = \det Y^{(2)}(\zeta). \tag{9.9}$$

Proof. We shall suppose for simplicity that $X^{(1)-1}(\zeta)$ and $X^{(2)-1}(\zeta)$ do not have singularities for $|\zeta| < 1$. Using Theorem 2.2, we construct a sequence of Blaschke products $\Pi_n(\zeta)$ such that $\Pi_n(\zeta) \longrightarrow X^{(1)}(\zeta)$ and $\Pi_n^{-1}(\zeta) \longrightarrow X^{(1)-1}(\zeta)$ $(|\zeta| < 1)$. Put $Z_n(\zeta) = \Pi_n(\zeta) X^{(2)}(\zeta)$. Obviously

$$Z_n(\zeta) \longrightarrow Z(\zeta) \equiv X^{(1)}(\zeta) X^{(2)}(\zeta) \ (|\zeta| < 1), \tag{9.10}$$

$Z_n(\zeta) \in C^{\mathfrak{S}}$, and $Z_n(\zeta)$ has for $|\zeta| < 1$ a finite number of poles. Splitting off from $Z_n(\zeta)$ on the right the corresponding elementary factors (Theorem 1.4), we find that

$$Z_n(\zeta) = \widetilde{X}_n^{(1)}(\zeta) \widetilde{\Pi}_n(\zeta), \tag{9.11}$$

where $\widetilde{\Pi}_n(\zeta)$ is a finite product of elementary factors. From the equation $\Pi_n(\zeta) X^{(2)}(\zeta) = \widetilde{X}_n^{(1)}(\zeta) \widetilde{\Pi}_n(\zeta)$ it follows that

$$\det \Pi_n(\zeta) \det X^{(2)}(\zeta) = \det \widetilde{X}_n^{(1)}(\zeta) \det \widetilde{\Pi}_n(\zeta).$$

Since $\det \Pi_n(\zeta)$ and $\det \widetilde{\Pi}_n(\zeta)$ are Blaschke scalar products, and $\det X^{(1)}(\zeta) \neq 0$ and $\det X^{(2)}(\zeta) \neq 0$ for $|\zeta| < 1$, we have

$$|\det \Pi_n(\zeta)| = |\det \widetilde{\Pi}_n(\zeta)|, \ |\det X^{(2)}(\zeta)| = |\det \widetilde{X}_n^{(1)}(\zeta)|. \tag{9.12}$$

(9.11) implies the inequality

$$0 \leq I - \widetilde{\Pi}_n^*(\zeta) \widetilde{\Pi}_n(\zeta) \leq I - Z_n^*(\zeta) Z_n(\zeta), \ 0 \leq \widetilde{\Pi}_n^{-1}(\zeta) \widetilde{\Pi}_n^{-1*}(\zeta) - I \leq Z_n^{-1}(\zeta) Z_n^{-1*}(\zeta) - I$$

so that, on the basis of Corollary 7.3 and (9.10), the sequences $\{I - \widetilde{\Pi}_n^*(\zeta) \widetilde{\Pi}_n(\zeta)\}$ and $\{\widetilde{\Pi}_n^{-1}(\zeta) \widetilde{\Pi}_n^{-1*}(\zeta) - I\}$ $(|\zeta| < 1)$ are $\mathfrak{S}$-compact.

Now we write out inequality (6.2) for the operator-function $\widetilde{\Pi}_n(\zeta)$, with $\lambda = \zeta$, $|\zeta| < 1$, $\mu = 0$:

$$\frac{|([\widetilde{\Pi}_n^{-1}(0) \widetilde{\Pi}_n(\zeta) - I]f, g)|^2}{|\zeta|^2} \leq \frac{([I - \widetilde{\Pi}_n^*(\zeta) \widetilde{\Pi}_n(\zeta)]f, f)([\widetilde{\Pi}_n^{-1}(0) \widetilde{\Pi}_n^{-1*}(0) - I]g, g)}{1 - |\zeta|^2}$$

---

[1] Other applications of Lemma 9.3 were indicated in [8].

for all f, $g \in \mathfrak{H}$. By Theorem 7.2 the sequence $\{\tilde{\Pi}_n^{-1}(0)\tilde{\Pi}_n(\zeta) - I\}$ is $\mathfrak{S}$-compact for each ζ with $|\zeta| < 1$. In exactly the same way, the sequence $\{\tilde{\Pi}_n^{-1}(\zeta)\tilde{\Pi}_n(0) - I\}$ is $\mathfrak{S}$-compact.

The operators $\tilde{\Pi}_n(0)$ may be regarded as Hermitian-nonnegative because of the choice of the unitary operators in the expression for the elementary factors (see § 6.2). Using Theorem 4.2, we obtain

$$0 \leq I - \tilde{\Pi}_n(0) \leq I - \tilde{\Pi}_n^*(0)\tilde{\Pi}_n(0), \ 0 \leq \tilde{\Pi}_n^{-1}(0) - I \leq \tilde{\Pi}_n^{-1}(0)\tilde{\Pi}_n^{-1*}(0) - I.$$

Corollary 7.3 then implies the $\mathfrak{S}$-compactness of the sequences $\{\tilde{\Pi}_n(0)\}$ and $\{\tilde{\Pi}_n^{-1}(0)\}$. In what follows we shall suppose these sequences $\mathfrak{S}$-convergent. From the $\mathfrak{S}$-compactness of the families $\{\tilde{\Pi}_n^{-1}(0)\tilde{\Pi}_n(\zeta) - I\}$ and $\{\tilde{\Pi}_n^{-1}(\zeta)\tilde{\Pi}_n(0) - I\}$ follows the $\mathfrak{S}$-compactness of the sequences $\{\tilde{\Pi}_n(\zeta)\}$ and $\{\tilde{\Pi}_n^{-1}(\zeta)\}$ for each ζ of the unit disk. Since $\|\tilde{\Pi}_n(\zeta)\| \leq 1$, by the generalized Vitali theorem [9] it is possible to select a sequence $\{\tilde{\Pi}_{n_j}(\zeta)\}$, $\mathfrak{R}$-converging uniformly in ζ inside the unit disk. In view of the $\mathfrak{S}$-compactness we have $\tilde{\Pi}_{n_j}(\zeta) \longrightarrow Y^{(1)}(\zeta)$ and $\tilde{\Pi}_{n_j}^{-1}(\zeta) \longrightarrow Y^{(1)-1}(\zeta)$ where $Y^{(1)}(\zeta)$ and $Y^{(1)-1}(\zeta)$ are holomorphic for $|\zeta| < 1$.

It follows from (9.10) and (9.11) that the sequence $\{\tilde{X}_n^{(1)}(\zeta)\}$ $\mathfrak{S}$-converges for $|\zeta| < 1$. Let $Y^{(2)}(\zeta)$ be its limit. Then

$$X^{(1)}(\zeta)X^{(2)}(\zeta) = Z(\zeta) = \lim Z_{n_j}(\zeta) = \lim \tilde{X}_{n_j}(\zeta)\lim \tilde{\Pi}_{n_j}(\zeta) = Y^{(2)}(\zeta)Y^{(1)}(\zeta).$$

$$(9.13)$$

Moreover

$$|\det X^{(1)}(\zeta)| = \lim |\det \tilde{\Pi}_{n_j}(\zeta)| = \lim |\det \tilde{\Pi}_{n_j}(\zeta)| = |\det Y^{(1)}(\zeta)|,$$

$$|\det X^{(2)}(\zeta)| = |\det \tilde{X}_n^{-1}(\zeta)| = |\det Y^{(2)}(\zeta)|.$$

Since (9.13) implies the equation

$$\det X^{(1)}(\zeta)\det X^{(2)}(\zeta) = \det Y^{(2)}(\zeta)\det Y^{(1)}(\zeta),$$

it follows replacing $Y^{(1)}(\zeta)$ by $UY^{(1)}(\zeta)$ and $Y^{(2)}(\zeta)$ by $Y^{(2)}(\zeta)U^{-1}$, where U is an appropriately chosen unitary operator with $U - I \in \mathfrak{S}$, that we can arrange that (9.9) is satisfied. Here (9.8) holds (see (9.13)). The lemma is proved.

Proof of Theorem 9.3. We show first of all that in the representation (9.1) the function $\theta(t)$ may be taken to be canonical. Suppose that

$$Y(\zeta) = \int_0^l \exp\{k[\zeta, \theta(t)]\,dE(t)\} \quad (\operatorname{tr} E(t) \equiv t),$$

where the function $\theta(t)$ is nondecreasing, continuous on the left, but not canonical (in fact, that $\theta(t) < 2\pi$ for $0 < \theta(t) < l - \epsilon$ and $\theta(t) = 2\pi$ for $l - \epsilon \leq t \leq l$). It is easy to reduce the general case to the case in which $\theta(t) > 0$ for $t > 0$. Next, put

$$X_1(\zeta) = \int_{l-\epsilon}^{l} \exp\left\{\frac{\zeta+1}{\zeta-1} dE(t)\right\}, \quad X_2(\zeta) = \int_{0}^{l-\epsilon} \exp\{k[\zeta, \theta(t)]\, dE(t)\}.$$

Then $Y(\zeta) = X_1(\zeta) X_2(\zeta)$. Using the lemma, we find that $Y(\zeta) = Y_2(\zeta) Y_1(\zeta)$, where

$$\det Y_2(\zeta) = \det X_1(\zeta) = \exp\left\{\epsilon \frac{\zeta+1}{\zeta-1}\right\}, \tag{9.14}$$

$$\det Y_2(\zeta) \det X_2(\zeta) = \exp\left\{\int_{0}^{l-\epsilon} k[\zeta, \theta(t)]\, dt\right\}. \tag{9.15}$$

It follows from (9.14) and (9.15) that

$$Y_1(\zeta) = U \int_{0}^{\epsilon} \exp\left\{\frac{\zeta+1}{\zeta-1}\, dF(t)\right\} \quad (\operatorname{tr} F(t) \equiv t),$$

$$Y_2(\zeta) U = V \int_{0}^{l-\epsilon} \exp\{k[\zeta, \theta(t)]\, dG(t)\} \quad (\operatorname{tr} G(t) \equiv t).$$

Here U and V are unitary operators. Here, as earlier, we shall use the following elementary fact:

(∗) *The function* $f(\zeta) (|f(\zeta)| < 1)$, *holomorphic for* $|\zeta| < 1$, *admits the unique representation*

$$f(\zeta) = e^{i\alpha} \prod_{j} \left[\frac{\zeta_j - \zeta}{1 - \bar{\zeta}_j\zeta} \cdot \frac{|\zeta_j|}{\zeta_j}\right]^{m_j} \exp\left\{\int_{0}^{l} k[\zeta, \theta(t)]\, dt\right\},$$

with $\operatorname{Im} \alpha = 0$, $\theta(t)$ *a canonical function, and* $\zeta_{j'} \neq \zeta_{j''}$ *if* $j' \neq j''$.

This proposition is an immediate consequence of the theorem on the uniqueness of the representation

$$f(\zeta) = e^{i\alpha} \prod_{j} \left[\frac{\zeta_j - \zeta}{1 - \bar{\zeta}_j\zeta} \cdot \frac{|\zeta_j|}{\zeta_j}\right]^{m_j} \exp\left\{\int_{0}^{2\pi} k(\zeta, \theta)\, d\sigma(\theta)\right\},$$

$\sigma(\theta)\,(\sigma(0) = 0)$ being a nondecreasing left-continuous function on $[0,\, 2\pi]$.

Thus

$$Y(\zeta) = Y_2(\zeta)Y_1(\zeta) = V \overset{l-\epsilon}{\underset{0}{\int}} \exp\{k[\zeta,\, \theta(t)]\,dG(t)\} \cdot \overset{\epsilon}{\underset{0}{\int}} \exp\left\{\frac{\zeta + 1}{\zeta - 1}\,dF(t)\right\}. \quad 9.16)$$

$$(9.16)$$

Put

$$\tilde{\theta}(t) = \begin{cases} 0 & (0 \le t \le \epsilon), \\ \theta(t - \epsilon) & (\epsilon \le t \le l) \end{cases}, \quad \tilde{E}(t) = \begin{cases} F(t) & (0 \le t \le \epsilon) \\ F(\epsilon) + G(t - \epsilon) & (\epsilon \le t \le l). \end{cases}$$

Since $0 < \theta(t) < 2\pi$ for $0 < t < l - \epsilon$, it follows that the function $\tilde{\theta}(t)$ is canonical. Moreover it is easy to see that $\operatorname{tr}\tilde{E}(t) \equiv t$. (9.16) implies the equation

$$Y(\zeta) = V \overset{l}{\underset{0}{\int}} \exp\{k[\zeta,\, \tilde{\theta}(t)]\,d\tilde{E}(t)\}.$$

The first part of Theorem 9.3 is proved.

For the proof of the second part it is sufficient to use formula (9.4) and proposition (*).

Theorem 9.4 [4]. *Suppose that $Y(\zeta) \in C^{\mathfrak{G}}$ and that (9.1) is its multiplicative representation with a canonical function $\theta(t)$. If t_0 is a point of growth for the function $\theta(t)$, then the operator $E(t_0)$ is determined by $Y(\zeta)$ uniquely.*

This theorem follows directly from the general proposition (Theorem 10.1) proved in the following section, from Theorem 9.3, and from the fact that on the basis of Theorem 9.2 the Blaschke product is defined up to a left unitary factor.

§ 10.   Proof of the theorem of uniquess
for a multiplicative integral

1. The basic objective of this section is the proof of the following theorem:

Theorem 10.1 [8]. *Suppose that*

$$X_j(t;\, \zeta) \equiv \overset{t}{\underset{0}{\int}} \exp\{k[\zeta,\, \theta(\tau)]\,dE_j(\tau)\} \quad (j = 1,\, 2;\ 0 \le t \le l,\ |\zeta| < 1), \quad (10.1)$$

where $\theta(t)$ is a canonical function and the $E_j(t)$, with $E_j(0) = 0$, are Hermitian operator-functions, continuous and of bounded variation in the $\mathfrak{R}$-metric. Then, if one of the two equations

$$X_2(l;\, \zeta) \equiv X_1(l;\, \zeta)\,U\ (|\zeta| < 1), \quad (10.2')$$

$$X_2(l; \zeta) \equiv UX_1(l; \zeta) \ (|\zeta| < 1) \tag{10.2''}$$

is valid, with U a unitary operator, then $U = I$, and for any point of growth t_0 of the function $\theta(t)$ the relations

$$X_2(t_0; \zeta) \equiv X_1(t_0; \zeta) \ (|\zeta| < 1), \tag{10.3}$$

$$E_2(t_0) = E_1(t_0), \tag{10.4}$$

hold.

Proof. We note first of all that the identity $(10.2')$ reduces immediately to the form $(10.2'')$. Indeed, if $(10.2')$ holds, then

$$X_2(l; \zeta) = U \overleftarrow{\int_0^l} \exp\{k[\zeta, \theta(t)] \, d[U^{-1}E_1(t)\, U]\}.$$

Thus it is sufficient to restrict ourselves to the case when X_1 and X_2 satisfy $(10.2'')$.

Suppose that t_0 is a point of growth of the function $\theta(t)$. Introducing the notation

$$Y_j(t; \zeta) = \overleftarrow{\int_t^l} \exp\{k[\zeta, \theta(t)] \, dE_j(t)\} \ (j = 1, 2),$$

we rewrite $(10.2'')$ in the following form:

$$Y_2^{-1}(t_0; \zeta)UY_1(t_0; \zeta) \equiv X_2(t_0; \zeta)X_1^{-1}(t_0; \zeta) \ (|\zeta| < 1).$$

Suppose that $Z(t_0; \zeta)$ is the common value of the left and right side of the last equation.

We suppose first that $\theta_0 = \theta(t_0) > 0$. Since the $Y_j(t_0; \zeta) \ (j = 1, 2)$ are holomorphic and unitary on the arc $\Gamma_{t_0} = \{e^{i\theta}, 0 < \theta < \theta_0\}$, and the $X_j(t_0; \zeta)$ have the same property on $\Delta_{t_0} = \{e^{i\theta}, \theta_0 < \theta < 2\pi\}$, the operator-function $Z(t_0; \zeta)$ is holomorphic in the entire extended plane, with the possible exclusion of the points $\zeta = 1$ and $\zeta = e^{i\theta_0}$, and takes on unitary values on the arcs Γ_{t_0} and Δ_{t_0}. We shall show that in fact $Z(t_0; \zeta)$ is holomorphic at those points.

Consider for example the point $\zeta = e^{i\theta_0}$ and a circular neighborhood $\mathfrak{U}$ of it which does not contain the point $\zeta = 1$ and in which $1 < |\zeta + e^{i\theta_0}|, \ |\zeta| < 2$. We shall estimate $\|Z(t_0; \zeta)\|$ first in the semineighborhood $\mathfrak{U}_1 = \{\zeta = \rho e^{i\theta} \in \mathfrak{U}, 0 < \theta < \theta_0\}$. Obviously for $\zeta \in \mathfrak{U}_1$, $j = 1, 2$, we have

$$\|Y_j^{\pm 1}(t_0; \zeta)\| \le \exp\left\{\frac{3}{|\zeta - e^{i\theta_0}|} \operatorname*{var}_{[0, l]} E_j\right\} \le \exp\left\{3|k(\zeta, \theta_0)| \operatorname*{var}_{[0, l]} E_j\right\},$$

since for $\zeta \in \mathfrak{U}_1$ and $\theta_0 < \theta(t) < 2\pi$ the inequality $|\zeta - e^{i\theta_0}| < |\zeta - e^{i\theta(t)}|$)
holds. Hence for $\zeta \in \mathfrak{U}_1$

$$\|Z(t_0; \zeta)\| = \|Y_2^{-1}(t_0; \zeta)UY_1(t_0; \zeta)\| \leq e^{\sigma|k(\zeta, \theta_0)|},$$

where $\sigma = 3(\mathrm{var}_{[0, l]}E_1 + \mathrm{var}_{[0, l]}E_2)$.

Consider the semineighborhood $\mathfrak{U}_2 = \{\zeta = \rho e^{i\theta} \in \mathfrak{U}, \theta_0 \leq \theta < 2\pi\}$. Using
the equation $Z(t_0; \zeta) = X_2(t_0; \zeta)X_1^{-1}(t_0; \zeta)$, we prove that for $\zeta \in \mathfrak{U}_2$ the
estimate

$$\|Z(t_0; \zeta)\| < e^{\sigma|k(\zeta, \theta_0)|} \tag{10.5}$$

holds. Since $\mathfrak{U} = \mathfrak{U}_1 \cup \mathfrak{U}_2$, then (10.5) holds for any $\zeta \in \mathfrak{U}$. We will now
produce an estimate of $\|Z(t_0; \zeta)\|$ on the diameter $d = \{\zeta = \rho e^{i\theta_0} \in \mathfrak{U}\}$ of the
neighborhood $\mathfrak{U}$. Since t_0 is a point of growth of the function $\theta(t)$, there
exists a neighborhood, say a right neighborhood $(t_0, t_0 + \delta_0)$, such that
$\theta(t) > \theta(t_0) = \theta_0$ for $t_0 < t < t_0 + \delta_0$.

Let $0 < \delta < \delta_0$. Since

$$Y_j(t_0; \zeta) = Y_j(t_0 + \delta; \zeta) \int_{t_0}^{t_0+\delta} \exp\{k[\zeta, \theta(t)]dE_j(t)\} \quad (j = 1, 2),$$

and since the quantity $\|Y_j(t_0 + \delta; \zeta)\|$ is bounded near $e^{i\theta_0}$ because the
operator $Y_j(t_0 + \delta; e^{i\theta_0})$ is unitary, then for $\zeta = \rho e^{i\theta_0} \in d$

$$\|Y_j^{\pm 1}(t_0; \zeta)\| \leq \mu_\delta \exp\left\{\left|\frac{1 + \rho}{1 - \rho}\right| \mathop{\mathrm{var}}_{[t_0, t_0+\delta]} E_j\right\},$$

μ_δ being a positive number (generally speaking, depending on δ). Since
the function $E_j(t)$ is $\Re$-continuous, for $\zeta \in d$ one has the estimate

$$\|Z(t_0; \zeta)\| \leq \mu_\delta^2 \exp\left\{\left|\frac{1 + \rho}{1 - \rho}\right| \epsilon(\delta)\right\}, \tag{10.6}$$

where $\epsilon(\delta)$ becomes small along with δ.

We turn to a new complex variable $\lambda = k(\zeta, \theta_0)$. Suppose that $F(\lambda) = Z(t_0; e^{i\theta_0}(\lambda + 1)/(\lambda - 1))$. Then there exists an $r_0 > 0$ such that $F(\lambda)$ is holo-
morphic for $|\lambda| > r_0$. (10.5) implies the validity of the inequality $\|F(\lambda)\| \leq e^{\sigma|\lambda|}$ for $|\lambda| > r_0$. Inequality (10.6) shows that

$$\varlimsup_{r \to \infty} r^{-1} \ln \|F(\pm r)\| \leq 0.$$

Moreover, the operators $(\pm ir)$ $(r \geq r_0)$ are unitary, since $Z(t; \zeta)$ is unitary for $\zeta \in \Gamma_{t_0} \cup \Delta_{t_0}$. Accordingly the operator-function $F(\lambda)$ is bounded on the curve Λ consisting of the rays $\lambda = \pm ir$ $(r \geq r_0)$ and the circumference $|\lambda| = r_0$.

Now we use the following theorem[1] of Phragmén-Lindelöf type:

Suppose that $f(\lambda)$ *is holomorphic in a region* $\mathfrak{D}$ *gotten from the upper halfplane by removing the disk* $|\lambda| \leq r_0$, *and satisfies the following conditions:*

1) *there exists a* $\sigma > 0$ *such that*

$$|f(\lambda)| \leq e^{\sigma |\lambda|} \ (\lambda \in \mathfrak{D});$$

2)
$$\varlimsup_{r \to \infty} r^{-1} \ln |f(r)| \leq 0 \ (r \geq r_0);$$

3) $|f(\lambda)| \leq a$ *for some* $a > 0$ *for any* λ *belonging to the boundary of the region* $\mathfrak{D}$.

Then $|f(\lambda)| \leq a$ *for any* $\lambda \in \mathfrak{D}$.

Suppose that x and y are arbitrary vectors of $\mathfrak{H}$. Applying the above theorem to the function $f(\lambda) = (F(\lambda)x, y)$ first in the region $\mathfrak{D}$, and then in $\mathfrak{D}^*$ symmetric to $\mathfrak{D}$ relative to the real axis, we show that $f(\lambda)$ is bounded in $\mathfrak{D} \cup \mathfrak{D}^*$. Accordingly, $f(\lambda)$ is regular at infinity. In view of the arbitrary choice of $x, y \in \mathfrak{H}$ the operator-function $F(\lambda)$ has the same property. Accordingly $Z(t_0; \zeta)$ is holomorphic for $\zeta = e^{i\theta_0}$. Analogously one proves that $Z(t_0; \zeta)$ is holomorphic at $\zeta = 1$. Therefore it is holomorphic in the entire extended plane, and hence a constant operator relative to ζ. This operator will still generally speaking depend on t_0, and it is obviously unitary. Thus $Z(t_0; \zeta) \equiv U(t_0)$ or

$$X_2(t_0; \zeta) = U(t_0) X_1(t_0; \zeta) \tag{10.7}$$

for any point t_0 of growth of the function $\theta(t)$ such that $\theta_0 = \theta(t_0) > 0$. On the basis of (10.2'') we may take (10.7) to be true also for $t_0 = l$. Here $U(l) = U$. Analogously one establishes the validity of (10.7) also for $\theta_0 = 0$.

Now suppose that (α, β) is an interval of constancy of $\theta(t)$. Then α and β are points of growth of that function. Using equation (10.7) for $t_0 = \alpha$ and $t_0 = \beta$, we get

$$X_2(\beta, \zeta) = U(\beta) X_1(\beta; \zeta) = U(\beta) \int_\alpha^\beta \exp\{k[\zeta, \theta(\beta)] dE_1(t)\} X_1(\alpha; \zeta)$$

$$= U(\beta) \int_\alpha^\beta \exp\{k[\zeta, \theta(\beta)] dE_1(t)\} U^{-1}(\alpha) X_2(\alpha; \zeta).$$

---

[1] For the proof see for example [10], Section III, problem 325.

Since

$$X_2(\beta, \zeta) = \overleftarrow{\int_\alpha^\beta} \exp\{k[\zeta, \theta(\beta)]\, dE_2(t)\}\, X_2(\alpha; \zeta),$$

we have

$$\int_\alpha^\beta \exp\{k[\zeta, \theta(\beta)]\, dE_2(t)\} = U(\beta) \overleftarrow{\int_\alpha^\beta} \exp\{k[\zeta, \theta(\beta)]\, dE_1(t)\}\, U^{-1}(\alpha).$$

Putting $\zeta = - e^{i\theta(\beta)}$, we get $U(\beta)\, U^{-1}(\alpha) = I$, or

$$U(\beta) = U(\alpha), \tag{10.8}$$

$$\overleftarrow{\int_\alpha^\beta} \exp\{k[\zeta, \theta(\beta)]\, dE_2(t)\} = U(\beta) \overleftarrow{\int_\alpha^\beta} \exp\{k[\zeta, \theta(\beta)]\, dE_1(t)\}\, U^{-1}(\beta). \tag{10.9}$$

We note that the functions $E_1(t)$ and $E_2(t)$ may without loss of generality be supposed to satisfy Lipschitz conditions:

$$\|E_j(t'') - E_j(t')\| \le |t'' - t'| \quad (j = 1, 2;\ 0 \le t',\ t'' \le l).$$

Indeed, we can arrange this by taking a new variable of integration according to the formula

$$s = t + \operatorname*{var}_{[0,\, t]} E_1 + \operatorname*{var}_{[0,\, t]} E_2.$$

Denote the weak derivatives[1] of the functions $E_1(t)$ and $E_2(t)$ by $M_1(t)$ and $M_2(t)$ respectively:

$$M_1(t) = \frac{d}{dt} E_1(t), \ M_2(t) = \frac{d}{dt} E_2(t) \ (t \in \mathfrak{M} \subset [0, l], \ \operatorname{mes} \mathfrak{M} = l);$$

$M_1(t)$ and $M_2(t)$ are normal operator-functions on $[0, l]$, and $\|M_j(t)\| \le 1$, $j = 1, 2$, for any $t \in \mathfrak{M}$.

Suppose that (α_j, β_j), $j = 1, 2, \cdots$ are intervals of constancy of the function $\theta(t)$, and $\mathfrak{N}$ the set of its points of growth: $\mathfrak{N} = [0, l] \setminus \mathbf{U}_j(\alpha_j, \beta_j)$. We introduce the auxiliary function

$$\widetilde{M}_2(t) = \begin{cases} M_2(t) & (t \in \mathfrak{M} \cap \mathfrak{N}) \\ U(\beta_j) M_1(t) U^{-1}(\beta_j) & (t \in \mathfrak{M} \cap (\alpha_j, \beta_j), \ j = 1, 2, \cdots) \end{cases}$$

Obviously $\widetilde{M}_2(t)$ is normal on $[0, l]$. Put

---

1) Here and below we use the results of §4, subsections 1 and 2.

$$\widetilde{E}_2(t) = \int_0^t \widetilde{M}_2(\tau)\, d\tau.$$

Extending $U(t)$ from $\mathfrak{N}$ to the intervals (α_j, β_j) as a constant $(U(t) = U(\alpha_j) = U(\beta_j)$ for $t \in (\alpha_j, \beta_j))$, we show that for any $t \in [0, l]$ the following equation holds:

$$\overleftarrow{\int_0^t} \exp\{k[\zeta,\, \theta(\tau)]\, d\widetilde{E}_2(\tau)\} = U(t) \overleftarrow{\int_0^t} \exp\{k[\zeta,\, \theta(\tau)]\, dE_1(\tau)\}. \qquad (10.10)$$

To this end we consider the functions

$$M_2^{(n)} = \begin{cases} M_2(t) & (t \in \mathfrak{M} \cap \{\mathfrak{N} \cup [\underset{j>n}{\mathbf{U}} (\alpha_j,\, \beta_j)]\}) \\[2mm] U(\beta_j)M_1(t)U^{-1}(\beta_j) & (t \in \mathfrak{M} \cap (\alpha_j,\, \beta_j),\ j = 1, \cdots, n), \end{cases}$$

$$\widetilde{E}_2^{(n)}(t) = \int_0^t \widetilde{M}_2^{(n)}(\tau)\, d\tau \quad (n = 1, 2, \cdots).$$

It is not hard to prove by induction that

$$\overleftarrow{\int_0^t} \exp\{k[\zeta,\, \theta(\tau)]\, d\widetilde{E}_2^{(n)}(\tau)\} = U(t) \overleftarrow{\int_0^t} \exp\{k[\zeta,\, \theta(\tau)]\, dE_1(\tau)\} \qquad (10.11)$$

for $t \in \mathfrak{N} \cup [\mathbf{U}_{j=1}^n (\alpha_j,\, \beta_j)]$. We shall verify for example that (10.11) holds for $n = 1$. For $t \leq \alpha_1$ this is obvious. Now suppose $\alpha_1 < t < \beta_1$. In this case

$$\overleftarrow{\int_0^t} \exp\{k[\zeta,\, \theta(\tau)]\, d\widetilde{E}_2^{(1)}(\tau)\}$$

$$= \overleftarrow{\int_{\alpha_1}^t} \exp\{k[\zeta,\theta(\tau)]\, U(\beta_1) M_1(\tau) U^{-1}(\beta_1)\, d\tau\} \cdot \overleftarrow{\int_0^{\alpha_1}} \exp\{k[\zeta,\, \theta(\tau)]\, dE_2(\tau)\}$$

$$= U(\beta_1) \overleftarrow{\int_{\alpha_1}^t} \exp\{k[\zeta,\, \theta(\tau)]dE_1(\tau)\} U^{-1}(\beta_1) U(\alpha_1) \cdot \overleftarrow{\int_0^{\alpha_1}} \exp\{k[\zeta,\, \theta(\tau)]\, dE_1(\tau)\}$$

$$= U(t) \overleftarrow{\int_0^t} \exp\{k[\zeta,\, \theta(\tau)]\, dE_1(\tau)\}. \qquad (10.12)$$

Here we have use equations (10.7) and (10.8). If now $t > \beta_1$, then

$$\overleftarrow{\int_0^t} \exp\{k[\zeta,\theta(t)]d\widetilde{E}_2^{(1)}(\tau)\}$$

$$= \overleftarrow{\int_{\beta_1}^t} \exp\{k[\zeta,\, \theta(\tau)] M_2(\tau)\, d\tau\} \cdot \overleftarrow{\int_0^{\beta_1}} \exp\{k[\zeta,\, \theta(\tau)]\, d\widetilde{E}_2^{(1)}(\tau)\}$$

$$= \overleftarrow{\int_{\beta_1}^t} \exp\{k[\zeta,\, \theta(\tau)]\, dE_2(\tau)\} \cdot U(\beta_1) \overleftarrow{\int_0^{\beta_1}} \exp\{k[\zeta,\, \theta(\tau)]dE_1(\tau)\} =$$

$$= \int_0^t \exp\{k[\zeta,\ \theta(\tau)]\, dE_2(\tau)\} = U(t)\int_0^t \exp\{k[\zeta,\ \theta(\tau)]\, dE_1(\tau)\}.$$

Here we have used relations (10.7) and (10.12).

Now suppose that t is any number on $[0,\ l]$. There exists an $n_0 = n_0(t)$ such that (10.11) holds for $n \geq n_0$. Since $\widetilde{M}_2^{(n)}(t) \Rightarrow \widetilde{M}_2(t)$ almost everywhere on $[0,\ l]$, and $\|\widetilde{M}_2^{(n)}(t)\| \leq \max\{\|M_1(t)\|,\ \|M_2(t)\|\} \leq 1$, we have $\widetilde{E}_2^{(n)}(t) \Rightarrow \widetilde{E}_2(t)$ for $t \in [0,\ l]$, and

$$\operatorname*{Var}_{[t',\ t'']} \widetilde{E}_2^{(n)} \leq |t'' - t'| \quad (0 \leq t' \leq t'' \leq l).$$

Using an analog of the second Helly's theorem for the multiplicative integral (5.11); compare also with Theorem 3.4) and passing to the limit as $n \to \infty$ in (10.11), we establish the validity of (10.10). Putting $\zeta = 0$ in (10.10), we obtain

$$\widetilde{W}_2(t) = U(t)\,W(t) \quad (0 \leq t \leq l), \tag{10.13}$$

where

$$W_1(t) = \int_0^t e^{-dE_1(t)}, \quad \widetilde{W}_2(t) = \int_0^t e^{-d\widetilde{E}_2(t)}.$$

Denote by $R_1(t)$ and $\widetilde{R}_2(t)$ the moduli of $W_1(t)$ and $\widetilde{W}_2(t)$ respectively: $R_1(t) = \sqrt{W_1^*(t)W_1(t)}$ and $\widetilde{R}_2(t) = \sqrt{\widetilde{W}_2^*(t)\widetilde{W}_2(t)}$. Equation (10.13) shows that $R_1(t) = \widetilde{R}_2(t)$ for $0 \leq t \leq l$. Using Theorem 4.1, on the basis of which

$$E_1(t) = \widetilde{E}_2(t) \quad (0 \leq t \leq l), \tag{10.14}$$

we get

$$U(t) \equiv I \quad (0 \leq t \leq l), \quad U = U(l) = I. \tag{10.15}$$

Suppose that t_0 is a point of growth for $\theta(t)$. Then

$$\widetilde{E}_2(t_0) = \int_0^{t_0} \widetilde{M}_2(t)dt = \int_{\mathfrak{R}\cap[0,\ t_0]} \widetilde{M}_2(t)dt + \sum_{\beta_j \leq t_0} \int_{\alpha_j}^{\beta_j} \widetilde{M}_2(t)dt.$$

But $\widetilde{M}_2(t) = M_2(t)$ for $t \in \mathfrak{M}\cap\mathfrak{R}$. Further, decomposing the left and right sides of (10.9) into powers of $\lambda = k[\zeta,\ \theta(\beta)]$ and taking account of (10.15), we get

$$\int_{\alpha_j}^{\beta_j} M_2(t)dt = \int_{\alpha_j}^{\beta_j} M_1(t)dt \quad (j = 1,\ 2,\ \cdots).$$

Since $\tilde{M}_2(t) = M_1(t)$ for $t \in \mathfrak{M} \cap (\alpha_j, \beta_j)$, we have

$$\tilde{E}_2(t_0) = \int\limits_{\mathfrak{R}\cap[0,\,t_0]} M_2(t)dt + \sum\limits_{\beta_j \le t_0} \int\limits_{\alpha_j}^{\beta_j} M_2(t)dt = \int\limits_0^{t_0} M_2(t)dt = E_2(t_0).$$

Hence and from (10.14) there follows the equation

$$E_1(t_0) = E_2(t_0) \tag{10.4}$$

for any point t_0 of growth of $\theta(t)$. As to equation (10.3), it directly follows from (10.7) and (10.15). The theorem is proved.

2. By the same method used for Theorem 10.1 we can prove a uniqueness theorem for multiplicative integrals of the form (10.1) with *J*-Hermitian-increasing functions $E_j(t)$ in the case of an arbitrary *J* as well. However we have to impose some additional conditions on the functions $E_j(t)$, which are automatically fulfilled when $J = I$.

We call the operator-function $E(t)$ $(0 \le t \le l)$ *diagonalizable*, if $E(t)$ may be represented in the form

$$E(t) = \int\limits_0^t S(\tau)\,dF(\tau)\,S^{-1}(\tau), \tag{10.16}$$

where $S(t)$ and $F(t)$ are operator-functions, $\mathfrak{R}$-continuous and of bounded $\mathfrak{R}$-variation, while $F(t)$ takes selfadjoint values on $[0, l]$, and $S(t)$ is a bounded inversible operator for each $t \in [0, l]$.

Theorem 10.2. *Suppose that*

$$X_j(t;\,\zeta) = \int\limits_0^t \exp\{k[\zeta,\,\theta(\tau)]dE_j(\tau)\} \quad (j = 1, 2;\; 0 \le t \le l,\; |\zeta| < 1),$$

where $\theta(t)$ *is a canonical function and the* $E_j(t)$ *are J-Hermitian, nondecreasing diagonalizable operator-functions. Then, if one of the two equations*

$$X_2(l;\,\zeta) \equiv X_1(l;\,\zeta)U,\; X_2(l;\,\zeta) \equiv UX_1(l;\,\zeta) \;(|\zeta| < 1)$$

holds, in which U *is a J-unitary operator, then* $U = I$, *and for any point* t_0 *of growth of the function* $\theta(t)$ *we have the relation*

$$X_2(t_0;\,\zeta) \equiv X_1(t_0;\,\zeta);\; E_2(t_0) = E_1(t_0).$$

The proof of this theorem will not be carried out here. It repeats verbatim the proof of Theorem 10.1, if one uses the following proposition:

Lemma 10.1.[1] *If* $E(t)$ *is a diagonalizable operator-function, and* $\theta(t)$

---

1) If $J = I$ the assertion of the lemma follows from the fact that the function $X(t_0;\,\zeta)$ is unitary on Γ_{t_0}.

is a canonical function, then the multiplicative integral

$$X(t_0; \zeta) = \overset{t_0}{\underset{0}{\overleftarrow{\int}}} \exp\{k[\zeta, \theta(t)]\, dE(t)\}$$

is bounded on the ray $\Gamma_{t_0} = \{e^{i\theta}; \theta_0 = \theta(t_0) < \theta < 2\pi\}.$

For the proof of this lemma we fix on some point $e^{i\phi} \in \Gamma_{t_0}$ and put

$$U(\phi) = \overset{t_0}{\underset{0}{\int}} \exp\{k[e^{i\phi}, \theta(t)]\, dE(t)\} = \overset{t_0}{\underset{0}{\int}} \exp\left\{- i\, \mathrm{ctg}\, \frac{\phi - \theta(t)}{2}\, dE(t)\right\}.$$

Suppose that $\Sigma: 0 = \tau_0 < \tau_1 < \cdots < \tau_{n-1} < \tau_n = t_0$ is some subdivision of $[0, t_0]$. Put

$$\Phi_\Sigma(\phi) = \overset{n}{\underset{j=1}{\overleftarrow{\prod}}} \exp\left\{- i\, \mathrm{ctg}\, \frac{\theta(\tau_j) - \phi}{2}\, S(\tau_j)\, [F(\tau_j) - F(\tau_{j-1})]\, S^{-1}(\tau_j)\right\}. \quad (10.17)$$

It is easy to prove that $\Phi_\Sigma(\phi) \Rightarrow U(\phi)$ when the diameter of the subdivision Σ tends to zero. For this it suffices to compare $\Phi_\Sigma(\phi)$ with the integral product corresponding to the subdivision Σ.

We now estimate the product $\Phi_\Sigma(\phi)$. To this end we introduce the notation $Q_j = S^{-1}(\tau_j)\, S(\tau_{j-1})$. Then

$$\prod_j \|Q_j\| = \prod_j \|I + S^{-1}(\tau_j)\, [S(\tau_{j-1}) - S(\tau_j)]\|$$

$$\leq \prod_j e^{\|S^{-1}(\tau_j)\|\, \|S(\tau_j) - S(\tau_{j-1})\|} \leq e^{\mu v}\left[\mu = \max_{0 \leq t \leq l} \|S^{-1}(t)\|,\; v = \underset{[0,\, l]}{\mathrm{Var}}\, S\right].$$

Now we write equation (10.17) in the following form:

$$\Phi_\Sigma(\phi) = S(\tau_n) \exp\left\{- i\, \mathrm{ctg}\, \frac{\phi - \theta(\tau_n)}{2}\, [F(\tau_n) - F(\tau_{n-1})]\right\}$$

$$\times \left[\overset{n-1}{\underset{j=1}{\overleftarrow{\prod}}}\, Q_{j+1} \exp\left\{- i\, \mathrm{ctg}\, \frac{\phi - \theta(\tau_j)}{2}\, [F(\tau_j) - F(\tau_{j=1})]\right\}\right] S^{-1}(\tau_1).$$

Since

$$\exp\left\{- i\, \mathrm{ctg}\, \frac{\phi - \theta(\tau_j)}{2}\, [F(\tau_j) - F(\tau_{j-1})]\right\}$$

are unitary operators, the following estimate, which does not depend on ϕ or on the subdivision Σ, holds:

$$\|\Phi_\Sigma(\phi)\| \leq \|S(\tau_n)\|\, \|S^{-1}(\tau_1)\|\, \prod_j \|Q_j\| \leq \mu v e^{\mu v}$$

$$\left[\nu = \max_{0 \leq t \leq l} \|S(t)\|\right].$$

Therefore it follows that $\|U(\phi)\| \leq \mu\nu e^{\mu\nu}$ $(0 < \phi < \theta(t_0))$. The lemma is proved.

We note that Theorems 10.1 and 10.2, which generalize and sharpen a result of L. A. Sahnovič [12], may be reformulated as theorems for the existence of certain inverse problems for systems of differential equations.

BIBLIOGRAPHY

[1] Ju. P. Ginzburg, *Multiplicative representations of J-contractive operator-functions*. I, Mat. Issled. 2 (1967), no. 2, 52–83; English transl., Amer. Math. Soc. Transl. (2) 96 (1970), pp. 189–221. MR 38 #1551a.

[2] M. S. Livšic, *On spectral decomposition of linear nonselfadjoint operators*, Mat. Sb. 34 (76) (1954), 145–199. (Russian) MR 16, 48.

[3] I. M. Gel'fand and N. Ja. Vilenkin, *Generalized functions.* Vol 4: *Applications of harmonic analysis*, Fizmatgiz, Moscow, 1961; English transl., Academic Press, New York, 1964. MR 26 #4173; 30 #4152.

[4] Ju. P. Ginzburg, *On J-contractive operator functions*, Dokl. Akad. Nauk SSSR 117 (1957), 171–173. (Russian) MR 20 #1203.

[5] V. P. Potapov, *The multiplicative structure of J-contractive matrix functions*, Trudy Moskov. Mat. Obšč. 4 (1955), 125–236; English transl., Amer. Math. Soc. Transl. (2) 15 (1960), 131–243. MR 17, 958; 22 #5733.

[6] I. C. Gohberg and M. G. Kreĭn, *Introduction to the theory of linear nonselfadjoint operators in Hilbert space*, "Nauka", Moscow, 1965; English transl., Transl. Math. Monographs, vol. 18, Amer. Math. Soc., Providence, R. I., 1969. MR 36 #3137.

[7] Ju. P. Ginzburg, *A maximum principle for J-contractive operator functions and some of its consequences*, Izv. Vysš. Učebn. Zaved. Matematika 1963, no. 1 (32), 42–53. (Russian) MR 26 #6793.

[8] ———, *Multiplicative representations of bounded analytic operator-functions*, Dokl. Akad. Nauk SSSR 170 (1966), 23–26 = Soviet Math. Dokl. 7 (1966), 1125–1128. MR 34 #611.

[9] E. Hille and R. S. Phillips, *Functional analysis and semi-groups*, rev. ed., Amer. Math. Soc. Colloq Publ., vol. 31, Amer. Math. Soc., Providence, R. I., 1957; Russian transl., IL, Moscow, 1962. MR 19, 664.

[10] G. Pólya and G. Szegö, *Aufgaben und Lehrsätze aus der Analysis*, 2nd ed., Die Grundlehren der math. Wissenschaften, Band 19, Springer-Verlag, Berlin, 1954; Russian transl., IL, Moscow, 1956. MR 7, 418; 15, 512.

[11]   M. S. Brodskiĭ and M. S. Livšic, *Spectral analysis of non-selfadjoint operators and intermediate systems*, Uspehi Mat. Nauk 13 (1958), 3–85; English transl., Amer. Math. Soc. Transl. (2) 13 (1960), 265–346. MR 20 #7221; 22 #3982.

[12]   L. A. Sahnovič, *On the reduction of nonselfadjoint operators to diagonal form*, Proc. Third All-Union Math. Congress (Moscow, 1956), vol. 1, Izdat. Akad. Nauk SSSR, Moscow, 1959, pp. 120–122. (Russian)

Translated by:
J. M. Danskin